# ZINC–SILVER OXIDE BATTERIES

# THE ELECTROCHEMICAL SOCIETY SERIES

# ZINC–SILVER OXIDE BATTERIES

Edited by
## ARTHUR FLEISCHER
*Consulting Electrochemist*

WITH

### JOHN J. LANDER
*Senior Scientist*
*Air Force Aero Propulsion Laboratory*

*Sponsored by*

## THE ELECTROCHEMICAL SOCIETY, INC.
New York, New York

## JOHN WILEY & SONS, INC.
NEW YORK · LONDON · SYDNEY · TORONTO

Library of Congress Catalog Card Number: 70-129658

ISBN 0 471 26350 8

Printed in the United States of America

10  9  8  7  6  5  4  3  2  1

# Contributors

F. C. ARRANCE, *Astropower Laboratory, McDonnell Douglas Corporation, Newport Beach, California*

D. BERNDT, *Varta AG, Forschungs- und Entwicklungszentrum, Kelkheim/ Taunus, Germany*

K. F. BLURTON, *Energetics Science, Inc., The Bronx, New York*

HANS H. BODE, *Varta AG, Forschungs- und Entwicklungszentrum, Kelkheim/ Taunus, Germany*

FREDERIC M. BOWERS, *U.S. Naval Ordnance Laboratory, White Oak, Silver Spring, Maryland*

W. N. CARSON, Jr., *Research and Development Center, General Electric Company, Schenectady, New York*

E. J. CASEY, *Defense Research Establishment Ottawa, Ottawa, Ontario, Canada*

A. M. CHREITZBERG, *ESB Incorporated, Exide Missile and Electronics Division, Raleigh, North Carolina*

VINCENT D'AGOSTINO, *RAI Research Corporation, Hauppage, New York*

GEORGE A. DALIN, *Elizabeth, New Jersey*

T. P. DIRKSE, *Calvin College, Grand Rapids, Michigan*

D. J. DOAN, *Eagle-Picher Industries, Inc., Joplin, Missouri*

FRED EDELSTEIN, *Grumman Aircraft Engineering Corporation, Bethpage, Long Island, New York*

S. UNO FALK, *Svenska Ackumulator Aktiebolaget Jungner, Oskarshamn, Sweden*

ARTHUR FLEISCHER, *Consultant, Orange, New Jersey*

M. G. GANDEL, *Lockheed Missiles and Space Company, Sunnyvale, California*

HARRY P. GREGOR, *Department of Chemical Engineering and Applied Chemistry, School of Engineering and Applied Science, Columbia University, New York*

SIDNEY GROSS, *The Boeing Company, Seattle, Washington*

v

ELMER GUBNER, *U.S. Naval Ordnance Laboratory, White Oak, Silver Spring, Maryland*

N. A. HAMPSON, *Department of Chemistry, Loughborough University of Technology, Leicestershire, England*

THOMAS J. HENNIGAN, *NASA/Goddard Space Flight Center, Greenbelt, Maryland*

PAUL L. HOWARD, *Consultant, Centreville, Maryland*

JOHN F. JACKOVITZ, *Westinghouse Electric Corp., Research and Development Center, Pittsburgh, Pennsylvania*

J. A. KERALLA, *Delco-Remy Division, General Motors Corporation, Anderson, Indiana*

R. L. KERR, *Air Force Aero Propulsion Laboratory, Wright-Patterson Air Force Base, Ohio*

J. J. LANDER, *Air Force Aero Propulsion Laboratory, Wright-Patterson Air Force Base, Ohio*

ALOIS LANGER, *Westinghouse Electric Corp., Research and Development Center, Pittsburgh, Pennsylvania*

D. LAX, *Chemistry Department, Constantine College of Technology, Middlesbrough, England*

JOSEPH LEE, *RAI Research Corporation, Hauppage, New York*

DAN LEHRFELD, *Grumman Aircraft Engineering Corporation, Bethpage, Long Island, New York*

RALPH LUTWACK, *Jet Propulsion Laboratory, Pasadena, California*

G. H. MILLER, *Air Force Aero Propulsion Laboratory, Wright-Patterson Air Force Base, Ohio*

G. MOE, *Astropower Laboratory, McDonnell Douglas Corporation, Newport Beach, California*

FRANCIS G. MURPHY, *Naval Underwater Weapons Research and Engineering Station, Newport, Rhode Island*

G. D. NAGY, *Defence Research Establishment Ottawa, Ottawa, Ontario, Canada*

P. NESS, *Varta AG, Forschungs- und Entwicklungszentrum, Kelkheim/Taunus, Germany*

V. A. OLIAPURAM, *Varta AG, Forschungs- und Entwicklungszentrum, Kelkheim/Taunus, Germany*

GUYLA ORBAN, *RAI Research Corporation, Hauppage, New York*

H. G. OSWIN, *Energetics Science, Inc., The Bronx, New York*

CHARLES F. PALANDATI, *NASA/Goddard Space Flight Center, Greenbelt, Maryland*

ROBERT E. POST, *Lewis Research Center, National Aeronautics and Space Administration, Cleveland, Ohio*

ALLEN H. REMANICK, *Whittaker Corporation, Research and Development/ San Diego, San Diego, California*

FRED E. ROBBINS, *Delco-Remy Division, General Motors Corporation, Anderson, Indiana*

PAUL RUETSCHI, *Leclanché S. A., Yverdon, Switzerland*

ALVIN J. SALKIND, *ESB Incorporated, Research Center, Yardley, Pennsylvania*

S. F. SCHIFFER, *Yardney Electric Corporation, Pawcatuck, Connecticut*

MANNY SHAW, *Whittaker Corporation, Research and Development/San Diego, San Diego, California*

H. R. THIRSK, *School of Chemistry, University of Newcastle, Newcastle upon Tyne, England*

NICHOLAS T. WILBURN, *Power Sources Division, Electronic Components Laboratory, USAECOM, Fort Monmouth, New Jersey*

# Foreword*

The Air Force Aero Propulsion Laboratory takes pleasure in cosponsoring with the Electrochemical Society and its Battery Division the symposium on silver-zinc batteries. Air Force interest in the development of this type of battery has been high over the past decade or so because this battery is used on aircraft, missiles, and space vehicles, and the Air Force has actively supported research and development projects aimed at performance improvement and increasing the battery's usefulness in military and space programs.

Most applied research and development on this electrochemical power system has been supported by several government agencies and, for this reason, the advances achieved, the negative results, and the elucidation of problem areas and their detailed definition have come in the open literature only slowly and in piecemeal fashion before the general scientific and engineering public. Cosponsorship of this symposium by the Air Force Aero Propulsion Laboratory has the purpose not only of airing as much of the available information as can be achieved in one concentrated week of conferences, participated in by that portion of the scientific and engineering populace most able and likely to contribute to further advances, but also has the aim of providing a permanent record of these proceedings in the form of an Electrochemical Society Monograph. It is expected that the monograph will be useful: *to the systems designer*, who sees the battery as but one component of an overall system or vehicle and who is responsible for the generation of performance specifications; *to the battery engineer*, who is responsible for design to meet a body of specifications; *to the manufacturing engineer*, who must translate the design engineer's handiwork into actual construction of a highly reliable end product; *to the researcher*, who is expected to contribute the basic information and understanding of the chemical and electrochemical functioning of the device in order to provide a firm scientific basis for continued improvement; and *to the general scientific and engineering public* for their further education.

* Opening address to the Silver-Zinc Battery Symposium.

We have attempted, in laying out the symposium content, to treat the silver-zinc battery from its early history and chemical and electrochemical theory through components function, design, and manufacture to final assembly, performance testing, and reliability assurance of the finished product in a cross-section of its more important applications. It is a part of the overall purpose to bring together in one conference a large group of those having expertise in all phases and aspects in the hope of encouraging a broader mutual understanding of the battery and its present capabilities and limitations so that not only will an up-to-date and excellent review be provided but also existing information gaps will be disclosed and fresh new ideas and approaches toward a much-improved product will be generated.

Communication and dissemination of information, to say nothing of its digestion and assimilation by the receiver, is always a difficult task. It seems particularly appropriate at this time, therefore, to undertake to bring together the available information on the silver-zinc battery which is so widely scattered through the scientific, technical, and patent literature and in project reports on government and government-sponsored work, in a more-or-less digested form before the literature becomes more unmanageably bulky than it is already. One may even look forward to updating the literature in perhaps ten years from now, an impossible development without the efforts of this symposium. Interaction among and within all groups, user, tester, manufacturer, engineer, and scientist, is vital to accomplishment of these goals.

It may be well to make a few general remarks about the performance capability of the silver-zinc battery, even though this will be considered in detail during the course of the symposium. The space program and weaponry programs, with their emphasis on economy of weight and volume and on reliability, resulted in putting the silver-zinc battery in a position of special significance because of its inherent high-energy and high-power capabilities. The battery industry, together with the systems groups and the cognizant government laboratories, was quick to respond, and development of the silver-zinc battery in its primary mode of operation advanced to a high degree of fruition, in terms of ultimate engineering capability, in a relatively few years after the inception of the military and space programs. Over 100 watt-hours per pound is obtained for some high-energy, low-rate applications. Reserve battery designs are especially remarkable, which, requiring as they do chemomechanical devices for rapid activation, high rates of energy delivery, and unusual reliability, have achieved a high degree of sophistication indeed since the time of their conception. This is not to say by any means that nothing remains to be done about them. Reduced costs are always welcome, and improved low-temperature performance as well as improved storageability to enable relaxation of control of the storage environment are desirable. Novel designs using lightweight materials offer promise of yet further performance increases for reserve batteries.

By comparison and perhaps partly in view of the rapid emergence of the primary battery, development of long-lived, high-energy secondary batteries has been disappointingly slow. Inadequate separator materials and the zinc electrode behavior have offered especially knotty problems, and their slow response to research, in terms of the needs, has prevented considerably more widespread use of the silver-zinc secondary battery. Elucidation and definition of the problem areas has, perforce, occupied a good deal of the research effort, while sealed-construction and sterilization procedures for the several space applications have additionally complicated the cycle-life issue. Increased attention to the separator problem in the past few years has resulted in signs that this portion of the problem will be successfully overcome, and quite likely the rapidity of development will be fairly in proportion to the availability of funding, from whatever source. The zinc electrode, on the other hand, appears to be a problem of a different nature. Although some progress has been made in extending its cycle life and minor gains are expected, there seems to be no indication at the present time that anything like an order of magnitude increase in the cycle life is near, based on approaches currently under investigation.

In spite of this, interest in the zinc electrode must continue because of its low cost and ready availability and high-energy density. In addition, anything that can be done to improve its cycle life in the silver-zinc battery should be applicable to its use in the nickel oxide-zinc and the zinc-air rechargeable batteries, which are of potential interest for vehicular propulsion as well as for military applications.

The Air Force Aero Propulsion Laboratory joins with the Electrochemical Society and its battery division in welcoming you to the silver-zinc battery symposium and wishes to thank all participants, especially those who have responded to invitations for papers and reviews. The Society and its Battery Division are understood to be departing somewhat from normal procedure in holding this particular symposium, and they are to be thanked for providing the vehicle by which this important subject can have a full and detailed treatment.

COLONEL H. A. LYON, Commander,
Air Force Aero Propulsion Laboratory

# Preface

The aims and scope of this book are stated clearly and elegantly in Colonel Lyon's Foreword, the opening address of the Symposium on Zinc-Silver Oxide Batteries at the 1968 Fall Meeting of the Electrochemical Society, Inc., at Montreal. The scientific and technological reviews of the present status of the zinc-silver oxide system should provide a source book of information to technologists of a wide range of interest as well as to specialists in the field. The book makes clear that the battery man will investigate every new laboratory technique, every new material, and every new theoretical or experimental approach or idea that might lead to an improved product.

The task of assembling this book has proceeded more slowly than I had anticipated. It was deemed necessary to reduce the total length of the manuscripts by about one-third to satisfy the probable economics of the market for this book. Moreover, the format was improved from a simple assemblage of papers by numbering the tables, graphs, and references in sequence. The references have been collected and placed at the end of the book; they are arranged alphabetically by authors' names. Most of the references were checked; it may surprise my colleagues to learn that there were inaccuracies in names, journals, and page citations in about one-sixth of the list. Reference numbers are indicated in parentheses, the numbers being preceded by the letter R. Temperatures are reported in Centigrade degrees unless otherwise noted. It was not always practical in the editing to follow through on The Electrochemical Society's instructions to authors.

The experience of editing the manuscripts was an enervating extension of activities in reviewing reports and abstracts for many years. At times it just seemed that the papers were getting longer as I moved further on in the book. It is not really so; the papers are interesting and fascinating all the way. My sense of proportion was restored on recalling one of Rabelais' tales. An early French king had devised a unique program for setting up milestones on the roads radiating from Paris with the result that the milestones were spaced further and further apart as one traveled away from the capital city.

The most difficult of all the decisions that had to be made was in the matter of acknowledgments. Each author had carefully appended names of those to whom they were indebted for advice and aid. These names have been mercilessly deleted from each paper and the rule applies here though it is not easy to forego the pleasure of acknowledging the author of the idea for this symposium and the many who have assisted in its promotion and with the manuscripts for the book. The editor thanks all on behalf of the authors and on his own behalf; the editor thanks the authors and many colleagues who responded to his requests for clarifying information.

*Orange, New Jersey*                                   ARTHUR FLEISCHER
*April 1970*

# Contents

xv

# ZINC–SILVER OXIDE BATTERIES

# PART I
## INTRODUCTION

# I.

# Milestones in the Electrochemistry of Zinc–Silver Oxide Batteries

## Paul L. Howard and A. Fleischer

Bereshith. In the beginning there was Volta (R619) whose silver-zinc pile set the stage, on the eve of the nineteenth century, for electrochemical science and technology. But it was not until the fifth decade of our century that the alkaline zinc-silver oxide system achieved recognition and use as primary and secondary batteries. This apparent demarcation of successful effort is vividly reflected by referring to well-known books on batteries. Vinal's third edition (R614), fourth printing 1947, and Drucker and Finkelstein (R182) make no reference whatever to silver oxide batteries. Crennell and Lea devoted one paragraph in their *Alkaline Accumulators* (R109) to silver electrodes. Jumau (R328), third edition, devoted seven pages to alkaline batteries with special emphasis on the many combinations suggested and studied in the last decade of the nineteenth century. The principal mention of the silver electrode is in connection with Jungner's copper-silver oxide cell as illustrative of the attempts to make a secondary system of the de Lalande-Chaperon cell.

Nevertheless, it is of interest to look into the early history in setting the stage for the turning point in the development of this system and for this symposium. The highlights shall be followed in an approximately chronological manner, thus intertwining the development and evolution of the two electrodes, zinc and silver.

Zinc has been a favorite material for many battery systems because of the favorable relationships technologically and economically. Its low equivalent weight and comparatively high voltage place it high on the list of energy densities for aqueous systems. It has always been available in many forms and in quantity. Its behavior in the Volta pile, the battery that dominated the scene in the first third of the nineteenth century, was improved by amalgamation, usually ascribed to Kemp and Sturgeon in 1828 (R194). The Volta pile was displaced from the scene by two types of cells which also used zinc anodes and which, in turn, were used extensively for many years. These were the Daniell cell invented in 1836 and the Grove cell of 1839 (R194).

During the middle third of the nineteenth century, there was great activity in

3

examining possible combinations for achieving practical cells. There emerged the two cell systems that became and remain dominant in the battery industry. The primary cell utilizing zinc as the anode was the Leclanché invention of 1866 (R393). The secondary cell was the lead-acid system invented by Plante (R491) in 1860. Pausing a moment, it may be noted that Plante selected the lead system after polarization studies on nine metals, including silver in dilute and strong sulfuric acid solutions. He sought currents of high power and one might speculate on the course of events if he had included strongly alkaline electrolytes as another variable.

The last third of the nineteenth century saw additional efforts on a wide variety of systems. It was also a time of intensive development of the lead-acid battery. The Faure (R194) pasted plate was patented in 1881, and the Tudor high-surface plates two years later. Concurrently, Gladstone and Tribe (R243) published their proposal of the double sulfate theory for the basic charge-discharge reaction. In 1889, Frankland (R227) proposed the use of electrolyte specific gravity to indicate the state of charge of the battery. The intimate history and background would suggest a growing technology in its infancy, with many problems and difficulties. It is in such a background that one can project back to a beckoning for searches to find a better battery.

The first complete alkaline zinc-silver oxide primary battery is described in a patent to Clarke (R97) in 1883. Also, the first secondary zinc-silver oxide battery seems to have been proposed in an 1887 patent to Dun and Hasslacher (R186). It is not known whether these cells were ever manufactured or used. At about the same time, the de Lalande-Chaperon (R194, 615A) cell came into use. Many efforts were made to convert this alkaline zinc-copper oxide cell into a rechargeable one. Boettcher (R54) in 1890 patented a procedure of depositing compact, coarsely cut crystalline zinc on the upper horizontal face of the electrode, otherwise insulated at the edges, bottom, and straps, in the presence of excess alkali zincate electrolyte. Schoop (R550) in 1894 reported on large batteries used by the Wadell-Entz Company to operate trams on their Second Avenue line in New York City. Recharging of the batteries required control of the zinc deposition from the alkali zincate electrolyte and was carried out in a central charging station. During recharge, the electrolyte was heated and circulated.

The recorded nineteenth century experience with the alkaline zinc electrode is indeed a gloomy heritage.

At the turn of the century, the two founders of the alkaline storage battery industry were busy in this field. Edison's approach was based on the deposition of zinc on a magnesium substrate (R187). Jungner developed the cadmium-silver oxide storage battery which he tested for the operation of an electric automobile in Stockholm (R530). Thus we may credit Jungner (R329-331) with the development of one type of silver electrode suitable for alkaline electrolyte.

Before making a complete transition to alkaline systems and the twentieth century, attention is directed to many efforts to combine the zinc electrode with lead dioxide in zinc sulfate-sulfuric acid electrolytes. This cell has a somewhat higher voltage than the usual lead-acid cell and was championed and studied by many. Pouchain (R498) obtained many patents in the twenties. At this time the system is used for some specialized and limited applications.

The establishment and growth of the nickel-iron and nickel-cadmium battery industry in the early part of this century probably slowed up a vigorous pursuit on the silver battery. Nevertheless, there are a number of interesting highlights deserving mention.

Luther and Pokorny (R401) in 1908 published a study of the electrochemical behavior of plated-silver electrodes in one *normal* electrolyte, clearly demonstrating the two-step oxidation and reduction with the known inequality of the two plateaus. Although their main purpose in the study was indicated to be related to certain problems in photochemistry, they mention a revived interest in the alkaline silver battery.

In the same year, Hubbell (R300) patented an electrode made by pressing a mixture of silver oxide and nickel hydroxide to a perforated plate and enclosing the tablet in an asbestos filter paper. The purpose of this envelope was to overcome the tendency of silver oxide to be displaced during charging and discharging. In 1925, Jirsa (R319) indicated that the tendency for colloidal silver oxide to form and migrate is decreased by the addition of potassium sulfate to the electrolyte. This statement has often been cited without a reminder that Jirsa based his conclusion on the use of one *normal* alkali electrolyte.

Morrison (R433), as disclosed in a series of 1909-1910 patents, attempted to eliminate problems with the zinc electrode. His most interesting proposals were the providing of compartments in a horizontal electrode to prevent mechanical washing and the purifying of the potassium hydroxide electrolyte by electrolysis at suitable current densities. In the 1930s Drumm (R598) in Ireland started out on his development of an electric train battery with the zinc-silver oxide system. His zinc electrode, indicated to be zinc deposited on a nickel substrate whose surface had been electroplated from an alkaline mercury cyanide bath, was claimed to be rechargeable without shape change. In 1935 Zimmerman reported on various alkaline systems including zinc-silver oxide, mainly with electrodeposited electrodes (R659). Also at this time, and with this we complete the early history, Kinoshita (R351) in Japan reported favorable results from his studies of pocket and pasted electrodes made from silver (I) oxide.

The giant step to practical batteries was provided by Henri André whose work truly ushered in the modern development of the zinc-silver oxide battery. He described his early work, extensive and impressive, in a 1941 paper (R15), "The Silver-Zinc Accumulator" (published in the *Bulletin Société française Electriciens*, sixth Series, Volume 1, pp. 132-146). His recognition of the problems, the scholarly analyses, and his practical solutions are classical and deserve our attention and praise.

First and foremost, he noted the solubility of silver oxide in the alkaline electrolyte and the migration of the silver species to the zinc electrode, with its deposition there. This action incubated a sequence of troubles that were self-accelerating. His remedy was to provide the semipermeable membrane, cellophane, to retard the migration of silver from the anode compartment created in applying this separator. In his early studies, it is reported that André made his own cellophane (R531).

The semipermeable membrane paved the way for a practical rechargeable cell. But this was only possible by an adjustment of the volume of the zinc electrode compartment and by the adjustment of the pressure on the entire element by suitable design. The compartment volume provided a reservoir of electrolyte and permitted a gentler liberation of hydrogen during charging without excessive disturbance of the spongy zinc.

André also established the optimum electrolyte concentration in the range of 40 to 45% by weight of potassium hydroxide for maximum capacity and performance. His work indicated a meticulous attention to details; here there will only be mentioned as a typical example, the need for insulating leads or tabs to the silver electrode to avoid waterline corrosion.

André's work inspired an examination of the potentialities of this system for military applications with the usual emphasis on higher power and energy densities. Research and development programs on the electrodes and their characteristic properties were undertaken at the National Bureau of Standards and at the U.S. Naval Research Laboratory. Under contract, Edison Storage Battery Company (R189) built a 60 kW battery (2000 amperes for 3 minutes) as well as mechanically activated reserve units provided with a storage compartment for the electrolyte.

From these beginnings there has grown a tremendous technology which this symposium will explore rather fully. The zinc-silver oxide system has been established as an important one for many military applications and for certain critical space projects. The record shows that many practical problems have been solved; we look ahead to progress toward the ultimate goal of long-life storage batteries of higher power and energy densities.

# 2.

# Thermodynamics of the Zinc–Silver Oxide Battery

Hans H. Bode with V. A. Oliapuram, D. Berndt, and P. Ness

The zinc electrode in alkaline medium is an electrode of the second kind for which the following reactions and the data of Table 1 apply:

$$Zn + 2H_2O = Zn(OH)_2 + H_2 \qquad (1)$$

$$Zn + H_2O = ZnO + H_2 \qquad (2)$$

TABLE 1.  SUMMARY[a] OF THERMODYNAMIC DATA FOR ZINC AT 25°

| Substance Formula | State and Description | $\Delta H$ kcal/mol | $\Delta G$ kcal/mol | $\Delta S$ kcal/mol | $pL$[b] |
|---|---|---|---|---|---|
| Zn | c | 0 | 0 | 9.95 | — |
| $Zn^{2+}$ | aq | −36.43 | −35.184 | −25.43 | — |
| ZnO | Inactive BET: 0.5 m$^2$/g | −83.17 | −76.63 | 10.5 | 16.83 |
| ZnO | Active BET: 16 m$^2$/g | | −76.40 | | 16.66 |
| $\epsilon$-Zn(OH)$_2$ | c (rhombic) | −153.5 | −132.83 | (19.9) | 16.47 |
| $\alpha$-Zn(OH)$_2$ | c (hexagonal) | | −132.1 | | 15.85 |
| Zn(OH)$_2$ | Amorphous | −150.5 | −131.54 | | 15.52 |
| Zn(OH)$_4^{2-}$ | aq | −261.4[c] | −208.6 | | — |

[a] Standard values of formation from the elements from References 388, 519, and 545 except for tetrahydroxyzinc ion, which was calculated for this paper.

[b] The solubility product $L$ is expressed as $pL$ or $-\log L$, $L$ being measured at ionic strength of zero.

[c] From the following chapter by Dirkse.

Taking into account the solubility product, $L$, there may be derived the following general expression for the potential:

$$E = -0.763 - 0.0295\ pL + 0.059\ pOH = E_0{}' + 0.059\ pOH \qquad (3)$$

The expression $E_0(Zn/Zn^{2+}) = -0.763$ is the standard potential on the

7

standard hydrogen scale. A number of different modifications of the hydroxides and oxides of zinc have been reported (R545). The standard potentials for three of the modifications have been calculated using Equation 3:

1. For the amorphous $Zn(OH)_2$ with $pL = 15.52$ as the most soluble, but the most unstable form.

2. For $\epsilon\text{-}Zn(OH)_2$ with $pL = 16.47$, the least soluble, but the most stable hydroxide.

3. For the "inactive" oxide having $pL = 16.83$, representing the oxides which distinguish themselves from each other by different reactivities (characterized by BET-surface area).

Substituting the corresponding $pL$ values, we have

$$\text{for } Zn(OH)_2 \text{ (amorphous)} \quad E = -1.221 + 0.059 \text{ pOH} \qquad (4a)$$

$$\text{for } \epsilon\text{-}Zn(OH)_2 \quad\quad\quad\quad E = -1.249 + 0.059 \text{ pOH} \qquad (4b)$$

$$\text{for } ZnO \text{ (inactive)} \quad\quad\quad E = -1.260 + 0.059 \text{ pOH} \qquad (4c)$$

(Note that the standard values of $-1.215$ and $-1.245$ given by DeBethune (R126) for the first two equations require $pL$ values of 15.32 and 16.35, respectively.)

The difference between the values $E_0{}'$ is comparatively small, but the temperature coefficients are appreciably different:

$$\left(\frac{dE_0}{dT}\right)_{\epsilon - Zn(OH)_2} = -1.001 \text{ mV/degree} \qquad \left(\frac{dE_0}{dT}\right)_{ZnO} = -1.161 \text{ mV/degree}$$

Because of their amphoteric character, the hydroxides dissolve in the alkali to form zincate; the values given above are valid only for solutions in which the respective modifications are present as the solid phase. This is evident for solutions saturated with zincate. But even if the solution is not saturated or is free from zincate, a covering layer is always formed on the surface of the electrode within a short time (R53). In such cases the potential of the electrode is determined by this surface film, which as a rule represents initially an unstable modification, but in course of time is converted to the stable modification. Thus the electrochemical behavior of the zinc electrode is independent of the concentration of zincate in solution.

The double negative tetrahydroxocomplex $Zn(OH)_4^{2-}$ has been postulated (Ref. 150 and see Jackowitz and Langer in this volume) to represent the composition of the zincate in 1 to $7N$ KOH solution, but a monovalent anion (R499), $HZnO_2^-$ or $Zn(OH)_3^-$ has also been reported. The solid phases in the system $ZnO-KOH-H_2O$ have not been investigated systematically. Only zincates of the type $[Zn(OH)_3]^-$ or $[ZnO(OH)]^-$ have been reported, with ZnO in dilute alkaline solutions, and KOH hydrates (R311) in more concentrated solutions.

A careful study (Figure 1) is available on the system $ZnO–NaOH–H_2O$ (R549). At room temperature and over a wide pH range, only a zincate of the $NaZn(OH)_3$ type appears as a metastable or stable solid phase besides $Zn(OH)_2$ and ZnO; $Na_2[Zn(OH)_4]$ appears as solid phase only in NaOH above 20 molar. For the formation of the zincate anion from the ions we have,

$$Zn^{2+} + 3OH^- = Zn(OH)_3^- \tag{5}$$

The equilibrium constant $K = 10^{14.36}$ (R283). The free energy of the reaction (Equation 5) is $\Delta G = -19.59$ kcal; the free energy of formation of the monovalent complex can be calculated to be $\Delta G = -167.6$ kcal/mol. The free energy of formation from the hydroxides according to Equation 6:

$$Zn(OH)_2 + OH^- = Zn(OH)_3^- \tag{6}$$

is only slightly positive, +2.8 kcal from $\epsilon$-$Zn(OH)_2$ and +1.5 kcal from amorphous $Zn(OH)_2$; hence this reaction should not proceed by itself. The same is the case for ZnO ($\Delta G = +3.3$ kcal). Consequently all alkali zincates are decomposed by water (R549).

The free energy of formation of the tetrahydroxo complex can be calculated with the help of the equilibrium constant of the reaction in Equation 7:

$$Zn^{2+} + 4OH^- = Zn(OH)_4^{2-} \tag{7}$$

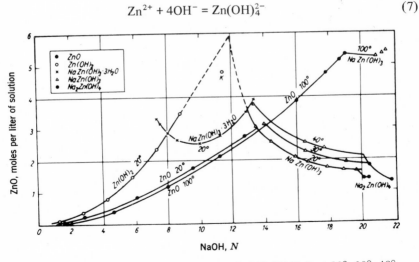

**Figure 1** Solubility isotherm for the system $ZnO\text{-}NaOH\text{-}H_2O$ at 20°, 30°, 40°, and 100°.

TABLE 2.    TETRAHYDROXOZINCATE ION, FREE ENERGY OF FORMATION

| pD | $\Delta G_R$ | $\Delta G_B$ | $E_0{}'$ | Author |
|---|---|---|---|---|
| −16.89 | −23.0 | −208.6 | 1.265 | M. v. Stackelberg and H. v. Freyhold (R574) |
| −15.45 | −21.1 | −206.7 | 1.225 | Latimer (R388) H. G. Dietrich and J. Johnston (R147) |
| −15.15 | −20.6 | −206.2 | 1.213 | Dirkse (R150) |

Several values of this equilibrium constant are available (Table 2). The $\Delta G_R$ of the reaction (Equation 7) is connected with the corresponding standard potential $E_0$ for Equation 8:

$$Zn + 4OH^- = Zn(OH)_4^{2-} + 2e^- \tag{8}$$

From this the solubility product of the modification which is in contact with the zincate solution, be it in the form of passive film or as surface layer, can be calculated (R335). This means that the potential of the zinc electrode depends not on the composition of the zincate solution but on the hydroxide or oxide formed on the surface of the electrode.

The transformation of the unstable, initially formed modification into the stable final product takes place via the zincate solution (R153). The velocity of the transformations taking place in the system $NaOH–H_2O–ZnO$ can be seen in Figure 2. The lower the $OH^-$ concentration, the longer is the stabilization time.

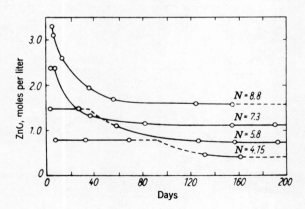

**Figure 2** Change in ZnO concentration with time for saturated solutions in various normalities of NaOH during the transformation to ZnO.

Among the values of the free energy for the $Zn(OH)_4^{2-}$ complex given in Table 2, the value for the least soluble, stable form should be $\Delta G = -208.6$.

The data obtained from solubility determinations and electrochemical measurements do not enable us to make any assertions about the "ionic weight," that is, the weight of the kinetic unit in the zincate solution. The dialysis and diffusion coefficients give some information on this. A zincate solution (in $4N$ and $6N$ KOH) has a dialysis coefficient of $\lambda = 0.258$ compared to $\lambda = 0.387$ for a chromate solution, measured under similar conditions (ionic weight (R66), $Zn(OH)_4^{2-} = 133.0$. Also there (R575) was determined the coefficient of diffusion, $D$, of a zincate solution in excess NaOH as conducting salt. This value may be compared with a $Fe(CN)_6^{3-}$ (ferricyanide) solution (ionic weight = 212.0) with added KCl to increase conductivity; for $1M$ concentration of the conducting salt

zincate (corr) $D = 6.5 \cdot 10^{-6}$
$Fe(CN)_6^{3-}$ $D = 7.63 \cdot 10^{-6}$
zincate $D = 6.86 \cdot 10^{-6}$ (5% KOH; 25°; Ref. 440)

With the simplifying assumption that these coefficients are inversely proportional to the square root of the molecular weight, we obtain from the dialytic coefficient the value 260 and from the diffusion coefficient the value 295. Comparing this with the ionic weight of 133 for $Zn(OH)_4^{2-}$, this value can be explained by a double nuclear ion (R66) $Zn_2(OH)_8^{4-}$. A complex of this type with more than one nucleus is not in contradiction with Dirkse's (R150) results. On the other hand, a strong hydration is also possible.

To summarize, a zincate solution in the NaOH-ZnO system can be described by the reaction:

Soluble phase: $Zn(OH)_4^{2-} \rightleftharpoons Zn(OH)_3^-$

| Solid phase | $Na_2[Zn(OH)_4]$ | $Na[Zn(OH)_3]$ | $Na[Zn(OH)_3] \cdot 3H_2O$ | $\epsilon\text{-}Zn(OH)_2$ | ZnO |
|---|---|---|---|---|---|
| Stable | $>19.5N$ | $19.5\text{-}13.5N$ | | | $13.5\text{-}0N$ |
| Metastable | — | $13.5\text{-}13N$ | $13\text{-}8N$ | $8\text{-}0N$ | — |

According to these results the composition of the soluble complex is not directly related to the solid phase. The stable final equilibrium is attained only after months, but the metastable composition is reached comparatively quickly.

In the system with KOH, besides ZnO only solid phases of the monovalent type have been observed; the solution contains probably the tetrahydrozincate ion.

The potential of the zinc electrode in alkaline medium is given by

$$E = E_0' + 0.059 \, pOH, \tag{9}$$

the $E_0$ values ranging from $-1.221$ to $-1.260$ V, depending upon the modification.

The more positive values $(-1.221$ V$)$ correspond to the unstable hydroxides which are formed on the zinc electrodes in freshly prepared solutions; in aged solutions, especially those with ZnO as solid phase, the more negative values are obtained. The different complex zinc ions present in the solution have no part in determining the electrode potential. The latter is only a function of the solid phase present on the electrode surface.

In the case of the temperature coefficients it is possible to differentiate only between $\epsilon$-Zn(OH)$_2$ and ZnO.

$$\left(\frac{dE_0}{dT}\right)_{Zn(OH)_2} = -1.00 \text{ mV/degree} \qquad \left(\frac{dE_0}{dT}\right)_{ZnO} = -1.16 \text{ mV/degree}$$

The values of the unstable hydroxides are not known. However, the entropy values for the different hydroxides ought not to be very different from each other, so that the value given for $\epsilon$-Zn(OH)$_2$ should be valid for the other hydroxides also.

## THE SILVER/SILVER OXIDE ELECTRODE

The characteristics of the silver/silver oxide electrode can be represented electrochemically by Equations 10 and 11 and related data in Table 3.

$$2Ag + H_2O = Ag_2O + H_2 ; E_0 = +0.345 \text{ V}; \frac{dE_0}{dT} = -1.337 \text{ mV/deg} \qquad (10)$$

$$Ag_2O + H_2O = 2AgO + H_2 ; E_0 = +0.607 \text{ V}; \frac{dE_0}{dT} = -1.117 \text{ mV/deg} \qquad (11)$$

$E_0$ and $dE_0/dT$ are the standard potentials and the isothermal temperature coefficients (R127).

TABLE 3.   SUMMARY[a] OF THERMODYNAMIC DATA FOR SILVER AT 25°

| Substance Formula | State of Aggregation | $\Delta H$ kcal/mol | $\Delta G$ kcal/mol | $\Delta S$ kcal/mol |
|---|---|---|---|---|
| Ag | c | 0 | 0 | 10.206 |
| Ag$^+$ | aq | 25.31 | 18.43 | 17.67 |
| Ag$_2$O | c | $-7.306$ | $-2.586$ | 29.09 |
| AgOH | c | | $-22.0$ | |
| AgO | c | $-2.769$ | 3.463 | 13.81 |
| AgO$^-$ | aq | | $-5.490$ | |

[a] Standard values of formation from the elements from Latimer (R388) except for AgOH which is from (R499) and $S$ value for AgO which is from (R56).

The behavior of the silver electrode may be described roughly. Charging and discharging reactions will take place at two distinct potential steps. In the following the process referring to the standard potential $E_0 = 0.345$ V* will be called the lower step and the other process characterized by $E_0 = 0.607$ V the upper step.

The reaction of Equation 10 represents an electrode of the second kind. The determining factor for the potential is the solubility product of the $Ag_2O$, which can be expressed in the following logarithmic form:

$$pL = pAg^+ + pOH^- - \tfrac{1}{2}pH_2O = 7.72 \qquad (12)$$

Introducing this expression into the following equation,

$$E = E_0(Ag/Ag^+) - 0.059pAg^+; E_0(Ag/Ag^+) = +0.799 \text{ V} \qquad (13)$$

one obtains:

$$E = 0.345 + 0.059pOH - 0.030pH_2O \qquad (14)$$

($E$ refers to the standard hydrogen electrode at pH $= 0$). For sake of simplicity the activity of water will be assumed to be 1.

Kinetic considerations do not fall into the realm of thermodynamics, and the subject as related to the silver electrode in alkaline solution is adequately covered in this symposium. Nevertheless it is well to recall the two different approaches, namely, the solid-state mechanism and the solution mechanism. In the former one, the oxide film grows by diffusion of $Ag^+$ ions to the surface or by the diffusion of $O^{2-}$ ions into the metal. The reaction path via the liquid phase (R493) requires $Ag^+$ ion migration from the metal surface in the form of a hydrated ion moving into the solution under an appropriate flow of electrons to the grid. As soon as the solution is saturated, a normal chemical precipitation of $Ag_2O$ will occur on the surface of the electrode.

It must be noted that part of the silver will stay in solution in the form of the argentous ion. For the formation of the $Ag(OH)_2^-$ ion from $Ag^+$ and $OH^-$ ions (equilibrium constant, $pK = 4.0$, Ref. 47), the following relation holds:

$$pAg(OH)_2^- = pAg^+ - 4.0 + 2pOH = 3.72 + pOH \qquad (15)$$

In highly concentrated alkali solutions, a multinuclear complex of the form $Ag_3O(OH)_2^-$ has been reported (R492), but this interpretation was not confirmed by other workers (R47).

---

*Editor's note: Some authors prefer to use the Hamer and Craig (R268) value of 0.342 V; however, a recent compilation (R126-7) has retained the 0.345 V value.

The reaction of Equation 11 is correlated to the potential of the upper step. However, some experimental data are not in agreement with this; for instance, an electrode which uses chemically precipitated AgO is discharged at the potential corresponding to the lower step. Only after few cycles can an upper step be observed on discharging. It seems that two reactions are probable.

On discharging an unaged, continuously cycled electrode with small or medium currents, the maximum capacity obtained at the potential of the upper step is equal to the excess of oxygen beyond the formula $AgO_{0.5}$. The usual oxidation state of such an electrode is $Ag_{0.8-0.9}$, the capacity corresponding to 30% to 40% of the calculated or 37.5 to 44% of the effective capacity can be taken out on the upper level.

## THE THERMODYNAMICS OF THE ZINC-SILVER OXIDE CELL

The fundamental thermodynamic data of both the silver-silver oxide electrode and the zinc-zinc hydroxide electrode are given in the previous sections. By combination of these data we obtain the thermodynamic characteristics of the zinc-silver oxide cell. The silver electrode displays its two potential plateaus which have been referred to as the upper and the lower potential step. The potential of the zinc electrode is determined by the type of the oxidation product that is being deposited on the electrode surface. The most stable solid phase is the so-called inactive form of zinc oxide followed by the stable modification of $\epsilon$-$Zn(OH)_2$, while amorphous $Zn(OH)_2$ is the most unstable, hence the most soluble compound. Depending on which of these solid phases is present, the standard potential may vary between the following values (Table 4).

TABLE 4.   STANDARD POTENTIALS[a] OF THE SILVER-ZINC CELL AT 25°

| Form of Zinc | $Zn(OH)_2$ Amorphous | $\epsilon$-$Zn(OH)_2$ | ZnO Inactive |
|---|---|---|---|
| Lower Ag plateau | 1.566 | 1.594 | 1.605 |
| Upper Ag plateau | 1.828 | 1.856 | 1.867 |

[a]Additional conditions: pOH = 0; $E_0$ in volts.

For the same reasons, the temperature coefficients of the standard potentials show the variations tabulated in Table 5. This range of variation caused by the different solid phases on the zinc electrode may account for the different and sometimes contradictory results obtained on thermal coefficient measurement reported in the literature. If, however, aged electrodes are used for such measurements, reproducible data are obtained which correspond closely to the stable form of the "inactive" zinc oxide. Hills (R289) has measured the lower potential step

$$E_0 = 1.604 \text{ V} \quad \text{and} \quad dE_0/dT = -0.169 \text{ mV/}°\text{C}$$

and the upper potential step

$$E_0 = 1.857 \text{ V} \quad \text{and} \quad dE_0/dT = +0.057 \text{ mV/}°\text{C}$$

which within the experimental errors correspond to the theoretical values for the inactive zinc oxide system.

TABLE 5.   TEMPERATURE COEFFICIENTS OF THE SILVER-ZINC CELL

|  | $dE_0/dT$ | $\epsilon$-$Zn(OH)_2$ | ZnO |
|---|---|---|---|
| Lower Ag plateau | mV/°C | −0.337 | −0.177 |
| Upper Ag plateau | mV/°C | −0.116 | +0.044 |

# PART II

## ZINC ELECTRODE, FUNDAMENTAL CHEMISTRY AND ELECTROCHEMISTRY

# 3.

# Chemistry of the Zinc/Zinc Oxide Electrode

## T. P. Dirkse

In the periodic classification of the elements, zinc is the first nontransition element following the first transition series of the elements. The third shell of electrons is full. Zinc has a valence of two in its compounds. Occasional references have been made to monovalent zinc (R45), but these are in connection with possible transient states in certain reaction mechanisms. No compounds containing monovalent zinc have been isolated.

Because of the filled third shell, the stereochemistry of zinc is determined only by size, electrostatic forces, and covalent bonding forces. The most common coordination numbers of zinc are 4 and 6. It is 4 in zinc oxide. It may be 6 in the zincate ion. (R150-151).

With respect to zinc-alkaline battery systems, the main items of interest are the oxides and hydroxides of the metal and the interaction of these with strongly alkaline solutions. The only compounds to be considered then are ZnO and $Zn(OH)_2$. However there is more than one form of each.

### ZINC OXIDE

Zinc oxide is a well-identified material. It can be prepared by heating zinc in air or by the thermal decomposition of zinc carbonate or other zinc salts. Normally it is a white substance which turns yellow on heating. Other colors of zinc oxide have been reported in connection with the anodization of zinc electrodes (R301). These colors result from various types of lattice defects or distortions. Some of these defects are responsible for the semiconducting properties of ZnO. ZnO crystallizes in the wurtzite structure in which each zinc is tetrahedrally bonded to four oxygens.

Although zinc oxide is a well-defined substance, its characteristics depend on the method of preparation, especially on the temperature of preparation (R207, 307). Thus, ZnO heated at $1000°$ has a free energy of one kcal less than that of ZnO prepared at room temperature (R207). The difference in enthalpy is about the same. Small differences have also been found between a ZnO prepared by heating at 940° and one prepared by heating at 600° (R520, 229). Some differences in solubility have also been reported. A ZnO prepared by heating $ZnCO_3$ below 400° dissolves in 0.28$N$ KOH to a greater extent and at a faster rate than a ZnO prepared by heating $ZnCO_3$ above 500° (R305). Furthermore, ZnO prepared by heating zinc oxalate at 400° is more soluble than a ZnO prepared by heating $ZnCO_3$ at 400°. However, these solubility limits were determined after 90 minutes. It is possible that a longer experimental time would have eliminated some of these differences in solubility. These differences were attributed to differences in density which could be related to differences in surface area (R305).

Because of these variations, it has been suggested that there are two forms of ZnO, an active and an inactive form (R306). The zinc oxides prepared by heating at lower temperatures have been referred to as "active" zinc oxides. It has been suggested that this activity is due to a lattice-defect structure (R307). Above 550° it loses its activity, probably due to an atomic rearrangement (R306). Values for the free energy and entropy are given in Table 1 (Bode's paper; see also R403).

The solubility of ZnO in NaOH and in KOH has been measured by many investigators, but the results show discrepancies. Some have noted that the solubility of ZnO in alkaline solutions undergoes a change on standing (R367, 277, 311). This may be related to the type of oxide that was used. Details are not always given. In spite of this, there is fairly good agreement in the older reported values, Figures 3 and 4.

More recently, an extended program has been carried out to determine the solubility of ZnO in KOH solutions, using five different samples of ZnO (R171). These varied in crystal type and in particle size. The solubility values were determined in a range of KOH concentrations and over a temperature range. The solubility was followed at 25° for a period of one year. It did not change after

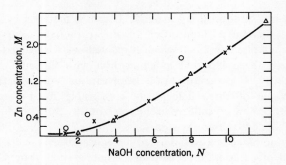

**Figure 3** Reported solubility values for ZnO in NaOH solutions: o from Reference 427; x from Reference 549, and △ from Reference 436.

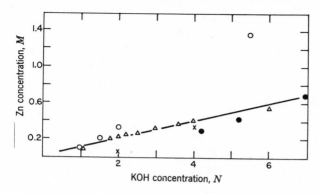

**Figure 4** Reported solubility values for ZnO in KOH solutions: ○ from Reference 427; × from Reference 436; △ from Reference 353; and ● from Reference 171.

the first few days in any of the KOH concentrations or for any of the ZnO samples. There was no decrease in solubility with time under these conditions. Nor was there any difference in the solubility of the various types of ZnO that were used. The variation in ZnO solubility with KOH concentration (Figure 5) is in agreement with that obtained in other work (Figure 4).

One of the noteworthy characteristics of the solubility of ZnO in KOH solutions is its independence of temperature (Figure 5), from $-3$ to $56°$. This temperature independence has been observed as low as $-62°$ (R26) and as high as $145°$ (R50). It indicates that the enthalpy change accompanying the dissolution of ZnO in KOH solutions is zero or very close to it. One explanation assumes that the so-called zincate solution is a colloidal mixture. However, the optical clarity of zincate solution and the reproducibility of the solubility values contradict the assumption of colloidal zincate. Assuming that the reaction accompanying dissolution is

$$ZnO_s + H_2O_{aq} + 2OH^-_{aq} = Zn(OH)_4^{2-}{}_{aq}$$

and that $\Delta H$ is zero for this reaction, then from the accepted values of enthalpy for $H_2O$ and $OH^-$, $\Delta H°_{298}$ for $Zn(OH)_4^{2-}{}_{aq}$ is $-261.4$ kcal/mol.

In a zinc-alkaline battery, rates of processes are fully as important as equilibrium values. So, too, with the dissolution of ZnO in alkaline solutions. However, here it is difficult, if not impossible, to set up realistic experimental conditions that may apply to a battery-type process or situation. The rate of dissolution will depend on conditions such as rate and type of stirring, and particle size. A little work has been reported in this area (R50, 381-382), but the results obtained are applicable only to the conditions under which these rates

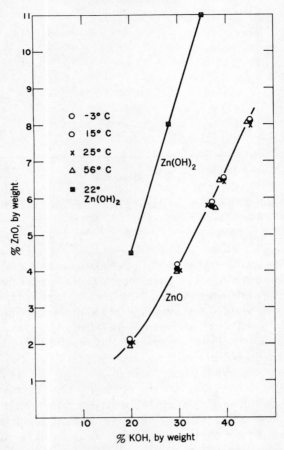

**Figure 5** Solubility of reagent grade ZnO in KOH solutions.

were measured. Using a rotating disc of ZnO (R381), it was concluded that the rate limiting process was the diffusion of zincate ions away from the interface.

## ZINC HYDROXIDE

Several methods have been suggested for preparing zinc hydroxide. One method calls for treating an aqueous solution of a zinc salt with the calculated amount of KOH or NaOH. An alternative approach uses $NH_4OH$. This is added in excess to dissolve the $Zn(OH)_2$ which is then subsequently recovered by evaporation of the solution. A third method consists of adding water to a zincate solution.

All these methods appear to be straightforward. Yet the preparation of $Zn(OH)_2$ is not an easy matter. Unless conditions are carefully controlled the addition of an alkali to an aqueous solution of zinc salt is likely to lead to the precipitation of a basic salt instead of, or in addition to, $Zn(OH)_2$. Furthermore, $Zn(OH)_2$ is not stable with respect to ZnO, and this instability increases with increasing NaOH or KOH concentration in solution (R307) and with increasing temperature. $Zn(OH)_2$ is stable at room temperature but appears to be unstable with respect to ZnO as the temperature is raised to about 35 to 45° (R307).

These factors undoubtedly account for the conflicting data reported in the literature about $Zn(OH)_2$. It is likely that in many cases a basic salt rather than the pure hydroxide was the solid phase being used. Solubility values that change with time may reflect the instability of the solid hydroxide phase.

A careful and detailed study (R204) of $Zn(OH)_2$ and its preparation yielded five crystalline forms as well as an amorphous form of $Zn(OH)_2$. These were obtained by varying the conditions under which the $Zn(OH)_2$ was precipitated from solution. The crystalline forms of $Zn(OH)_2$ have been designated as $\alpha$, $\beta$, $\gamma$, $\delta$, and $\epsilon$. The identification was apparently based entirely on X-ray diffraction patterns. About the only analytical information given—and that for only some of the forms—is the loss of weight during heating. Assuming that this loss is water, the weight loss corresponds to that theoretically expected from $Zn(OH)_2$, that is, 18.1%. The various forms of $Zn(OH)_2$ have different degrees of stability. The stable form is the $\epsilon$-modification (R207).

The amorphous $Zn(OH)_2$ (R204) was formed by carefully controlled precipitation from an aqueous zinc salt solution. The precipitate had a blue color. This blue material gave a very poor X-ray diffraction pattern. Later (R205) it was described as a substance having incompletely ordered layers in which zinc is tetrahedrally coordinated as in ZnO. Such a blue color has at times been associated with a defect type of structure. A blue material has also occasionally been noted on or at a zinc electrode which has been operating in an alkaline solution.

Not all the modifications of $Zn(OH)_2$ have been thoroughly characterized. The $\beta$ form appears to have two modifications, $\beta1$ and $\beta2$ (R206). The one form is precipitated from a neutral solution, the other from an alkaline solution. The differences are slight but real. The X-ray diffraction patterns of the two forms are similar but vary in the intensities of the lines. Both types have a layered lattice with a distance of 5.67 Å between layers.

All forms of $Zn(OH)_2$ are decomposed thermally to ZnO; the rate of this decomposition appears to depend on the crystal structure of the original hydroxide (R242). The free energies of formation of the various crystalline forms of $Zn(OH)_2$ vary from 132.1 to 132.83 kcal/mol at 25° (R147, 545).

The solubility of $Zn(OH)_2$ in NaOH and in KOH solutions has been measured several times. The results obtained are subject to some uncertainty because of the uncertain nature of the solid phase in some cases and because of the instability of $Zn(OH)_2$ in contact with alkaline solutions. As a result, the solubility has not been measured in the higher alkali concentrations. Some representative results are given on Figures 5 and 6. The results in KOH are shown in Figure 5 to provide a direct comparison with the solubility of ZnO in KOH solutions. The results in NaOH indicate that the solubility of $Zn(OH)_2$ is also temperature independent. However, others (R143) have observed a change in solubility with temperature. For example, in 6$N$ NaOH the solubility increased with temperature up to $55°$ and then decreased. In 10$N$ NaOH the solubility increases with temperature and then becomes temperature independent above $55°$. Certainly the solubility of $Zn(OH)_2$ is definitely larger than that of ZnO. Solutions of $Zn(OH)_2$ in the alkalies show a decrease of zincate content with time (R228) and the values approach those for saturated solutions of ZnO. This decrease in solubility with time has been called an ageing process (R218). It was found that the presence of $Li^+$ and silicate ions reduced the rate at which this decay occurred, the effect being dependent on the concentration of these ions and on the temperature. From solutions containing these ions the precipitate that formed was $Zn(OH)_2$ rather than ZnO.

The $\alpha$-$Zn(OH)_2$ has been observed as a corrosion product formed on zinc kept in distilled water at room temperature (R240). It has also been found on the surface of a zinc anode in alkaline solutions (R310).

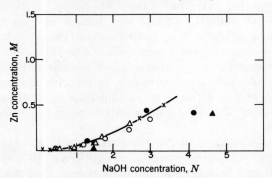

**Figure 6** Solubility of $Zn(OH)_2$ in NaOH solutions: ○, 0°, from Reference 436; x, 25°, from Reference 436; △, 35°, from Reference 436; ●, 40°, from Reference 549; and ▲, 100°, from Reference 549.

## ZINCATES

Zinc hydroxide is amphoteric and therefore its reaction with NaOH or KOH should produce zincates. Many attempts to isolate these zincates have been reported, yet unequivocal evidence for the existence of solid zincates is sparse.

There has been a difference of opinion with respect to the nature of the solute species. The titration of aqueous $ZnSO_4$ with NaOH was said to give no evidence for the existence of zincates (R67). It was also said to give evidence for an acid zincate, $HZnO_2^-$ (R288). Other work (R153, 367), however, provides rather good evidence for the presence of both $Zn(OH)_3^-$ and $Zn(OH)_4^{2-}$ in solutions which are strongly basic. As the alkalinity of the solution increases, more and more of the dissolved zincate is in the form $Zn(OH)_4^{2-}$. The formation constant at $25°$ for $Zn(OH)_3^-$ is $6 \times 10^{-4}$ while that for $Zn(OH)_4^{2-}$ is $10^{-2}$ (R152).

A compound corresponding to $Na_2ZnO_2 \cdot 4H_2O$ was isolated from concentrated NaOH solutions (R259). Directions for preparing $NaHZnO_2$ have been given (R436). However, there (R548) have been obtained precipitates corresponding to $NaZn(OH)_3$ and $Na_2Zn(OH)_4$; on studying the thermal decomposition, it was suggested that these were more likely $Zn(OH)_2 \cdot NaOH$ and $Zn(OH)_2 \cdot 2NaOH$. A similar conclusion was reached by others (R417). Later, directions (R549) were given for preparing $NaZn(OH)_3 \cdot 3H_2O$, $NaZn(OH)_3$, and $Na_2Zn(OH)_4$.

There is no conclusive evidence reported in literature for the existence of potassium zincates as a separate solid phase. Crystals corresponding to the composition of a potassium zincate appeared rather to be the compound $Zn(OH)_2 \cdot 3KOH \cdot H_2O$ (R417).

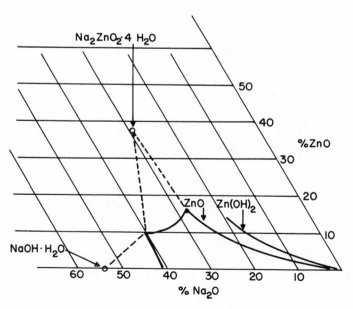

**Figure 7** Phase diagram for $ZnO-Na_2O-H_2O$ system at $30°$ (R259).

The results of all these solubility studies can best be summarized in terms of phase diagrams. The most extensively reported phase diagrams in NaOH are given in References 259, 436, 549. An attempt to summarize this information showed that comparisons are difficult because results are reported differently and density data for solutions are omitted. Figures 1 and 7 summarize these data. There is some disagreement as to numerical values and there is also disagreement as to the solid phases which are present, that is, whether and which zincates precipitate from some of these solutions.

There is much less information available in the literature on the phase diagrams in KOH solutions. What there is (R154), is summarized on Figure 8. Here the dashed lines represent regions in which no experimental data are available. The shape of the phase boundaries is different from that of the NaOH–ZnO–H$_2$O system, Figures 1 and 7. There is no evidence on Figure 8 for the existence of potassium zincates as stable solid phases. The only solid phases here were ZnO and KOH · 2H$_2$O. Only that part of the diagram in which KOH · 2H$_2$O is the solid phase shows a temperature dependence.

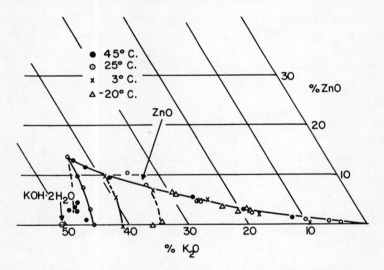

Figure 8 Phase diagram for ZnO-K$_2$O-H$_2$O system (R154).

## BATTERY ELECTROLYTES

On many occasions a white precipitate of ZnO appears in the electrolyte of a zinc-alkaline battery during discharge. Analysis of such an electrolyte shows a zincate content different from and higher than that for a solution in equilibrium with ZnO as the solid phase (R153, 219, 311). The zincate content may easily be twice that of the values shown on Figures 4 and 5. This zincate content arises as a product of the anodic charge transfer reaction. On standing, solid ZnO slowly precipitates from such solutions. The solute species in such solutions is the same as that for solutions prepared by dissolving ZnO in KOH (R153). The question has been raised as to whether this is a supersaturated solution. The slow precipitation of ZnO argues against this. The formation of crystals of the solid product does not hasten the further precipitation of the excess zincate. It has been suggested that some of the excess zincate may be in the colloidal form (R219). A small amount of this may be present but the solutions are optically clear and certainly most of the excess zincate is not in the form of a colloid. Neither seeding nor shock causes or hastens the precipitation of ZnO (R219). The ZnO is formed slowly and resembles a decomposition reaction (R153).

The results on the solubility of $Zn(OH)_2$ as shown on Figure 6 suggest that the anodic charge transfer product may be $Zn(OH)_2$, which is much more soluble than any known form of ZnO. Such solutions are unstable with respect to solid ZnO and consequently a precipitate of ZnO appears and the zincate content of the solution decreases and approaches the equilibrium value for saturated solutions of ZnO.

At present this phenomenon can be best accounted for as follows.

$$Zn + 4OH^- = Zn(OH)_4^{2-} + 2e \tag{1}$$

$$Zn(OH)_4^{2-} = ZnO + H_2O + 2OH^- \tag{2}$$

The reaction of Equation (1) is the overall electrode reaction and may proceed with $Zn(OH)_2$ as an intermediate. Its rate is governed by the current density and the rate of dissolution of $Zn(OH)_2$. The reaction of Equation 2 takes place at a rate which is independent of the current density. The amount of $Zn(OH)_4^{2-}$ in solution then, depends on the relative rates of these reactions. The rates of dissolution of $Zn(OH)_2$ and the rate of decomposition of $Zn(OH)_4^{2-}$ are unknown.

However, some work has been reported (R172) which attempts to determine the rate of the reaction of Equation (2). Figure 9 illustrates the type of experimental data obtained. A qualitative analysis of the results suggests that the rate for the reaction of Equation 2 can be represented by the equation,

$$\text{Rate} = k_1 \, [Zn(OH)_4^{2-}]^a - k_2 \, [OH^-]^b \tag{3}$$

Values for $a$ and $b$ could not be obtained. Equation 3 suggests that the cause for the decrease in rate of decomposition with time is due to two factors: the decrease in excess zincate concentration and the consequent increase of $OH^-$ ion concentration.

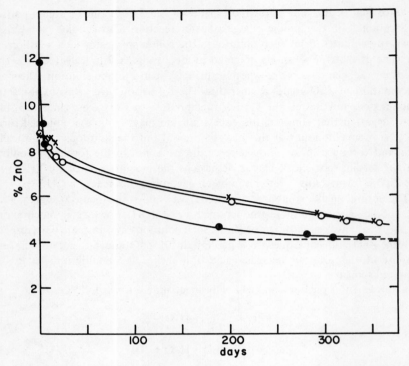

**Figure 9** Decomposition of electrolytically generated zincate ion in 30% KOH, room temperature.

The fact that the electrolytically generated zincate is soluble to a much greater extent than is ZnO and that the resultant solution decomposes very slightly is significant in understanding zinc-alkaline battery behavior. It allows for a high discharge rate at a zinc electrode. It also gives rise to the transport of zinc throughout such a cell. In a steadily cycling battery, it is likely that the electrolyte has excess zincate content most of the time while it is discharging. Much of this will be subject to electrochemical reduction on charge, but some of it will also undergo chemical decomposition to solid ZnO. The details of these phenomena are, as yet, not understood. Further work is needed to determine the rates of the processes involved and the possibility of stabilizing the electrolyte.

# 4.

# A Spectroscopic Investigation of the Zinc–Hydroxy System

## John F. Jackovitz and Alois Langer

The nature of the species formed when zinc oxide is dissolved in excess potassium hydroxide has been subject to considerable investigation for many years. The preceding paper (R178) emphasized disagreements on the structure of zincates and the many attempts to isolate solid zincates. The comprehensive studies of this system (R152) provide rather good evidence for the presence of both $Zn(OH)_3^-$ and $Zn(OH)_4^{2-}$ in solutions which are strongly alkaline. As the basicity of the solution increases, most of the dissolved zincate is in the form $Zn(OH)_4^{2-}$.

A spectroscopic investigation of the $ZnO-KOH-H_2O$ system (R224) combining infrared and Raman determinations, established the zincate species to have tetrahedral symmetry in agreement with the tetracoordinated species proposed by Dirkse. Unfortunately, high background noise did not allow a precise determination of frequencies or reliable polarization measurements.

Recently, we have studied zincate solutions using a laser-source Raman spectrometer and a modified infrared attenuated total reflectance (ATR) solution accessory. Our main objective was to locate, precisely, all of the Raman lines which are present in the aqueous zincate solutions over the 10 to 40% KOH concentration range.

In addition, a small silver-zinc electrochemical cell was constructed in a polished quartz cell, and spectra of the electrolyte were recorded during both charge and discharge. Infrared spectra, using a metal-coated reflection element, were obtained on small aliquots drawn from the cell at full charge and discharge.

It is a well-known phenomenon that the formation of oversaturated solutions of Zn(II) is favored by anodic action. If one places a metallic zinc electrode in a saturated ZnO–KOH solution, it is found that very significant amounts of Zn(II) can be anodically dissolved without precipitation. It was our aim to study this "supersaturation" effect and to identify the species involved.

## EXPERIMENTAL

Solutions were prepared by shaking excess zinc oxide (Fisher Certified) in the appropriate carbonate-free potassium hydroxide solution for several weeks at room temperature. The approximate zinc content of each solution was obtained from the data reported by Langer (R386). The upper limits of concentration 40% KOH, 6.3% ZnO) were fixed by the solubility of zinc oxide.

The small silver-zinc cell used for Raman studies was constructed using a quartz jacket, polished optically flat on all four sides. The half cell consisted of three flat zinc sheets (Fisher Certified) each with 24 cm$^2$ total area, counting both sides. Two large capacity, sintered silver electrodes of the same dimension were wrapped in cellophane and placed in an alternate parallel arrangement with the zinc sheets. About 30 ml of electrolyte completely covered the electrodes. A Teflon cap containing a tiny hydrogen escape vent covered the cell top. Charging of the cell was accomplished at a rate of 20 mA for 50 hours. Before discharging the cell, the electrolyte was drained from the cell and replaced with fresh saturated ZnO−KOH solution. Since the cell contained nonretentive separators, an insignificant amount of electrolyte remained after drainage.

Raman spectra were observed using an argon-ion laser for excitation. The laser was of the graphite capillary type and delivered about one watt in the 4880 Å line. The spectra were dispersed by a double tandem monochromator (Spex 1400). Because of the severe background scattering from the viscous solutions, a ¼ meter grating monochromator was employed, with a fixed band pass, as a prefilter to the double monochromator to further discriminate against the effect of scattered laser light. The signal was photoelectrically detected using a cooled S-11 Photomultiplier (EMI 9502S) having a dark count of about 1 sec$^{-1}$. The laser beam was chopped within the optical cavity at 90 Hz and the signal from the photomultiplier was amplified and synchronously detected. Further discrimination against elastic scattering from the sample was attained by using a long path length between the laser and sample position. In addition, the ZnO−KOH solutions were centrifuged and passed through an ultrafine filter before use.

ATR spectra were recorded using a Beckman IR-12 grating spectrophotometer. The reflection accessory was a Barnes Model 131 equipped with a 2 mm thick KRS-5 crystal. The three-inch length of the crystal allows a maximum of 25 reflections at a constant attenuation angle of 45°. The larger face of the trapezoid-shaped crystal allows 13 reflections and these were found sufficient for the present study. Since the ZnO−KOH solutions readily attack KRS-5, the exposed face of the crystal was coated carefully with a 40-Å-thick palladium film. This particular thickness of palladium produced distortion-free spectra without requiring a change in the attenuation angle and served to protect the crystal face from immediate attack by the electrolyte. To compensate for sluggish instrument response, the slit openings were adjusted to twice their normal widths. Spectra were obtained as quickly as possible after filling the cell. Successive runs indicated that the spectra were reproducible and reliable at least down to 350 cm$^{-1}$.

## RESULTS AND DISCUSSION

SOLUTION SPECTRUM

The observed Raman spectra for two solutions of zinc oxide in potassium hydroxide are shown in Figure 10. Instrument settings were maintained constant so that observed intensities show linearly with concentration. The nearly constant baseline and favorable signal-to-noise ratio for these scans allow positive identification of four bands. The inconsistency of the baseline in the region 250-330 cm$^{-1}$ relative to that between 330-550 cm$^{-1}$ is caused by strong background scattering. Transfer to a constant baseline reveals that the two low frequency bands are nearly equal in intensity with the $\Delta$430 cm$^{-1}$ band.

A qualitative measurement of polarization was accomplished by insertion of a Polaroid at the entrance slit of the Spex monochromator. Orientation of the Polaroid did not affect the intensity ratios of the three lowest frequency bands but did have a drastic effect on the intense band at $\Delta$480 cm$^{-1}$. The ratio $I\perp : I\parallel$ is nearly 1 : 3 for this band, identifying it as belonging to a totally symmetric optical mode. Extensive measurements on the three low-frequency

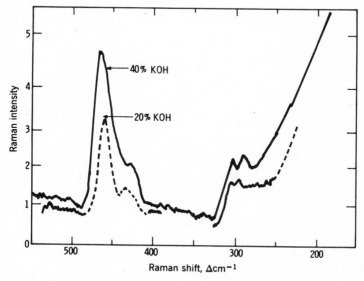

**Figure 10** Raman spectra of potassium hydroxide solutions saturated with chemically prepared zinc oxide at 26°C.

bands indicate strongly that all three are depolarized in agreement with the suggestion of Fordyce (R224).

A correlation between the areas under the $\Delta 480 \text{ m}^{-1}$ line for the two solutions in Figure 10 and their zinc content strongly support the existence of only one zinc-hydroxy species over this concentration range. The area ratios are 1 : 3.5 whereas the zinc concentration ratio established from analytical data is 1 : 3.6. Also, the favorable background for these scans would certainly allow significant concentrations of other zinc-hydroxy geometries to be readily observed.

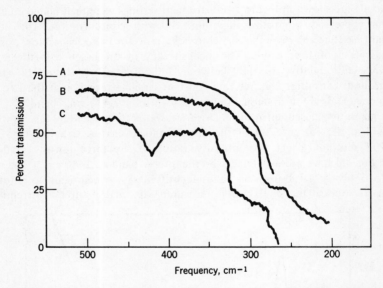

**Figure 11** Infrared ATR spectra of potassium hydroxide solution saturated with chemically prepared ZnO. (A) KRS-5; (B) 40% KOH; (C) saturated with ZnO.

To exclude any configurations containing a strict center or inversion center, the infrared spectrum must show mutual inclusion with the Raman. Perhaps the most reliable technique for studying difficult systems in the infrared is ATR. In Figure 11, the ATR spectra of both aqueous potassium hydroxide and the ZnO–KOH system are shown. An obvious difference occurs ca. 435 cm$^{-1}$, while an absorption increase near 330 cm$^{-1}$ indicates the possibility of another band.

An attempt was also made to obtain infrared data by transmission through a thin solution film. The technique for sample preparation involved sealing a few drops of the ZnO–KOH solution between two high density polyethylene films (Marlex, 0.02 mm thick). The film was flexible enough so that the optical path length could be adjusted by simply tightening the sample holders until reasonable transmission was obtained in a region of nonabsorption. The spectra obtained using this technique tended to be more diffuse than those obtained using the ATR method but nevertheless showed a definite absorption centered ca. 435 cm$^{-1}$, almost coincident with the Raman band at 430 cm$^{-1}$.

The appearance of one and perhaps two coincidences between the Raman and infrared spectra rules out the possibility of an octahedral (O ) or square-planar (D$_{4n}$) arrangement of hydroxy groups around the zinc ion. Also, the appearance of only four bands eliminates the possibility of low symmetry structures such as bridged or polymeric series.

The observed data best fit a tetrahedral zinc hydroxy arrangement. Assignments on the basis of the T$_d$ point group are given in Table 6. The intense , polarized band at $\Delta 480$ cm$^{-1}$ is undoubtedly $\nu_1(A_1)$, the totally symmetric stretching mode. Since this vibration does not involve a dipole change, it is absent in the infrared. The Raman and infrared active mode at 430 cm$^{-1}$ must necessarily be $\nu_3(F_2)$, the asymmetrical stretch. The remaining two frequencies at 313 and 286 cm$^{-1}$ are assigned to deformation modes according to a scheme for other $MO_4^{2-}$ anions, where more clearly defined spectral lines are reported (R454).

TABLE 6.    OBSERVED VIBRATIONAL FUNDAMENTALS FOR Zn(OH)$_4^{2-}$
BASED ON POINT GROUP T$_d$ (NEGLECTING THE HYDROGEN ATOMS)

| Raman Shift ($\Delta$ cm$^{-1}$) | Infrared ATR (cm$^{-1}$) | Infrared Transmission (cm$^{-1}$) | Assignment |
|---|---|---|---|
| 480 s, p |  |  | $\nu_1(A_1)$ |
| 450, w | 435 | 435 | $\nu_3(F_2)$ |
| 313, w | 330? |  | $\nu_4(F_2)$ |
| 286, w |  |  | $\nu_2(E)$ |

ANODIC DISSOLUTION OF ZN(II)

In Figure 12, the intensity of the $\Delta 480$ cm$^{-1}$ Raman band of a ZnO–40% KOH solution is shown versus that of the same solution from which zinc was deposited on a zinc plate by charging the cell. Since instrument settings remained fixed, the relative intensities of this band indicate the approximate fraction of zinc removed from solution. The overall spectrum remains the same, indicating that the remaining zinc in solution is present as Zn(OH)$_4^{2-}$.

At this point, the partially zinc-depleted electrolyte was removed and replaced with a fresh saturated ZnO–40% KOH solution. Thus, consequent discharge of the zinc electrode should produce the expected oversaturation effect and permit identification of any new species. It must be noted that, after complete discharge of the cell, the electrolyte remained clear except for minute amounts

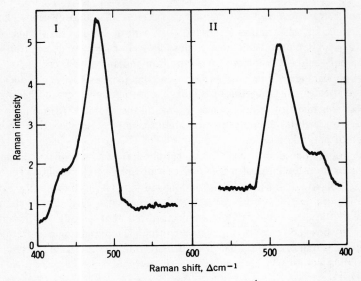

**Figure 12** Raman intensity of the $\Delta 480$ cm$^{-1}$ band of: (I) the chemically prepared ZnO-40% KOH electrolyte of a Ag-Zn cell; (II) the electrolyte after charging the cell at 20 mA for 50 hours.

of gray flakes from the zinc electrode. It was apparent under magnification that a pronounced refractive index change occurred around the zinc electrodes during discharge.

Raman spectra of the cell electrolyte were recorded both at the calculated half-discharge point and at full discharge. During discharge, the laser beam was focused on a volume immediately under the electrodes to allow identification of species in the concentration gradient. The final spectrum shown in Figure 13 hopefully represents the equilibrium electrolyte composition of the discharged cell.

The intensities of the $\Delta 480$ cm$^{-1}$ and $\Delta 430$ cm$^{-1}$ bands of the discharged cell relative to those of a standard ZnO$-$40% KOH solution are shown in Figure 13. The peak heights show only a minor increase but this certainly must be real, considering the stable baseline and favorable signal-to-noise ratio. Integration of the total area under the $\Delta 480$ cm$^{-1}$ band for both solutions shows that the concentration of Zn(OH)$_4^{2-}$ increased about 12% relative to the original electrolyte. Analytical data on these solutions indicate a zinc concentration of 77 g/liter for the standard ZnO$-$40% KOH electrolyte and 127 g/liter for the electrolyte after discharge, an increase of about 70%

Careful repetition of these experiments yielded excellent reproducibility and substantiated our conclusion that only a small fraction, about one-sixth, of the anodically dissolved zinc was converted to Zn(OH)$_4^{2-}$. It was also clear from

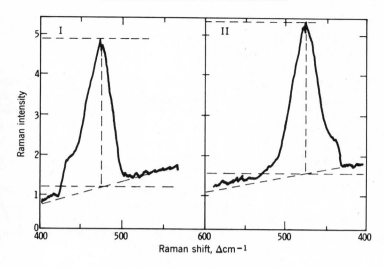

**Figure 13** Raman intensity of the $\Delta 480$ cm$^{-1}$ band of: (I) the saturated ZnO-40% KOH electrolyte contained in a Ag-Zn cell and having 77 g Zn per liter; (II) the electrolyte after discharge of the cell, now containing 137 g Zn per liter.

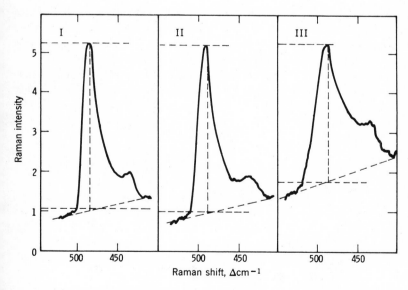

**Figure 14** Raman intensity of the $\Delta 480$ cm$^{-1}$ band of: (I) saturated ZnO-40% electrolyte; (II) saturated ZnO-40% KOH with 20 g per liter additional Zn as ZnI$_2$; (III) 74 g per liter Zn as ZnI, in 40% KOH.

these data that the $Zn(OH)_4^{2-}$ initially present in the electrolyte was unchanged by the anodic discharge process. A complete scan of the electrolyte from $100-600$ cm$^{-1}$ showed no bands in addition to the four observed for $Zn(OH)_4^{2-}$.

At a concentration of 40% potassium hydroxide, the solubility of zinc as zinc oxide is 77 g/liter. To this solution was added an additional 20 g/liter of zinc as zinc iodide. Dissolution was very slow, but the solution remained clear after two weeks. On cooling in an ice bath, crystals separated from solution and were identified as potassium iodide. As shown in Figure 14, the intensity of the $\Delta 480$ cm$^{-1}$ band does not increase on further addition of zinc to a saturated ZnO$-$40% KOH solution.

Solutions prepared by dissolving only ZnI$_2$ in 40% KOH show similar effects. As shown by curve III of Figure 14, a solution containing 74 g of zinc/liter (as ZnI$_2$) has approximately the same concentration of $Zn(OH)_4^{2-}$ as has the stock ZnO$-$40% KOH solution. The apparent broadness of the band is most likely associated with the high iodide content of the solution.

## CONCLUSIONS

The spectral data indicate that addition of large quantities of zinc, either anodically or as ZnI$_2$, to an already saturated ZnO$-$KOH solution does not appreciably increase the concentration of $Zn(OH)_4^{2-}$ already present. Careful scanning of all solutions showed no new spectral bands corresponding to other zinc-hydroxy configurations. Also, no significant band contour changes were noted in any of the Raman spectra of the "oversaturated" solutions. It is possible that the broad intense Raman bands between $400-500$ cm$^{-1}$ may overlap bands due to other species but this seems tenuous considering the above evidence.

# 5.

# Kinetics of the Zinc Electrode

## N. A. Hampson

Studies of exchange reactions (the processes of deposition and dissolution) at amalgam electrodes have provided the physical chemist with the vast majority of experimental data (R585) on which modern electrode kinetic theory rests (R140, 611). Progress in the study of exchange reactions at solid metal electrodes has been made only during the past 20 years. The need for more stringent electrolyte purification than is required for amalgams, where the use of the dropping mercury electrode reduces the purification necessary, has been made a relatively easy matter by the use of charcoal, so that electrolytes of ultrapurity are now available experimentally (R31, 41, 273).

Although electrodes of very high purity can be produced, the study of solid electrodes offers difficulties that are not encountered with liquid amalgams because the latter provide smooth, structureless surfaces of readily determinable area. The solid metal surface area may differ from its superficial area because of a multiplicity of such factors as crystal planes, steps, or dislocations, which can contribute to surface topography. Moreover each exposed crystal plane is, potentially, a different surface; for example, close-packed planes would be expected to be less reactive than are planes in which atoms are farther apart. Thus an apparently smooth metal surface could possess a range of reactivities at equilibrium. The process of inclusion or release of atoms into or from the ordered lattice is believed (R71, 226) to take place only at a limited number of sites (kink sites); but an amalgam surface is able to incorporate an atom at any point because all points are equivalent. Single-crystal solid metal surfaces would provide fairly homogeneous electrodes but handling and preparing perfect single-crystal surfaces requires considerable skill and it is doubtful if even a single crystal would retain its homogeneity during exchange.

## METAL/ELECTROLYTE SYSTEM

A metal electrode in contact with an electrolyte forms a double layer (R140) consisting of two charged layers of opposite sign but equal magnitude, separated by a distance of a few angstroms. If a double layer can be set up by transport of

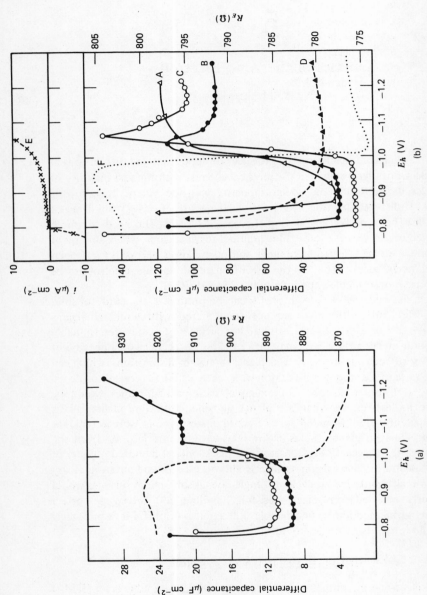

**Figure 15** Differential capacitance curves for single crystal zinc electrodes in aqueous KCl electrolytes, 25°; 1 kc/sec; chemical etch (10% HClO₄); superficial area, 11.3 × 10⁻² cm². ○, 0.1N KCl; ●, 0.01N KCl; dashed line = $R_E$ for 0.01N KCl. (b) Differential capacitance curves for polycrystalline zinc electrodes in aqueous KCl electrolytes, 25°; 1 kc/sec; chemical etch (10% HClO₄); superficial area, 9.1 × 10⁻² cm². (A) 1.0N KCl; (B) 0.1N KCl; (C) 0.01N KCl; (D) 1.0N KCl, electrode polarized at potentials more anodic than –0.8 V; (E) faradaic current curve corresponding to (C); (F) $R_E$ for electrode corresponding to (C).

38

charge through an external circuit without charge crossing the interphase, the electrode is called "ideally polarized." An example is a liquid mercury electrode in contact with an aqueous electrolyte free of a potential-determining ion. Electrically the interphase behaves as a condenser of capacitance $C_L$, the double layer capacitance, and measurements of this quantity as a function of potential have been of considerable use in investigating the structure of the interphase. The solid metal electrode/electrolyte interphase rarely constitutes an ideal polarized electrode even in a relatively small potential region.

## THE DOUBLE-LAYER CAPACITANCE OF THE ZINC ELECTRODE IN AQUEOUS SOLUTION

Electrometric and impedance measurements (R68) at both single crystal (0001) plane and polycrystalline electrodes in a number of aqueous electrolytes show the zinc electrode not to be ideally polarizable (Figure 15b upper curve), with the experimentally polarizable region being larger for single crystal ($-0.8$ V to $-1.2$ V) than for polycrystalline electrodes ($-0.8$ V to $-0.98$ V; compare Figures 15(a) and (b)). The close-packed (0001) plane is less effective for hydrogen-ion discharge, visible hydrogen evolution occurring from the (0001) plane at $-1.47$ V but at only $-1.3$ V from the polycrystalline surface. Increase in the resistive component, $R_E$, of the electrode impedance as the electrode potential is swept anodically is taken as evidence of an anionic (probably OH$^-$) interaction with the electrode surface. Direct experiments (R68) indicate strong adsorption of OH$^-$ ions at zinc electrodes, while films are produced by anodic polarization beyond $-0.8$ V.

## THE EXCHANGE REACTION

At electrode/electrolyte systems in equilibrium the electrochemical potential of each charged component in one phase is equal to that in the other. Altering the phase boundary potential changes the concentration of ion in one phase until equilibrium is again reached with a new concentration of charged component at a new value of potential. Alternatively, a change in concentration in one phase results in a change of phase boundary potential.

The equilibrium between a reduced form $R$ (Zn) and an oxidized form $O$ (Zn(II)) involving Z electrons (2 in the case of Zn$^{2+}$/Zn)

$$O + Z\,e = R \qquad (1)$$

requires that equal current $i_0$ (the exchange current) flows for both anodic and cathodic reactions. When the equilibrium is disturbed by the application of an overpotential, $\eta$, net current flows. The relationship between the faradaic current, $i$, and the charge-transfer overpotential (activation overpotential), $\eta_D$, is (R140)

$$i = \underset{\rightarrow}{i} - \underset{\leftarrow}{i} = i_0 \left( \exp \frac{-\alpha ZF\eta_D}{RT} - \exp \frac{(1 - \alpha)ZF\eta_D}{RT} \right) \qquad (2)$$

The constants, $i_0$ and $\alpha$ can be obtained from $\eta_D - i$ data (or impedance data). Also, $\alpha$ can be obtained from the variation of exchange current with concentration of the electroactive species, since combination of Equation 2 with the Nernst equation for equilibrium conditions gives (R140)

$$i_0 = ZFk^\circ C_R^\alpha C_0^{1-\alpha} \qquad (3)$$

Thus for the charge-transfer process (Equation 1), a knowledge of $k^\circ$ and $\alpha$ completely define the kinetics of the electrode process.

## THE Zn(II)/Zn(Hg) EXCHANGE REACTION

In the case of the simple charge-transfer reaction at an amalgam electrode

$$Zn^{2+} + 2e = Zn(Hg) \qquad (4)$$

Most of the reliable data up to 1963 have been tabulated (R585). They indicate that $k^\circ$ and $\alpha$ exhibit some dependence on the nature and concentration of the supporting electrolyte, but generally $k^\circ$ is $\sim 5 \times 10^{-3}$ cm sec$^{-1}$ and $\alpha$ is $\sim 0.25$. A detailed study (R308) showed $k^\circ$ to increase sharply with increasing ionic strength below $\mu = 2.0$ with $\alpha$ being dependent on ionic strength. Some evidence was provided for a consecutive mechanism:

$$Zn^{2+} + e = Zn^+ \qquad (5)$$

followed by

$$Zn^+ + e = Zn(Hg) \qquad (6)$$

It is noteworthy that the enthalpy of activation (10 kcal mol$^{-1}$) is independent of ionic strength.

## THE Zn(II)/Zn(Hg) EXCHANGE REACTION
## IN ALKALINE SOLUTION

In alkaline solution the exchange reaction is more complicated because of the formation of a complex. Overall the reaction is the following one, written conveniently as the oxidation or discharge of the zinc negative in the remainder of this paper for the exchange reaction:

$$Zn(Hg) + 4OH^- = Zn(OH)_4^{2-} + 2e \qquad (7)$$

Studies of the effect of concentration of electroactive species on the exchange current (R238), keeping the appropriate intensity factors constant, gave the equations:

$$\left( \frac{\partial \log i_0}{\partial \log C_{Zn(Hg)}} \right) = 0.5 \tag{8}$$

$$\left( \frac{\partial \log i_0}{\partial \log C_{Zn^{2+}(aq)}} \right) = 0.5 \tag{9}$$

$$\left( \frac{\partial \log i_0}{\partial \log C_{OH^-}} \right) = 0 \tag{10}$$

This was interpreted as indicating that the overall reaction occurred in two stages, involving charge transfer,

$$Zn(Hg) + 2OH^- = Zn(OH)_2 + 2e \tag{11}$$

followed by complex formation,

$$Zn(OH)_2 + 2OH^- = Zn(OH)_4^{2-} \tag{12}$$

For reaction 11 we have, using Equation 3,

$$i_0 = 2Fk^0 C_{Zn(Hg)}^{\alpha} \cdot C_{OH^-}^{2\alpha} \cdot C_{Zn(OH)_2}^{1-\alpha} \tag{13}$$

and for the equilibrium 12,

$$\beta = \frac{C_{Zn(OH)_4^{2-}}}{C_{OH^-}^2 \cdot C_{Zn(OH)_2}} \tag{14}$$

where $\beta$ is a constant. Since most of the Zn(II) is present as the complex $Zn(OH)_4^{2-}$

$$C_{Zn(OH)_2} = \frac{C_{Zn^{2+}(aq)}}{\beta C_{OH^-}^2} \tag{15}$$

Combining Equations 13 and 15 gives

$$i_0 = \frac{2Fk^0}{\beta^{(1-\alpha)}} \cdot C_{Zn(Hg)}^{\alpha} \cdot C_{OH^-}^{2(2\alpha-1)} \cdot C_{Zn(aq)}^{(1-\alpha)} \tag{16}$$

With $\alpha = 0.5$ the experimentally observed relationship, Equations 8, 9, and 10 result. The value of $\alpha$, 0.5, has recently been confirmed by radiochemical techniques (R503) and the faradaic impedance method (R200). The value of the experimentally determinable constant, $2Fk^\circ/\beta^{(1-\alpha)}$, is reported to be $\sim 1 \times 10^{-3}$ cm sec$^{-1}$ (R200) at 22$^\circ$ for electrolytes of $3M$ ionic strength (added NaClO$_4$). Equilibrium constants (R556) measured at $3M$ ionic strength (added NaClO$_4$) yield a calculated value of $\beta$ of $10^{15.38}$. The value of the rate constant $k^0$ is therefore $\sim 5 \times 10^4$ mol cm$^{-2}$ sec$^{-1}$. This magnitude expresses the ease with which the neutral Zn(OH)$_2$ species crosses the double layer.

## THE Zn(II)/Zn EXCHANGE REACTION AT A
## SOLID ELECTRODE IN ALKALINE SOLUTION

Ionic exchange at a solid zinc electrode in alkaline solution is more complicated than at amalgams because of the process of lattice disruption. Investigations of the exchange reaction (R198) by the a-c faradaic impedance method in the frequency range up to 4 kHz were made at single crystal (0001 face), polycrystal, and heavily cold-worked zinc electrodes in alkaline solution. The method involves measurement of the electrode impedance as a function of a-c frequency, analysis of the data, and "matching" of the data with electrical analogues representing the various possible reaction mechanisms. Conclusions can usually be drawn where one or two processes in the reaction mechanism contribute impedance components which predominate in the electrical analogue.

The magnitude of the adatom flux $(V_g^\circ)^3$ (with $ZFV_g^\circ$ expressed in A cm$^{-2}$; $Z = 2$ for zinc adatoms using the nomenclature of the main references), which is the extent of flow of adsorbed zinc atoms on the surface, was shown to control the rate of exchange, with $2FV_g^\circ$ dependent only on the metal characteristics. Figure 16 shows typical plots of in-phase and out-of-phase components of the faradaic impedance data matched to a simple model representing diffusion in solution and a single rate controlling process (R507) using a computer technique (R198). The analytical procedure also yields a frequency independent resistance, $R_0$, and frequency independent capacitance, $C_0$, characteristic for each system. For the range of frequencies investigated, it was shown that the difference between the numerical magnitude of the in-phase and out-of-phase components,

$$R_R - \frac{1}{\omega C_R} = \Delta,$$

is characteristic of the adatom flux

$$\Delta = \frac{RT}{Z^2 F^2 V_g^0}$$

as shown in Figure 16, and is approximately constant throughout the series of

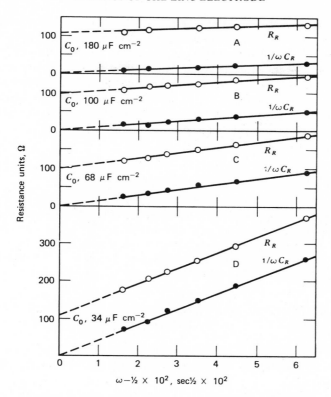

**Figure 16** Faradaic impedance curves for single crystal zinc electrodes (0001). Electrode area $2.5 \times 10^{-2}$ cm$^2$, electrolytically etched in HClO$_4$ (10% v/v); electrolyte ionic strength $3N$ with NaClO$_4$ at $0.001M$ Zn$^{2+}$; temperature 22°C; 2 hr stabilization period; (A) $3N$ OH$^-$; (B) $0.5N$ OH$^-$; (C) $0.14N$ OH$^-$; (D) $0.05N$ OH$^-$. Data from Reference 198 by permission of The Faraday Society.

hydroxide concentrations. Values of electrode impedance and therefore, $C_0$ and $V_g^\circ$, are independent of zincate concentration. The magnitude of the adatom flux at 22° (R198):

| Electrode Type | Single Crystal | Polycrystal | Heavily Cold-Worked |
|---|---|---|---|
| $2FV_g^\circ$, mA cm$^{-2}$ | 5.4 | 32.1 | 92.0 |

increases with lattice disorder and kink density as would be expected if kinks provide the essential points for lattice dissolution.

An important feature of the analysis of electrode impedance data is the necessity of ascribing a large value of capacitance, $C_0$ ($100\ \mu$F cm$^{-2}$ for $0.5N$ NaOH), as the frequency independent capacitance shunting the faradaic impedance. At the potential at which the experiments are conducted the double layer capacitance on zinc is expected to be $\sim$20 $\mu$F cm$^{-2}$ (R68,599). The large value of $C_0$ is due to the presence of adsorbed species (OH$^-$ and adatoms) and has recently been confirmed by direct differential capacitance measurements (R68). It is to be concluded that adatoms are stabilized by hydroxide as a result of reactions of the type

$$Zn_{ad} + OH^- = Zn(OH)^- \tag{17}$$

The enthalpy of activation for the adatom flux, $\Delta H_k$, estimated from the temperature dependence of log $2FV_g^\circ$ (Figure 17) is:

|                                  | Single Crystal (0001) | Polycrystal |
|----------------------------------|:---------------------:|:-----------:|
| $\Delta H_k$, kcal mol$^{-1}$    | $11.0 \pm 1.0$        | $9.7 \pm 1.0$ |

The enthalpy of adsorption $\Delta H_{ad}$ results from the temperature dependence of concentration of adsorbed species illustrated in Figure 18;

|                                   | Single Crystal (0001) | Polycrystal |
|-----------------------------------|:---------------------:|:-----------:|
| $\Delta H_{ad}$, kcal mol$^{-1}$  | $7.9 \pm 1.0$         | $8.7 \pm 1.0$ |

The activation energy for adatom diffusion on the electrode follows from differences between $\Delta H_k$ and $\Delta H_{ad}$, being about 3 kcal mol$^{-1}$ for the single crystal (0001) and about 1 kcal mol$^{-1}$ for the polycrystal, so that the difference is considered to be barely significant.

Investigation (R199) has been carried out of the charge transfer process in the microsecond time range whereby adatom diffusion effects would not interfere. It is possible to differentiate between the adatom diffusion step and the charge transfer step since complete relaxation of the former occurred at frequencies less than $\sim$100 kHz. In other words the technique depends upon observing the necessary $i - \eta_D$ data within the space of a few microseconds after the beginning of the experiment, that is, before adatom diffusion can "get going." The double-impulse method (R239) that was used involves the application of an L-shaped pulse of current to the system at equilibrium. An initially high double-layer charging pulse is followed by a much lower faradaic current pulse. The duration $t_1$, of the initial pulse is $\sim$1 $\mu$sec and charges the double layer to the overpotential, $\eta_D$, necessary to drive the charge transfer reaction at a rate

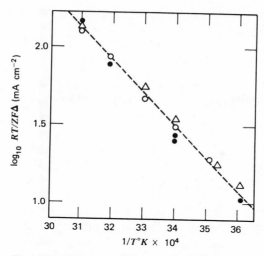

**Figure 17** Temperature dependence of equilibrium adatom flux ($RT/ZF\Delta = 2FV_g^\circ$) on polycrystalline zinc electrodes: ($\bullet$) electrolyte, $3N$ NaOH + $0.015M$ $Zn^{2+}$, run 1; ($\circ$) electrolyte, $3N$ NaOH + $0.015M$ $Zn^{2+}$, run 2; $\triangle$, electrolyte, $5N$ NaOH + $0.48M$ $Zn^{2+}$.

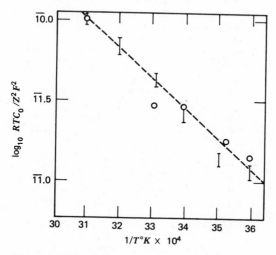

**Figure 18** Temperature dependence of concentration of adsorbed species ($\Gamma = RTC_0/Z^2F^2$) for polycrystalline zinc electrodes. (I) Electrolyte, $3N$ NaOH; ($\circ$) electrolyte, $5N$ NaOH. Data from Reference 198, by permission of The Faraday Society.

45

equivalent to the current density, $i$, the amplitude of the secondary pulse. At $t_1$ for correct balance,

$$\left(\frac{\partial \eta}{\partial t}\right)_{t\,=\,t_1} = 0$$

Experiments were limited to the low overpotential region and $i_0$ was calculated from

$$i_0 = \frac{RT}{2F(-\partial \eta_D/\partial i)} \tag{18}$$

Figure 19 shows typical overpotential-current density data and also the variation of the resulting exchange current with hydroxide concentration. The exchange current was found to be independent of zincate concentration as shown in Table 7. The highest exchange currents were obtained in KOH electrolytes. The exchange current for heavily cold-worked electrodes was about the same as for the polycrystalline material; however, the exchange current

TABLE 7.   EXCHANGE CURRENTS AT $25°$
IN HIGH ELECTROLYTE CONCENTRATION[a]

| $N$ | $M$ | A cm$^{-2}$ |
|------|------|------|
| [NaOH] | [Zn$^{2+}$] | $i_0$ |
| 7.0 | Nil | 0.140 |
| 7.0 | 0.0014 | 0.160 |
| 7.0 | 0.0047 | 0.120 |
| 7.0 | 0.0094 | 0.130 |
| 7.0 | 0.470 | 0.138 |
| [KOH] | [Zn$^{2+}$] | $i_0$ |
| 7.0 | 0.016 | 0.238 |
| 7.0 | 0.040 | 0.241 |
| 7.0 | 0.080 | 0.231 |
| 7.0 | 0.64 | 0.224 |
| [LiOH] | [Zn$^{2+}$] | $i_0$ |
| 4.95 | Nil | 0.185 |
| 4.95 | 0.056 | 0.167 |
| 4.95 | 0.140 | 0.175 |
| 4.95 | 0.280 | 0.166 |

[a] Data from Reference 199 by permission of Elsevier Publishing Company.

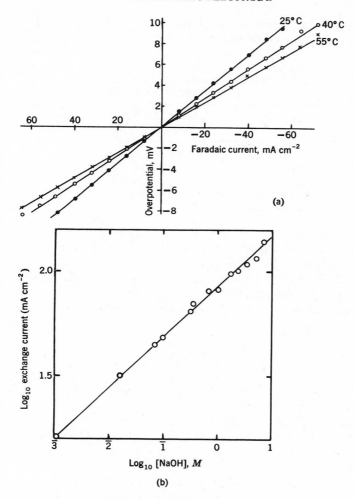

**Figure 19** Results of double-impulse experiments. Polycrystalline zinc electrode; electrolyte at constant ionic strength 7 (with NaClO$_4$). (a) Typical faradaic current-overpotential curves at 0.35$N$ NaOH + 0.001$M$ Zn$^{2+}$. (b) Variation of exchange current with hydroxide concentration, 25°. Data from Reference 199, by permission of Elsevier Publishing Company.

estimated at a single crystal (0001) face is only about one-third of the value on the polycrystalline electrode. The value of $i_0$ is considerably greater than $2FV_g^\circ$ and confirms that adatom dislocation and diffusion is the rate-determining step this may not be the case of course at very low [OH$^-$].

Enthalpies of activation in Table 8, estimated from the temperature dependence of $i_0$, confirm that the charge transfer step is energetically relatively easy.

TABLE 8.   ENTHALPY OF ACTIVATION
FOR CHARGE TRANSFER[a]

| Polycrystal | Single Crystal (0001) | Heavily Cold-worked |
|---|---|---|
| NaOH 3.1 | 3.4 | 3.0 |
| KOH  2.9 | | |
| LiOH 2.65 | | |

[a] $\Delta H$, kcal mol$^{-1}$, all values $\pm 0.2$

These enthalpies of activation may be compared with the enthalpy of activation for the exchange (charge transfer control) at an amalgam electrode, 10 kcal mol$^{-1}$ (R200). The occurrence of the reaction in the adsorbed state obviously lowers the activational enthalpy.

The variation of $i_0$ with hydroxide concentration

$$\left( \frac{\partial \log i_0}{\partial \log [OH^-]} \right)$$

is reported to be 0.2 (R199) and appears to indicate a value of $\alpha$ of 0.1 although an iterative analysis of the $\eta_D - i$ data indicates a value of $\alpha \sim 0.5$. The reason for this is considered to be either that the electron transfer step is taking place in stages (Equation 19) or that there is very strong adsorption of OH$^-$ (Equation 20) or both.

$$Zn_{kink} + OH^- = Zn_{ad} OH^- \tag{19a}$$
$$Zn_{ad} OH^- = Zn(OH)_{ad} + e \tag{19b}$$
$$Zn(OH)_{ad} + OH^- = Zn(OH)_2 + e \tag{19c}$$
$$Zn(OH)_2 + 2OH^- = Zn(OH)_4^{2-} \tag{19d}$$
$$Zn_{kink} + OH^- = Zn_{ad} OH^- \tag{20a}$$
$$Zn_{ad} OH^- + OH_{ad}^- = Zn(OH)_2 + 2e \tag{20b}$$
$$Zn(OH)_2 + 2OH^- = Zn(OH)_4^{2-} \tag{20c}$$

The insensitivity of the reaction kinetics to the zincate concentration of the electrolyte indicates that strong adsorption of species based on Zn(II) occurs at the electrode. The reaction sequences (Equation 19) seem to be the most likely mechanism.

## ANODIC BEHAVIOR OF THE ZINC ELECTRODE IN
## ALKALINE SOLUTION (DISCHARGE PROCESS)

### OVERPOTENTIAL DURING DISSOLUTION, "THE TAFEL REGION"

As the potential of the zinc electrode is forced increasingly from the equilibrium, it is expected on both experimental and theoretical grounds (R49) that the adatom diffusion step should decrease in importance in rate control (the number of active kink sites increases), although it is stressed that it remains a part of the overall process. Charge transfer becomes rate controlling.

Kinetic measurements at high overpotential are much more scarce than at low overpotential. From the available data it seems certain that charge transfer is rate controlling because overpotential-current density data (R274) are adequately represented by a Tafel plot as shown in Figure 20. Value for $\eta_D$ was determined by extrapolation of overpotential-time curves to zero time. In the presence of Zn(II) the Tafel lines are curved in the intermediate overpotential region, and it is noteworthy that in the Zn(II) concentrated range investigated changes in [Zn(II)] had no effect on the overpotential. The values of $\alpha$

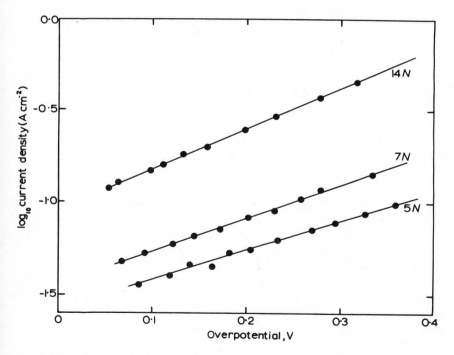

**Figure 20** Tafel lines for smooth zinc electrodes anodically polarized in KOH, 23°C. Data from Reference R274.

calculated from the slopes of these Tafel plots (Figure 20) at high overpotentials are surprisingly large (Table 9). The discrepancy between these values of $\alpha$ and those calculated previously from iterative matches of $\eta_D - i$ data in the low overpotential region, and also from the concentration dependencies of $i_0$, has yet to receive an unequivocal explanation. There is no doubt that $OH^-$ and $Zn(OH)_2$ (or intermediates) are both adsorbed at the electrode. Adsorption of $OH^-$ is probably potential dependent; accordingly, a more complete adsorption layer of $OH^-$ might be present at high overpotential. Differing extents of adsorption might well explain changes in $\alpha$. A change in charge transfer mechanism from a simultaneous two-electron transfer in the equilibrium region to successive one-electron steps at high overpotential might provide an explanation also.

TABLE 9.    PARAMETERS $i_0$ AND $\alpha$ CALCULATED FROM
ANODIC TAFEL CURVES[a]

| [KOH] $N$ | $i_0$ mA cm$^{-2}$ | $\alpha$ |
|:---:|:---:|:---:|
| 1 | 12 | 0.93 |
| 3 | 22 | 0.93 |
| 5 | 30 | 0.92 |
| 7 | 40 | 0.96 |
| 10 | 62 | 0.96 |
| 14 | 91 | 0.99 |

[a]Data from Reference 274.

ACTIVE–PASSIVE TRANSITION FOR THE DISSOLUTION OF
ZINC IN ALKALINE SOLUTION.
THE ANODIC PASSIVATION OF ZINC

Duration of the discharge process in alkaline solution is determined by the ability of the electrode to remain "active" for the dissolution process. The active region is terminated when passivation occurs and Zn ceases to be converted into Zn(II). The process of transition from active to passive has been studied by a number of workers: References 150, 153, 179, 190, 195, 211, 232, 269-70, 301-304, 311, 334, 379-380, 463, 495-496, 501, 537, 539, 588, 591-592). Reference 325 gives a particularly useful summary. It is generally agreed that passivated electrodes are physically blocked for the dissolution of Zn by the presence of a passive film.

*Nature of the Passivating Film.* Some (R311, 463) consider that passivation is due to films of the orthorhombic modification of zinc hydroxide. "Aging" of hydroxide films on the surface of zinc crystals has been suggested (R220). Direct electron diffraction measurements (R270) and X-ray diffraction methods (R270) have shown that the passivating layer on the surface of the anode consists essentially of zinc oxide of grain size 1-3 $\mu$. In 0.1$N$ KOH the passive layer is equivalent to ~20 atomic layers of ZnO (R463).

Overall stoichiometry, by chemical analysis of the oxide film, rarely corresponds exactly to ZnO. Dark, zinc-rich films have been observed with passivation at low current densities (R270, 304). The presence of $ZnO_2$ has been postulated, but X-ray and electron diffraction studies (R270, 302) do not confirm this, evidence for $ZnO_2$ resting on the development of a sky-blue color when a fragment of the film is treated with strong KI solution (R211). Films examined by direct transmission X-ray and electron diffraction (R270, 304) have shown a minor part of $\gamma$-$Zn(OH)_2$. It has been concluded (R302) that $ZnO_2$ does not form a part of the stable, passivating film even if it is formed; and, it has been considered (R496) from careful rotating disk, potentiostatic, and a-c measurements that $ZnO_2$ may participate in ZnO dissolution at high anodic overpotential. When the passive electrode, polarized at very high anodic potentials ($-0.3$ V N.H.E. or above), is cathodically activated, plateaus occur in the voltage curves, suggesting the presence of compound(s) more electropositive than is ZnO. Excess oxygen in the film was explained by adsorption (R588), but others (R496) concluded that if oxygen was bound to the oxide layer the bond was as strong as that in metal peroxides.

Experimental conditions have varied considerably among investigators; it is clear that the major part of the passivating film is ZnO, although other constituents may be present depending on experimental conditions.

*Passivation Time Relationships for Galvanostatic Polarization.* Experimentally, the simplest method of following electrode behavior is by galvanostatic polarization, which involves observing the electrode potential as a function of time at constant current density (conforming to the most used testing method in battery engineering practice). Attainment of the passive condition at time, $t_p$, is marked by a rise in the electrode potential to values required to cause oxygen evolution at a rate equivalent to the current density. A typical curve is shown in Figure 21.

It is generally accepted that passivation occurs when the capacity of the electrolyte layers immediately surrounding the electrode for soluble anodic products of Zn(II) is exceeded, ZnO being produced at the anode by processes of the type

$$Zn(OH)_2 = ZnO + H_2O \qquad (21)$$

or

$$Zn(OH)_4^{2-} = 2OH^- + ZnO + H_2O \qquad (22)$$

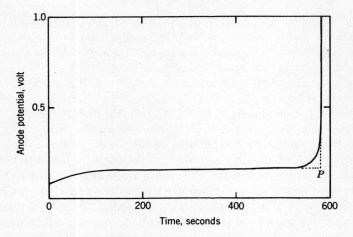

**Figure 21** A typical passivation curve ($1N$ KOH, 0.0233 A cm$^{-2}$, 25°C). $P$ marks the passivation time. From Reference R270, by permission of The Electrochemical Society.

A number of investigations of the relationship connecting current density and passivation time have been reported. A common feature has been a marked irreproducibility of the results, depending on electrode orientation and passivation times. For vertical electrodes and long passivation times, inconsistent passivation times resulted (R537). In the intermediate time region replicates differed by as much as 100% especially for the longer passivation times (R190). Good reproducibility was obtained (R380) for very short passivation times of 30 sec or less (most being less than 2 sec).

Experimental passivation time-current density data are conventionally expressed in the form (R179, 190, 269, 334, 380)

$$(i - i_1)\, t_p^{1/2} = k \tag{23}$$

where $i_1$ = current below which the anode will not passivate, and $k$ = constant for a particular system.

*Interpretation of Equation 23.* See References 190-191, 269-271. For systems in which transport of faradaic products away from the anode is by diffusion, Sand (R536) has shown that using the laws of non-steady-state diffusion the time taken for the development of a concentration difference $\Delta C$ between the bulk of the electrolyte and at the electrode is given by

$$\Delta C = \frac{i}{ZF}\, (4t/D\pi)^{\frac{1}{2}} \tag{24}$$

where $D$ is the diffusion coefficient.

If $\Delta C$ attains its critical value, $\Delta C_i$, so that any increase exceeds the capacity of the electrolyte immediately surrounding the electrode for Zn(II), then the further flow of charge must result in the anode's being blocked with products based on Zn(II). The development of the passivating layer results on transition from the zinc dissolution reactions

$$Zn + 2OH^- = Zn(OH)_2 + 2e \tag{25}$$

$$Zn(OH)_2 + 2OH^- = Zn(OH)_4^{2-} \tag{26}$$

to the discharge of $OH^-$ ion

$$2OH^- = H_2O + \tfrac{1}{2}O_2 + 2e \tag{27}$$

The chemical reaction by which the passive layer is laid down is probably described by Equations 21 and 22 but at the instant of transition the electrode may be directly oxidized to ZnO

$$Zn + 2OH^- = ZnO + H_2O + 2e \tag{28}$$

When diffusion is the only mode of mass transport away from the anode, Equation 24 should describe the kinetics of passivation at constant current. Figure 22 shows that, at electrodes carefully orientated in the horizontal plane so that the mass of the electrolyte was above the anode, the relationship, Equation 23, holds for times up to at least 1000 sec (R269).

To a first approximation, when convection occurs, Equation 24 may be modified by adjusting $i$ to take into account the effect of convection, that is, $i$ is replaced by $(i - i_l)$ where $i_l$ represents the magnitude of transport process other than diffusion. Extrapolation of $i$ versus $t_p^{-1/2}$ curves to $t_p^{-1/2} = 0$ (infinite passivation time) yields the value of $i_l$ as the intercept on the current density axis. The value of $i_l$ in the experiments with horizontal electrodes is very small, confirming that the contribution of nondiffusional processes to the available modes of mass transport is relatively unimportant.

Investigations at high current densities with vertical electrodes (R380, 270) demonstrated that Equation 23 holds for relatively short passivation times. Significant values of $i_l$ were reported. For systems subject to convection, the current necessary for maintaining the critical concentration across the diffusion layer may be calculated (R190), and reasonable results have been obtained for similar systems (R272). This theoretical approach is justified if diffusion remains the main mode of mass transport (R191). Theoretically, times of the order of 20 sec mark the limit of diffusion as the main controlling mechanism (R190); however, practical studies have shown Equation 23 to be applicable far beyond

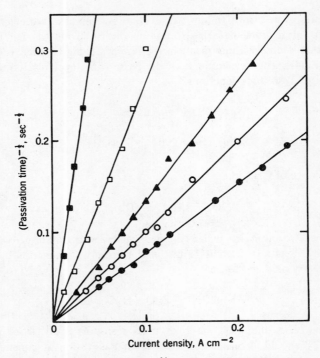

**Figure 22** (Passivation time)$^{-1/2}$ versus current density. KOH electrolytes at 20$^\circ$C: (■) $N$; (□) 2$N$; (▲) 3.5$N$; (○) 13.8$N$; (●) 10$N$. Data from Reference 269, by permission of the Electrochemical Society.

this limit (R179, 272) although this is a matter for convenience without theoretical justification (R270, 271). Figure 23 shows passivation time for three extreme orientations for smooth anodes in 30% KOH (R190).

Vertical orientation is usually the most, and often the only orientation, convenient for practical systems. Consequently vertical electrodes have received most experimental attention (see particularly R179, 190, 270, 380). For such experimental arrangements Figure 24 shows the variation of $i_l$ with KOH concentration at three temperatures. Figure 25 shows the variation of $k$ with KOH concentration: in the range 20-40$^\circ$, $k$ was independent of temperature. Combination of the results of Figures 24 and 25 indicates that the maximum passivation time occurs in concentrations 7-8$N$ KOH (~30% KOH) at temperatures in the range of 20-40$^\circ$.

Hydrodynamic difficulties associated with static electrodes have been considerably simplified by Russian workers (R496) using a rotating disk technique. In agreement with other investigations (R270, 380), mass transport in the electrolyte was shown to control the passivation process and the process is somewhat reversible, that is, depassivation occurs if the current density is reduced; total depassivation occurs if anode products are able to completely leave the vicinity of the electrode. A further quantitative conclusion (R496) is that the molar ratio of $Zn(OH)_4{}^{2-}$ to $OH^-$ at the electrode at the point of

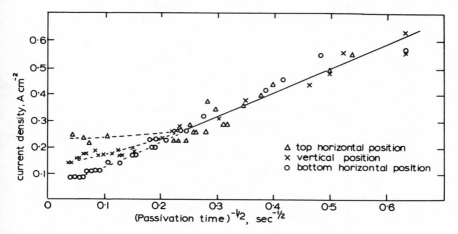

**Figure 23** (Passivation time)$^{-\frac{1}{2}}$ versus current density for three anode positions in a small rectangular sectioned cell, 30% KOH electrolyte at 24°C. Data from Reference 190, by permission of The Electrochemical Society.

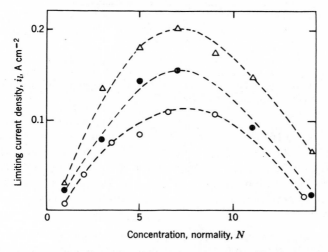

**Figure 24** The variation of limiting current density with electrolyte concentration, vertical anodes: (○) 20°; (●) 25°; (△) 40°.

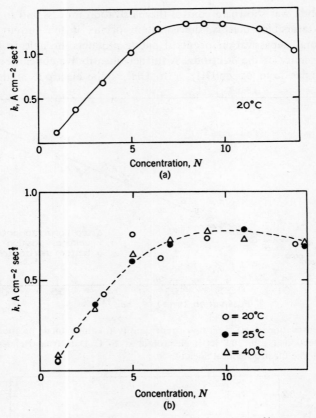

**Figure 25** Variation of $k$, from the equation $(i - i)t^{1/2} = k$, with concentration: (a) horizontal and (b) vertical anodes in KOH electrolyte. Data from References 269 and 270, by permission of The Electrochemical Society.

passivation is ~0.16, being considerably higher than that corresponding to the dissolution of ZnO in alkaline solution (R150, 153). Saturation of the electrolyte with ZnO causes remarkably little shortening of $t_p$ (R269, 380). At the instant of passivation, therefore, considerable supersaturation of the solution with respect to Zn(II) exists in the layers of electrolyte immediately surrounding the electrode.

*Potentiostatic Polarization.* A few applications of the potentiostatic method have been made (R304, 496, 538). The potentiostatic curve for zinc in alkaline solution (R538) indicated that at the start of passivation an insoluble film of ZnO forms on the anode which later passes into a soluble form. Potentiostatic measurements and simultaneous measurements of the a-c impedance of the active and passive electrode have been made. The passivation of Zn occurs at a potential of $\sim-1.0$ V N.H.E. From measurements of reduction (cathodic activation) curves carried out on passive electrodes, it was considered (R496) that the passive layer itself may give rise to a small active region. This region occurs at potentials more positive than $-0.3$ V N.H.E. and is caused by the reactions:

$$ZnO + 2OH^- = ZnO_2 + H_2O + 2e \qquad (29)$$

$$ZnO_2 + 2H_2O + 2OH^- = Zn(OH)_4^{2-} + H_2O \qquad (30)$$

An elegant voltammetric study of the anodic behavior of zinc in $2N$ and $5N$ KOH was made (R620) resulting in conclusions generally in agreement with those of earlier workers (R304, 496, 538). In addition the blackening of the oxide film, due to the deposition of finely divided zinc, is given a formal mechanism involving disproportionation

$$Zn + 2OH^- = Zn(OH)_2^- + e \qquad (31)$$

$$2Zn(OH)_2^- = Zn + Zn(OH)_4^{2-} \qquad (32)$$

## DEPOSITION OF ZINC FROM ZINCATE ELECTROLYTES
## (CHARGE PROCESS)

Electrolysis of zincate electrolytes at rates of sufficient magnitude to be practically significant have been studied in some detail particularly in connexion with the preparation of zinc dust (R246, 429), production of zinc sponge (R430, 32), and the extraction of zinc from alkaline electrolytes (R202, 203, 535, 641). The limits for the formation of dense and spongy deposits have been investigated (R655). In general, at commercially desirable current densities, the deposit is laid down in unsatisfactory dendritic form, becoming more dense, however, at low current densities. The causes of the differing qualities remain uncertain and are variously attributed to codeposition of hydrogen (R25, 350), formation and subsequent deposition of hydrides (R192), adsorption of zincate on growing crystals (R222, 554, 593), or to the discharge of complex ion (R358). Spongy deposits are favored at low concentration of zincate and high cathodic current density (R400). The product of the charge transfer reaction may be colloidal particles (ultramicrons) of zinc in which case a random crystal growth process causes a sponge to be formed (R219, 361).

Results of galvanostatic polarization measurements at a rotating cathode (R580) led to considerations that the cause of sponge formation lies in the nature and supply of the species being discharged at the cathode. It is suggested that the shortage of $Zn(OH)_2$ at the cathode engenders the formation of a spongy deposit.

Compact electrolytic deposits of zinc are favored by the addition of $Hg^{2+}$, $Pb^{2+}$, and $Sn^{2+}$ to the electrolyte (R359, 362). Additions (usually $Pb^{2+}$) are of the order of 0.05 g liter$^{-1}$ and are reported to have practically no effect on the current efficiency and cathode potential; the deposit, however, contains a small amount of the added metal. Other additions made in order to achieve the same effect (e.g., surfactants, $Cd^{2+}$, $Cu^{2+}$) are not so beneficial (R362, 656).

Overpotential-current density data for the process of zincate (30 g liter$^{-1}$ ZnO, 3$N$NaOH) discharge at a rotating solid zinc electrode have been reported (R656). Log $i - \eta_D$ data form a curve; at medium current densities (10 mA cm$^{-2}$) the Tafel slope is low, indicating a high value for the charge transfer coefficient; at higher current densities the Tafel slope increases steeply. It is not possible therefore to calculate a unique value of the charge transfer coefficient. The high value of $\alpha$ at medium current densities agrees with the value observed in the anodic experiments (R274). The low value of $\alpha$ at high current densities appears not to be due to codeposition of $H_2$ since the current efficiency is 100%. In agreement with the anodic behavior of the system at high overpotentials (high current densities), differences in the experimentally observed charge transfer coefficient strongly suggest a change in mechanism (from a single two-electron transfer step to two consecutive one-electron transfer steps) as the overpotential increases.

## THE HYDROGEN OVERVOLTAGE ON ZINC

### ACID ELECTROLYTES

The hydrogen overvoltage at a clean zinc electrode has been studied by a number of workers. Accurate measurements (R230) of the hydrogen overvoltage in acid solution for a single crystal in 2$N$ H$_2$SO$_4$ at 25$^\circ$ led to the equation

$$\eta_h = -1.24 - 0.118 \log_{10} i \tag{33}$$

where $\eta_h$ is the hydrogen overvoltage and $i$ is the current density in A cm$^{-2}$. A similar equation was obtained (R521) for polycrystalline zinc in H$_2$SO$_4$.

### ALKALINE ELECTROLYTES

The overpotential at a clean zinc cathode in KOH solutions (0.1$N$–14$N$) was determined (R312). The variation of $\eta_h$ with $\log_{10} i$ is linear from ~10$^{-5}$ to ~5 x 10$^{-2}$ A cm$^{-2}$ at 20$^\circ$. Table 10 (R312) records the constants in the equation

$$\eta_h = a - b \log_{10} i \tag{34}$$

TABLE 10.    HYDROGEN OVERVOLTAGE ON ZINC IN KOH SOLUTIONS

| [KOH] | 0.1N | 0.2N | 1N | 5N | 7N | 10N | 13N | 14N |
|---|---|---|---|---|---|---|---|---|
| a | −1.394 | −1.375 | −1.312 | −1.230 | −1.189 | −1.164 | −1.230 | −1.235 |
| b | 0.128 | 0.130 | 0.130 | 0.120 | 0.118 | 0.118 | 0.120 | 0.125 |

The Tafel slope is ~120 mV per decade of current density. On increasing the alkali concentration the overvoltage at constant current density decreases by about 90-100 mV per pH unit in solutions of intermediate concentration. This behavior agrees with the equations of the slow discharge theory

$$H_2O + e = H_{ad} + OH^-$$    (35)

if account is taken of the effect of the structure of the double layer.

In concentrated solutions of KOH the overvoltage shows a sharp increase at concentrations 10-14N amounting to 160 mV. This cannot be explained by changes in the double layer structure or in the activity of the water molecules. It is suggested, however (R312), that the effect may be due to the increased screening effect of the alkali metal cations on the water molecules undergoing electron transfer in the double layer.

Experiments on single crystal faces of zinc in alkaline solution are as yet unrecorded in the literature, and further work on this topic would be of considerable interest.


## AMALGAMATION OF THE SOLID ZINC ELECTRODE SURFACE

It is often found expedient to attempt to modify the behavior of an electrode by physical means. Two processes have been (separately) widely applied in practice, namely, the preparation of microporous electrodes and amalgamation. Microporous zinc increases the effective area of the electrode and involves no fundamental change in kinetics.

Amalgamation is a change of a more fundamental kind. Its purpose is to modify the surface of the electrode so that its behavior as a cathode for the discharge of hydrogen ion resembles mercury rather than zinc. Mercury has a much smaller exchange current for the hydrogen evolution reaction than has zinc (R140, 611). Thus for a given current density, $i$, Equation 2 shows that the $H_2$ overvoltage at a zinc anode will be much less than that at a mercury electrode. Amalgamation "raises the hydrogen overvoltage" and has the effect of considerably reducing the rate of the open circuit reaction (Equation 35) and, consequently, corrosion.

Amalgamation can also affect the discharge. Polarization experiments at both smooth and amalgamated zinc electrodes (R380) showed that amalgamated electrodes were rather more reactive (as measured by the ability to resist passivation) than solid metal electrodes. A recent investigation (R177) has confirmed that amalgamated electrodes have a lower overpotential than have solid electrodes and a slight increase in limiting current density was also found, confirming the earlier work (R380). The results can be interpreted in terms of the absence of the adatom diffusion step in the process occurring at the amalgamated electrode. Because adatoms can be incorporated at any point on the amalgam surface, the total overpotential at any current density is thereby modified.

The electrode capacitance for a zinc amalgam drop approaches that of a mercury drop at the same potential whereas the electrode capacitance of a solid zinc electrode in alkaline solution is high because of adsorption (R68, 198, 200). The "normal" capacitance (R177) of an amalgamated zinc electrode in alkaline solution ($\sim$40 $\mu$F cm$^{-2}$) indicates the absence of adsorption of OH$^-$(or adatoms) which might explain the marginal improvement in its ability to withstand passivation.

## CONCLUSIONS AND FURTHER WORK

The common factor to emerge from the results of recent investigations is the pronounced effect of adsorption of OH$^-$at the electrode suface. The impression obtained is that the effect of OH$^-$ is both important and fundamental. At overpotentials not far removed from the equilibrium there is considerable evidence that the charge and discharge reactions occur mainly as surface reactions between adsorbed species with the nature of the electrode surface determining both adatom flux and exchange current density. At higher overpotentials the situation is less certain but what evidence there is points to considerable adsorption of OH$^-$ with charge transfer as the controlling mechanism. Therefore, mechanistic change as the overpotential is increased from the equilibrium seems to be indicated. Further work is clearly required in the higher overpotential region ($>$10 mV about the equilibrium).

The mechanism of the active/passive transition which marks the limit of useful discharge of the negative electrode is fairly well understood. The process is controlled by mass transport in solution. Production of perfectly dense deposits of zinc from alkali zincate electrolytes is an extremely desirable goal; although progress has been made, a detailed description of the mechanisms of the processes by which dendritic deposits of zinc are laid down is still lacking.

The kinetics of the hydrogen evolution reaction from metals has been under constant investigation for the last 70 years. It is only relatively recently that single crystal faces of various orientations have been studied. Zinc with a very convenient cleavage plane, the basal plane (0001) provides an ideal metal for the study of the hydrogen evolution reaction. A little exploratory work has been done (R68, 599) but much remains. The main problem involved in studying the hydrogen evolution reaction in alkali at specific crystal faces will be to maintain the identity of the electrode under conditions of exchange. Even in the absence of soluble zinc ion, sufficient $OH^-$ is adsorbed at the electrode to provide for some exchange between adsorbed species. Experimental work in which measurements are made immediately the surface is born at the instant of electrode/electrolyte contact would be extremely interesting.

At an amalgam (or amalgamated) electrode the exchange reaction and the hydrogen evolution reaction are simpler. The structure of the electrode is removed by amalgamation and atoms can be incorporated at any point. The extent of the adsorption of $OH^-$ ion at the surface is very much reduced if indeed not completely removed. The effect is predictable: the overpotential at a fixed current density is modified by the removal of adsorption and the adatom diffusion step. Amalgamation only marginally affects the time taken to reach the onset of the active/passive transition, caused, possibly, by changes in the structure of the interphase.

# 6.

## The Morphology of Zinc Electrodeposited from Alkaline Electrolyte

### H. G. Oswin and K. F. Blurton

Zinc use as the anode in batteries is based on its stability, its energy- and power-density characteristics and its low cost per watt-hour. Low cycle life has limited its use as the negative active material in storage batteries. It is also necessary to operate with a concentrated alkaline electrolyte so that the anodic oxide film formed on zinc during discharge can be rapidly dissolved to form the zincate ion thus preventing electrode passivation. However it is this high zincate solubility that limits the cycle life of the zinc electrodes in an electrochemically rechargeable cell, permitting the zinc to be deposited in a dendritic or nonadherent form during charging.

To prevent passivation and electrolyte concentration polarization at the current densities required from modern storage batteries, it is necessary to use a relatively large quantity of electrolyte in the cell and to accommodate this volume within a highly porous zinc electrode. During discharge this low-density zinc matrix becomes less dense as zinc oxidizes and dissolves. In order to replate the zinc during the charge cycle back onto the matrix in its original form at least two conditions must be met; namely, the zinc matrix must not be distorted when in its weaker, discharged state, and the soluble and colloidal zinc must not be permitted to migrate from the region where it was formed during discharge. In present-day batteries these requirements are partially achieved by incorporating electrode design features (thickness, porosity, and rigidity) and by using separators that must maintain an even pressure on the electrodes and restrict diffusion of zincate ions away from the negative plate. Presently available separators only limit zincate diffusion at the expense of reduced ionic mobility, which causes increased ionic resistance in the cell. Separators also suffer from chemical instability in the presence of KOH and oxidizing species. Operating methods and cell modification are sought to partly compensate for the lack of a perfect separator material in order to maintain the original, physical form of the zinc electrode.

This paper reviews the studies reported on the charging of zinc electrodes and relates phenomena observed on smooth (i.e., nonporous) zinc electrodes to the reported modes of cell failure. It is difficult to isolate and identify the causes of cell failure during cycling. However, at least three modes of cell failure have been observed, namely, zinc penetration of the separator (shorting to the positive), gross changes of the physical dimensions of the zinc electrodes (shape change and "shedding"), and incomplete reduction of the zinc oxide formed during the electrode discharge.

The first section of this review attempts to relate the first two failure modes to phenomena observed in experiments conducted in relatively simple systems, where the effect of the separator and counter electrode necessarily have been ignored. In a subsequent section, the application of these fundamental data to practical cells will be discussed, and finally further studies on the secondary zinc electrode are suggested.

## FUNDAMENTAL STUDIES

### DEPOSIT MORPHOLOGY

On deposition from an alkaline zincate solution, a zinc deposit is either mossy or dendritic unless charging is performed under special conditions. The former is black and porous while the latter is needlelike and conforms to the usual concept of dendrites. Other characteristic differences between these two deposits are that the diameter of the mossy whiskers is considerably less than that of the dendrites (see Figure 26 and R572) and that the density of the mossy deposit is one-seventh that of the dendritic deposit (R571).

Other types of zinc deposits have been observed on foil electrodes but these morphologies are less important in battery technology. Thus smooth, bright zinc deposits have been obtained by deposition from an alkaline electrolyte onto a rotating disk electrode at low overpotentials and at high Reynolds numbers and by deposition onto a stationary electrode with an intermittent charging technique (R38, 39). Fine black deposits have been obtained at low current densities (less than 5 mA/cm$^2$) on a stationary electrode (R475), granular deposits have been obtained by charging at a current density intermediate between that for moss and dendrite formation (R458) and a heavy dendritic zinc sponge has been produced (R144, 571-572) at overpotentials more cathodic than that for the formation of dendrites.

*Current Density.* Mossy deposits are formed at low current densities (4-20 mA/cm$^2$) (R474-475, 477, 458, 21) from concentrated zincate electrolyte, that is, under conditions contrary to those usually favoring the formation of powder electrodeposits (R309), while dendritic deposits are produced at higher current densities from zincate-depleted electrolyte. The transition from

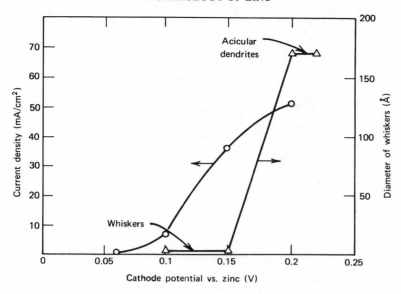

**Figure 26** Current density and whisker diameter as a function of potential, from Reference 572.

mossy to dendritic growth is characterized by a temperature-dependent, critical current density (R475-478; see also Figure 27). The physical form of deposits obtained by charging at constant current will depend on the value of the current. At a current less than a critical current only moss will grow. At a current greater than this critical current, dendrites will first be formed and then, because of the decrease in overpotential corresponding to the increase in real surface area, there is a transition in morphology leading to the deposition of moss.

The characteristic potential-time curves for moss are given in Figure 28 and for dendrite formation in Figure 29. The overpotential of the zinc electrode at a small, constant current has been claimed to be the same for a compact (i.e., coherent and high density) deposit as for a mossy deposit (R360, 474). However further work (R21-22) has shown that once moss is formed there is a small, but significant, decrease in the overpotential (Figure 28). This decrease in the overpotential is presumably caused by the decrease in the true current density resulting from the increased electrode surface area. The equilibrium potential of the mossy deposit is 5-10 mV more cathodic than that of a zinc foil, (R21, 38-39).

During zinc deposition at constant, high current densities, the surface concentration of zincate ion approaches zero so that the electrode overpotential approaches that for hydrogen evolution. When the hydrogen bubbles are formed the electrolyte is stirred and the surface concentration of zincate ions is no

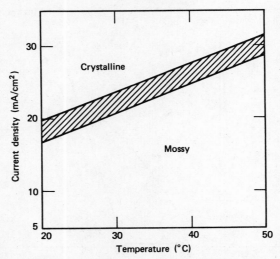

**Figure 27**   Variation of critical current density with temperature within experimental limits, 43% KOH, 1.13*M* Zn(II); from Reference 475.

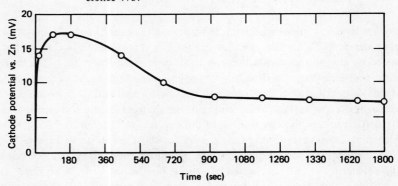

**Figure 28**   Potential versus time curve for mossy zinc deposition at a constant current density of 10 mA/cm², 43% KOH, 1.13*M* Zn(II), 30°.

longer zero and, hence, the potential increases to near its original value. This accounts for the oscillations in the potential-time curve during dendrite formation shown in Figure 29. Ultimately, however, the electrode roughness increases as a result of dendrite growth and, since the diffusion layer thickness does not exceed the dimensions of the surface irregularities, the transition time is no longer attained (R474).

*Potential.* Mossy electrodeposits are formed at low overpotentials (R21, 144, 471-473, 571-572) while dendrites are formed at high overpotentials. The transition from one morphology to the other was characterized by the half-wave

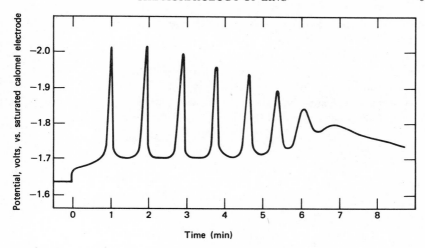

**Figure 29.** Potential versus time curve for dendritic zinc deposition at a constant current density of 30 mA/cm$^2$, 43% KOH, 1.13$M$ Zn(II), 30°; from Reference 474.

potential for zincate reduction (R571-572). In view of this transition in deposit morphology, deposition at constant potential results in the propagation of a particular morphology and, after an initial current decrease, deposit growth is accompanied by an increase in the total current because of the increase in the real surface area of the electrode (R572).

The same change (with changing potential) in the morphology of zinc deposited from an alkaline solution with zinc oxide is observed on porous and smooth electrodes (R472). The rate of propagation of each morphology appeared to be lower on the porous zinc electrodes, but no quantitative measurements were reported (R472).

*Role of Zincate Mass Transport.* Electrodeposited metal dendrites propagate under conditions of mass-transport control (R49), and several effects noted in zinc deposition studies suggest that zinc dendrites grow under similar conditions. The following reported observations indicate that in the zinc/zincate system, mass transfer of the zincate ion controls zinc dendrite propagation:

1. The dependency of the critical current density corresponds to the change in deposit morphology on temperature (R22, 476-477; see Figure 27).

2. Dendrite growth is reduced by intermittent charging at current densities greater than the critical value (R22).

3. Dendrites are formed at a lower current density from the higher viscosity zincate-saturated NaOH solution than from the zincate-saturated KOH solution (R474).

4. Dendritic deposits are formed at a potential corresponding to a limiting current condition (R572; see Figure 26).

5. Measurement of dendrite propagation rates (R476-477) is also an indication.

Thus the factors promoting rapid transport of the zincate ions, such as high zincate concentration, high temperature, or low electrolyte viscosity, tend to suppress zinc dendrite growth.

The numerical value of the critical current density depends on the particular design of the electrochemical cell used since this determines the contribution of

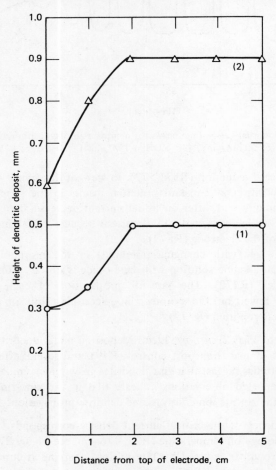

**Figure 30**  Height of mossy zinc deposit versus distance from top of electrode; current density 15 mA/cm$^2$, 43% KOH, 1.13$M$ Zn(II), 30°; (1) 27 coulombs/cm$^2$; (2) 54 coulombs/cm$^2$; from Reference 474.

convection to the transport of the zincate ion. Thus the equation developed empirically (R286) is not general and is applicable only to results obtained in their apparatus. The overpotential is controlling only to the extent that it determines whether the deposition is mass-transfer controlled.

Experimentally it was found (R287, 474) that mossy deposits form preferentially at the electrode base (i.e., in concentrated zincate electrolyte, Figure 30) while dendrites reach their maximum height at the top of the

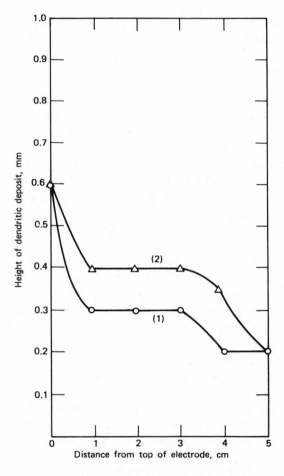

**Figure 31**    Height of dendritic zinc deposit versus distance from top of electrode; current density 30 mA/cm$^2$; 43% KOH; 1.13$M$ Zn(II); 30°; (1) 27 coulombs/cm$^2$; (2) 54 coulombs/cm$^2$; from Reference 474.

electrode (i.e., in zincate-depleted electrolyte, Figure 31). At higher current densities, the zincate concentration gradient will cause electrolyte density differences and, if free convection can occur, zincate concentration will be larger at the bottom of the electrode. The same effect should occur in cells and batteries although the presence of a separator and the restricted electrolyte movement will diminish the magnitude of this phenomenon.

## DEPOSIT ADHERENCY

Nonadherent material will tend to fall from the plate if subjected to vibrations or other physical stresses and will result in a lower available negative capacity.

A portion of the mossy and dendritic deposits are nonadherent (defined as that portion of the deposit easily removed by wiping, (R515), but both the adherent and nonadherent portions appear to have similar forms when viewed at 250 times magnification. In an early study (R360-361), a mossy deposit was not obtained on the cathode during the electrodeposition of zinc from an alkaline solution if the anode was platinum or nickel, but a subsequent study showed that moss was formed even with these anodes and that the deposit adherency was the same for nickel, gold, platinum, or zinc anodes (R22).

*Charging Rate.* Initial studies found (R475) that, for a constant quantity of charge passed, the deposit adherency decreased from 80% to a minimum of 40% as the current density was increased from 4 mA/cm$^2$ to 10 mA/cm$^2$ and then increased to a maximum with further current increase. However a subsequent study (R21) found that the mossy deposit was more adherent at a charging rate of 10 mA/cm$^2$ than at 4 mA/cm$^2$. The adherency of dendrites was found to be independent of charging rate up to a current density of 30 mA/cm$^2$ (R475), but measured dendrite deposit adherency is probably dependent on the cell geometry since dendrite formation is controlled by the transport of the zincate ion.

*Quantity of Charge.* The mossy deposit adherency as a function of the number of coulombs of charge passed at selected temperatures is given in Figure 32 (R475). The characteristic initiation time before the formation of non-adherent deposits and the decrease in deposit adherency with increasing quantity of charge passed was also reported (R515). Regardless of the total number of coulombs, only a weight corresponding to approximately 6 coulombs/cm$^2$ deposits adherently after the appearance of moss (R21). This additional adherent zinc deposits during the period when both compact and mossy deposits form simultaneously and it is a function of the current density. Thus the time required before appearance of moss is a better means of deposit characterization than the less easily measured "percent nonadherent deposit."

The initiation time for the formation of a nonadherent deposit corresponds to the time of appearance of moss and is dependent on electrolyte purity

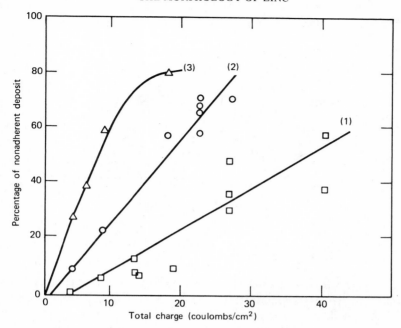

**Figure 32** Variation of mossy zinc deposit adherency with total charge at selected temperatures. Current density 15 mA/cm$^2$; 43% KOH; 1.13$M$ Zn(II); (1) 25°; (2) 30°; (3) 50°; from Reference 475.

(R500B) and temperature (R475). The increase of temperature also causes a decrease in mossy deposit adherency.

The effect of the quantity of charge passed on the adherency of the dendritic deposit at selected temperatures is shown in Figure 33 (R474-475). Again there is an initiation time during which all the zinc is deposited adherently. As deposition proceeds an increasing percentage of the dendritic zinc is deposited nonadherently. Both the dendritic adherency and the initiation period are independent of temperature. The initiation time is less when the zinc is deposited from NaOH solution rather than from KOH solution (R474). These initiation times are probably similar to those observed (R37) before the appearance of dendrites in the Ag/Ag NO$_3$ system and have been reported during zinc deposition by other investigators (R144).

*Intermittent Charging.* The effects of three modes of intermittent charging of zinc deposits from alkaline solution have been studied: (1) alternating current superimposed on a direct current at low current densities (R38, 515); (2) pulse charging with selected ON and OFF times at low (R21-22, 39, 477, 517) and high current densities, (R22, 473-474, 458); (3) periodic reversal of current with selected anodic and cathodic periods at constant current (R22, 39) and at constant potential (R500B).

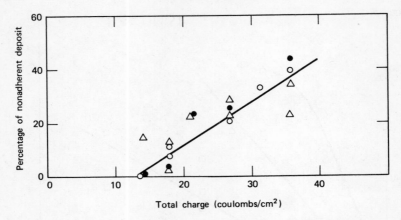

**Figure 33**    Variation of dendritic zinc deposit adherency with total charge at selected temperatures. Current density 30 mA/cm$^2$; 43% KOH; 1.13$M$ Zn(II); (●) 25°; (○) 30°; (△) 40°; from Reference 475.

By charging with an alternating current of frequency less than 1000 cps superimposed on a direct current (R38,515), the time of appearance of moss was delayed and the deposit adherency increased by charging with this technique at low current densities. This delay in moss formation occurred when a definite peak value of the anodic current was reached. The peak value depended on the electrolyte temperature and the rate of stirring; the optimum a-c frequency was between 1 and 250 Hz. An improvement (R38) in the deposit was also obtained when the alternating current was briefly increased at the commencement of the electrolysis.

Pulse charging at low current densities as a means of obtaining adherent zinc deposits has been studied by many workers (R21-22, 39, 477, 515, 517-518). Charging with this technique delays the appearance of moss and increases the deposit adherency provided that relatively short ON and OFF periods are used. Thus no improvement was found in the deposit adherency at the charge rate of 15 mA/cm$^2$ when the ON and OFF times were 45 and 120 sec, respectively, but there was a slight improvement in the deposit adherency with ON and OFF times of 1 sec, respectively. It has been claimed (R39, 515, 517) that the adherency of the deposit depends on the ratio of the ON to OFF times but a later study showed (R22) that, at constant current, the adherency of the deposit depended not only on the ratio of the ON and OFF times but also on the actual values of the ON and OFF times. The deposit adherency obtained by pulse charging depended on the current density in the same fashion as with direct current (R22), i.e., the deposit formed at 10 mA/cm$^2$ was more adherent than that formed at 4 mA/cm$^2$. Using pulse-charging techniques, it was found that the tendency to moss formation increased with increasing temperature (R39).

Pulse charging (R473-4) at currents greater than the critical current density resulted in a more adherent deposit than was obtained with direct current charging when the ON and OFF times were 45 seconds and 120 seconds, respectively. In contradiction (R458), it is claimed that, if the magnitude of the pulsed current was adjusted so that the mean charging current density was the same as that with direct current charging, no significant improvement in the deposit was obtained when charging with cycles from 10 seconds ON and 0.1 second OFF up to 10 seconds ON and 20 seconds OFF.

However the effect of pulse charging on the nature of the deposit obtained at high current densities in more extensive studies (R22) showed that the four factors decisive in determining the deposit characteristics with pulsed current charging at constant bath composition and constant temperature are current density, the quantity of charge passed, the ON time, and the OFF time. As expected for a constant ON time, the deposit is smoother the longer the OFF time. However, the ratio of ON to OFF time was not by itself decisive whereas the actual values of the ON and OFF time were very important. For a constant ON time to OFF time ratio, ON and OFF times of the order of 10 msec produced smoother and more adherent deposits than those produced with ON and OFF times of the order of 1 sec. In contrast (R458), a smoother deposit was obtained with a charging rate of 80 mA/cm$^2$ and ON and OFF times of 10 msec than with direct current at half the charge rate.

The other intermittent charging technique used in studies of zinc deposition is periodic reversal of the electrode polarity. The initiation of dendrites was delayed by switching from a constant cathodic to anodic current when the ratio of cathodic to anodic periods was only slightly in excess of one (R22) and by switching from a constant overpotential of −100 mV to +100 mV or +150 mV when the anodic period was less than one percent of the total duty cycle (R500B). The formation of moss is also significantly delayed by switching from a constant cathodic to anodic current providing the ratio of the cathodic to anodic periods is less than 6 (R39), although the tendency for moss formation increases with increasing temperature.

Two reasons have been given in the literature for the effect of intermittent charging on the quality of the zinc deposit at low current densities: On one hand (R22, 38-39) the beneficial effect of these techniques is due to the dissolution of the active mossy deposit during the anodic or open circuit period. On the other hand (R518), the effect of the anodic or open circuit period is to allow the zincate to be transported to the cathode. Although this latter explanation is probably correct at the higher current densities where zincate ion mass-transport controls, at the lower current densities the beneficial effect of intermittent charging is probably caused by the dissolution of mossy deposits during the OFF period. This view is supported by the facts that the equilibrium open-circuit potential of a mossy deposit is 5-10 mV more negative than that of a smooth foil electrode (R21-22, 38-39) and that hydrogen evolution occurs during the OFF period because of zinc dissolution (R22).

The beneficial effect of pulse charging at high current densities is that the zincate-ion concentration at the electrode/electrolyte interface approximates to the bulk concentration. During the current-OFF period, the surface concentration of the zincate ion is replenished by diffusion and convection and the repulsive electrostatic forces between the cathode and zincate ion diminish (R22). The importance of the latter effect is demonstrated by the fact that a smoother deposit was obtained with a charging rate of 80 mA/cm$^2$ and ON and OFF times of 10 msec than with direct current at half the charge rate (R22).

*Effect of Substrate.* Moss formation (R360-361, 363-365) was suppressed when the charge rate was 3 mA/cm$^2$ by the addition of mercury, tin, selenium, arsenic, antimony, or lead salts to the alkaline zincate electrolyte. Similar results were reported when these metals were incorporated into the zinc anode. These inorganic additives had no influence on the deposit when the charge rate was 10 to 20 mA/cm$^2$. Using zinc amalgam, cadmium, copper, lead, silver, and tin (R477) substrates, it was shown that at 15 mA/cm$^2$ an almost totally adherent mossy deposit formed on Pb while a more adherent deposit was formed on Zn amalgam, Ag, Cd, and Sn than on Zn (Table 11). Similarly, it was found that inorganic additives in the electrolyte increased the dendritic deposit adherency when charging at 30 mA/cm$^2$ and more adherent deposits were obtained on tin, cadmium, and lead than on zinc (Table 11). Further experiments showed that the lead substrate continued to affect the deposition process even when completely covered by a thick layer of zinc.

TABLE 11.   VARIATION OF ZINC DEPOSIT
ADHERENCY WITH METAL SUBSTRATE (R477)[a]

| Substrate | Mossy Zinc Adherency[b] | Dendritic Zinc Adherency[b] |
|---|---|---|
| Copper | 74 | 89 |
| Zinc | 69 | 23 |
| Silver | 63 | 86 |
| Cadmium | 54 | 12 |
| Tin | 30 | 1 |
| Zinc amalgam | 13 | 39 |
| Lead | 2 | 6 |

[a]Conditions: total current = 27 coulombs/cm$^2$; current density = 15 mA/cm$^2$; electrolyte = 43% KOH, 1.13$M$ Zn; temperature = 30°.
[b]Figures in percentage of nonadherent deposit.

This suggested that soluble lead is the controlling species since any effect of the substrate lattice would be obscured after deposit of a few layers of Zn, and depositions were carried out on zinc with lead strip inserted in the electrolyte but not connected to the cathode. The lead still had a beneficial effect showing that the presence of lead ions dissolved in the electrolyte is the important factor. The deposit adherency also increased when lead acetate was added to the electrolyte and the critical concentration range for the best deposit was 0.5 g/liter to 1.0 g/liter, that is, the lead is deposited under conditions of mass transport control. If the ratio of Pb to Zn in the deposit is small, zinc dendrites may propagate whereas if the ratio is large, lead dendrites may form. The reason for the influence of lead on the deposit adherency is not clear.

*Height of the Electrolyte.* Pulse-charging at a current density of $10 \text{ mA/cm}^2$ with ON and OFF times of 2.5 sec, the percentage adherency of the deposit of a zinc electrode with the electrolyte height 4.4 cm was approximately 10% whereas it was approximately 21% with the electrolyte height 1.0 cm (R21). This is probably a result of the change in the vertical zincate-ion concentration gradient.

## RATES OF PROPAGATION OF MOSSY AND DENDRITIC DEPOSITS

The rate of propagation of noncompact deposits is an important parameter in the cycle life of secondary zinc electrodes. The reported data are usually average values of linear growth in one direction, reflecting the difficulties in studying the growth of individual dendrites.

Essentially similar results on the rate of propagation of moss and dendrites have been reported (R472, 571) with the difference being in the duration of the experiment. At constant potential, the initial growth rate of moss is low (R472) and the thickness of the deposit does not increase linearly with time, but accelerates (R571). The propagation rate of dendrites remains constant after the passage of a few coulombs of charge (R571, 472). The rate (R144) of zinc dendrite propagation increases as the initial concentration of the zincate ion increases while the initiation time for dendrite formation decreases with increasing overpotential, temperature, and zincate concentration.

The claim (R361) that organic additives have no influence on the mossy zinc deposit is contradicted by evidence (R571) that surfactants reduce the rate of growth at constant potential and also change the morphology of the deposit from mossy to dendritic. Similarly, surfactants reduce the propagation rate of dendrites (R571), and this is presumably caused by deposition occurring preferentially at points where no surfactant is adsorbed. Some surfactants also increase the cycle life of the zinc electrode (R345), and this improvement appears to be caused by the adsorption of surface active material on the electrode surface (R177). The effect of separators in reducing the rate of dendritic growth is ascribed (R357) to dissolved surfactants but later work (R115, 571) suggests that the major effect of separators is merely to reduce the rate of zincate mass transport.

## MECHANISM OF ZINC MOSS AND DENDRITE FORMATION

*Mechanism of Zinc Moss Formation.* The conditions favoring zinc moss formation are in contrast to those favoring the formation of many other metal powders and hence the mechanisms of these depositions are probably different (R309, 38-39, 360-361, 363-365). The mechanism of zinc moss formation is based on the following experimental findings: (1) no moss is formed at low current densities if purified solutions and an insoluble anode (Pt or Ni) instead of zinc is used; (2) moss is formed when zinc dust is added to the purified electrolyte; (3) moss appears only after a long time and is formed, initially, at the air/electrolyte interface; (4) the initial overpotential is small and is the same during the initial period, when there is a compact deposit, as during moss formation; and (5) addition of small amounts of ions of nobler metals (Hg, Pb, Sn, Se, As, Sb) inhibit moss formation. It was assumed that moss is formed at points on the cathode where colloidal zinc particles are deposited, postulating that the colloidal zinc particles are introduced into the electrolyte owing to the inhomogeneity of the anode surface and are transported to the cathode by electrophoresis. The metallic zinc particles may also be formed by corrosion of the zinc cathode by differential aeration and hence there is preferential moss formation near the electrolyte surface. The action of the noble metals is ascribed to the dissolution of zinc particles by electrochemical replacement.

However, later work has disputed these experimental results and the interpretation. Thus several workers have found that moss is formed with Ni (R515, 518), Pt (R21-22, 472-475, 477), and Au (R21, 22) anodes in unpurified (R21-22, 472-475, 477, 515, 518) and purified (R515, 518) electrolytes and that the time of appearance of moss was approximately the same with a nickel as with a zinc anode (R22, 518). Preferential moss growth has also been observed (R21, 474, 515) at the base of the electrode and not at the air/electrolyte interface and, at constant current, the overpotential has been found to decrease on the appearance of moss (R21-22).

Moss formation has been postulated to be caused by (R518) insufficient electroreducible species, for example, $Zn(OH)_3^-$ and $Zn(OH)_4^{2-}$, at the electrode surface. These were visualized as forming at the electrode/electrolyte interface by the decomposition of the complexes $Zn(OH)_6^{4-}$ and $Zn_2(OH)_6^{2-}$. However, for this mechanism to be correct, the dissociation of the complex ions into the simpler electroreducible ion must be rate limiting whereas kinetic studies and measurement of the relationship between the limiting current at a dropping mercury electrode and the square root of the mercury pressure indicate that complex dissociation is not the rate-determining step.

Moss formation may involve deposition at screw dislocations for, under magnification, it appears that helical spirals are formed with a minimum of side branches (R479). Thus the mechanism of moss formation is most likely connected with the relationship between the kinetics of zincate deposition and the nature of the electrode surface rather than with the transport of ionic or colloidal zinc species to the electrode surface. However the latter process may contribute to moss formation since smooth deposits are produced on a rotating disk electrode under conditions of exclusive activation control (R571). To elucidate the mechanism of the formation of this noncompact zinc deposit more information is required on the mechanism of electrodeposition processes in general and, in particular, on the incorporation of adsorbed zinc atoms into the lattice and the effect of potential on this process.

*Mechanism of the Initiation of Dendrites.* Dendrites of metals and compounds may be formed during electrodeposition (R642), solidification from melts (R590), vapor deposition, and crystallization from solution. Only two studies on the nucleation of zinc dendrites from an alkaline electrolyte on selected crystal faces have been reported (R458-459, 500C) although the nucleation of other metal dendrites and of zinc dendrites from acid electrolyte has been extensively studied (R46, 201, 340, 605 and references therein).

Protrusions (R500C) propagate more rapidly on the single crystals than on polycrystalline zinc. This fact was explained by assuming that there was competition between dendrite propagation and the formation of new dendrite growth sites and, since there are fewer defects in zinc single crystals than in polycrystalline zinc, the dendrites grow on single crystals whereas on polycrystalline zinc new nucleation sites are preferentially formed.

The major effect of substrate orientation was the lower density of small growths on the basal plane compared to the other planes (R458-459). This may be caused either by a variation in the number of impurity sites or to a difference in the uniformity of the surfaces or to a difference in the chemical activity of the plane. For copper deposition (R605), the nucleation sites were probably impurity centers; physical defects acted as nucleation sites primarily because such defects are preferred sites for impurity segregation. However, the number of surface defects (R458-459) on a zinc crystal was less than the number of growth sites and it was concluded that impurity effects were not causing the formation of dendrites. It was also shown by interference microscopy that the basal plane and other planes were of equal smoothness and hence the primary reason for the difference in nucleation densities of the faces was the variation of the exchange current density for zinc deposition on each face. This explanation is similar to that advanced for the deposition of copper (R117). In the latter case the exchange current density is smallest on the close packed plane and, since in zinc the basal plane is the closest packed plane, the exchange current density will probably be lowest on the basal plane of zinc and hence there will be fewer active centers on that plane (R458). A fourth explanation, not previously considered, is that a dissolution of the zinc surface on open circuit caused by local anodic currents results in additional surface defects over the number determined by measurement of the etch pit densities. This explanation can be checked since the degree of surface disorder, and hence the number of nucleation sites, will depend on the time the zinc electrode stands on open circuit.

Dendrite growth centers may be developed on growth planes of zinc either by screw dislocations or by foreign particles or by two-dimensional nucleation (R500C). The first mechanism is unlikely since the growth patterns formed during electrodeposition onto the basal plane of zinc consisted of closed loops and no spirals were observed (R500C). New growth centers are formed by periodic anodic pulses (R500C) and this may arise from the formation of zinc oxide particles or from two-dimensional nucleation. Verification of this point could improve the understanding of dendrite nucleation on practical, discharged secondary zinc electrodes.

MECHANISM OF MOSS AND DENDRITE PROPAGATION

Both noncompact deposit morphologies are affected by zincate ion mass-transport. However moss forms only when the zincate concentration gradient is minimized while for gross dendrite formation the reverse is true.

Three regions have been postulated (R571) along a growing whisker of moss: the tip where plating is almost entirely activation controlled (the reaction is probably controlled by surface adatom diffusion rather than charge transfer (R198) in this region); the intermediate region where the deposition reaction is activation- and mass-transport controlled; and a third region closest to the substrate where the process is entirely mass-transport controlled. Whiskers of moss grow away from each other because of mutual screening of neighboring whiskers and this results in the complicated network of dendritic growth. A semiquantitative analysis showed that the critical overpotential for the change in deposit morphology from moss to dendrites should occur at 100 mV, which is approximately the experimentally determined value (R144, 471-473, 571).

The theory of dendrite propagation (R37) of silver dendrites from a molten nitrate melt has recently been reviewed (R49). The assumptions of this theory are that deposit growth occurs mainly at the dendrite tip and that the tip of the dendrite is spherical. The quantitative relationship between the radius of curvature of the tip, $r$, the observed local current density, $i$, and the overpotential is such that (R49)

$$\text{diffusion overpotential} = \eta_d = \frac{RT}{(zF)^2 Dc} \, ir = K_d ir \tag{1}$$

$$\text{activation overpotential} = \eta_a = \frac{RT}{i_0 2F} \, i = K_a i \tag{2}$$

$$\text{curvature overpotential} = \eta_k = \frac{2\gamma v}{2F} \frac{1}{r} = \frac{K_k}{r} \tag{3}$$

where $D$ is the diffusion coefficient of the metal ion, $c$ is the concentration of the metal ion, $v$ is the molar volume of the metal, and $\gamma$ is the surface tension of the metal-electrolyte interface. The curvature overpotential arises from the shift of the reversible potential of a curved (higher energy) surface toward cathodic values with respect to a flat electrode. The terms in $r$ occur instead of the diffusion layer thickness since diffusion is spherical rather than planar. The total applied overpotential ($\eta$) is given by

$$\eta = \text{constant} = \eta_d + \eta_a + \eta_k = K_d ri + K_a i + K_k \frac{1}{r} \tag{4}$$

The first term results in an increase of local current density while the third term results in a decrease of local current density for a decrease in the tip radius of curvature. Thus the rate of dendrite propagation increases to a maximum value as $r$ increases and thereupon decreases with further increase in $r$. The initiation time for the formation of dendrites arises from the delay time before the growing surface extends outside the overall diffusion layer so that spherical diffusion can occur.

This theory has been used to explain the propagation of zinc dendrites. An extension of the theory to encompass mixed activation and diffusion control showed that the critical overpotential for dendrite formation is $-60\,\text{mV}$ and that the critical current density depends solely on the conditions of diffusion to the flat surface. From the relationship between the rate of tip propagation and the tip radius, the individual zinc dendrites were shown to propagate slowly and the formation of large numbers of dendrites on the electrode surface was explained (R145).

This model for dendrite propagation is consistent with the observation (R144-145) that, at constant overpotential, the current increases exponentially with time. However, at constant overpotential, the current initially decreases with time due to concentration polarization and then it increases (R471-472,

571). The current will increase exponentially with time only during the growth of the dendrites and the relationship will be modified if another electrochemical process, such as hydrogen evolution, occurs simultaneously at the electrode surface.

The theories of moss and dendrite propagation involve the transport of zincate ion but differ in one important respect. During the formation of moss, it is postulated (R571) that zinc is deposited on the sides and tip of a growing mossy whisker whereas, during the propagation of dendrites, zinc is deposited only on the dendrite tip. Thus, the two different noncompact morphologies formed during zinc electrodeposition from an alkaline electrolyte may arise either because of a difference in the propagation of the deposit or a difference in the initiation of the deposit under the experimental conditions.

The formation of a heavy dendritic sponge at overpotentials more cathodic than those for the formation of dendrites was attributed (R144) to a secondary process occurring during zinc deposition. The possible secondary electrode reactions postulated were (R144) either hydrogen evolution or cation codeposition. The former has been shown (R233) to affect copper, silver, and lead deposit morphology during deposition from an aqueous solution while cation codeposition has been shown (R423) to affect the deposit morphology during deposition of titanium from alkali-metal salt melts.

## APPLICATION OF FUNDAMENTAL DATA TO CELLS AND BATTERIES

The experiments performed with foil electrodes have characterized the deposit morphology obtained by electrodeposition of zinc from an alkaline solution. The results suggest further experiments to understand the mechanism and controlling parameters of the electrodeposition reaction and also suggest techniques for improving the cycle life of silver/zinc batteries. Some of the possible improvements have already been incorporated into model experimental systems either as a result of or independently from the fundamental studies. We shall only discuss how the results obtained in fundamental studies might be used to improve silver/zinc battery technology.

To obtain long cycle life from a secondary zinc electrode, it is essential to deposit the dissolved zinc adherently and uniformly over the surface of the substrate. Thus the cell design and charge conditions must be consistent with those for adherent deposition of zinc and minimum zincate ion concentration gradients within the cell.

### ZINC ELECTRODE SHAPE CHANGE

The cycle life of silver/zinc batteries is limited by a variety of phenomena. Of six conventionally assembled silver/zinc cells, three failed because of short circuits, two failed because of loss of zinc from the zinc electrode, and one failed because of disintegration of the separator (R483). Of six cells with loosely assembled elements, two failed because of shorting of dendritic zinc while four failed because of shedding from the zinc electrode. All six cells with tightly assembled elements failed because of the separator. However it is possible that, in all these experiments, the cycle life of the battery was limited by the shape change of the zinc electrode and that the apparent faults occurred as a result of the slowly changing electrode configuration. Since cell plates are usually only 1 mm thick, small changes can be very significant.

During the cycle life of a secondary zinc electrode, zinc migrates from the top and sides to the bottom of the electrode. Cells examined (R439) in the charged state after 100 deep cycles showed that the geometric area of the zinc electrode had decreased by 50%, that the thickness of the center of the electrode was approximately 15% greater than the initial value, that the electrode thickness was less at the edge than at the center of the electrode, that dendrites grew preferentially at the top and edges of the plate, and that the deposit at the center of the plate consisted of a layer of mossy zinc over a layer of zinc oxide.

Preferential dendrite growth was also observed (R473, 477) at the top and edges of foil electrodes during the electrodeposition of zinc from an alkaline zincate solution. It was considered that the preferential growth at the top of the electrode was the result of the zincate-depleted electrolyte in that region whereas the dendrite growth at the edges of the electrode was caused by spherical diffusion at the edge of the plate resulting in a faster dendrite growth at those points. Faster dendrite propagation was also observed at the edges of a porous zinc electrode (R472). This phenomenon has been given some attention recently and cells made using negatives with modified edges have shown a significant improvement in cycle life (R282).

The transfer of material from the top to the bottom of the electrode is a result of the greater zincate ion concentration at the bottom. This vertical concentration gradient occurs as a result of the density difference between KOH solution and KOH solution saturated with zinc oxide. There is no adequate explanation of the migration of material from the sides to the plate center but the presence of unreduced ZnO may be a contributing factor. Thus there are two problems, which need to be overcome in order to eliminate zinc electrode shape change; firstly the vertical concentration gradient must be eliminated and secondly all the zinc oxide must be reduced during the charge process. Solution of these problems may also eliminate or modify dendrite growth on a practical zinc electrode.

An attempt to overcome electrode shape change has been made (R254) by allowing space for the expansion of zinc at the center of the electrode and by providing excess zinc at the edges. The convex area of the electrode was filled with a nylon fiber mesh. It would seem, however, that the dimensions of the space might be critical and particularly the dimensions of the inserted spacer. If the dimensions permit convection in the electrolyte, vertical zinc transfer will occur.

Studies on electrode-shape change indicate that zincate ion transport occurs within a practical cell although, of course, to a lesser extent than in the free electrolyte studies reviewed earlier in the paper. Zinc transport throughout a silver/zinc battery has also been demonstrated by Palagyi (R482-483). In these studies the center electrode of a regular silver/zinc cell was a labeled $^{65}$Zn electrode. Measurement of the radioactivity level of the originally nonlabeled electrodes showed that both the zinc and silver electrodes were contaminated after 40 cycles with labeled $^{65}$Zn and that the degree of contamination was greater when the cells were assembled loosely.

It is unlikely that this zinc transport occurs only via zincate ion diffusion for this mode of transport is relatively slow and hence it is probable that convective flow is significant in transporting material through the cell. This is a result of forced convection (e.g., shaking) or of natural convection. The latter originates close to the zinc electrode surface resulting from electrolyte density differences and hydrogen evolution. Its importance in transporting zincate ions to foil electrodes has been demonstrated (R472, 477) by eliminating convection in a capillary. It is of course known that tight packing of plates in a cell is essential and closer attention to plate flatness and separator flatness may prove worthwhile.

MODE OF CHARGING

*Charge Rate.* The rate of charge of a secondary zinc electrode will affect both the zinc deposit adherency and the concentration gradients within the cell. Studies (R458, 477) with foil electrodes suggest that dendrite formation may be eliminated on a practical electrode by charging at a current density less than the critical current density since the mossy deposits are more even with fewer protrusions. However the even growth of moss is not necessarily an advantage since it is largely nonadherent and leads to a loss of capacity of the electrode. Another advantage of low charge rates is that the zincate ion concentration gradient within the cell is minimized; this must tend to limit the redistribution of zinc over the electrode surface when charging at low overpotential (R475). Ultimately however at low overpotential, poor zinc adherency will become a limiting factor which will be more apparent at higher depths of discharge. Higher overpotentials are necessary for good adherency but this must be applied in a discontinuous manner to prevent dendrite propagation.

*Intermittent Charging.* Deposition from alkaline electrolyte onto foil electrodes either with a pulsed current source (R21-22, 39, 458, 473-474, 477, 517) or with an alternating current superimposed on a direct current (R38, 515) or with periodic reversal of the current (R22, 39, 500B) increases the zinc adherency and produces a smoother deposit compared to that obtained with a direct current source. It is expected that a similar beneficial increase in deposit adherency will be obtained with porous electrodes and that the zincate concentration gradients will be minimized as a result of zincate mass transport during the OFF or anodic period. However, few studies of the intermittent charging of practical cells have been reported in the literature. It was claimed (R515) that charging Ag/Zn batteries either with an asymmetric alternating current source or with a pulsed current source with ON and OFF times of 20 msec, respectively, decreases the formation of dendrites on porous zinc electrodes but no data on cycle life or polarization curves are given nor is any comment made on zinc electrode shape change.

Pulse charging (R255) with the characteristics of the cell determining the cycle regime was suggested as the best way to apply the intermittent charging technique. However no data on the cycling of silver-zinc cells with this charging mode are available.

### CHANGES IN THE DESIGN OF ELECTRODES AND CELL

*Nature of Zinc Electrode.* Two practical modifications of zinc electrodes have been made which are consistent with findings on foil electrodes. Cells made from square electrodes have better cycle life than cells with tall, narrow electrodes (R253). This improvement is presumably attributable to smaller, vertical zincate concentration gradients in square cells. Secondly, the cycle life of a silver/zinc cell was reportedly increased from 500 to 800 cycles when 1% lead oxide was added to the zinc electrode (R291). These two factors should be largely independent of each other.

*Nature of Electrolyte.* It has been suggested that dendrite formation on a practical electrode will be prevented by limiting the solubility of the zinc oxide. The addition of aluminum hydroxide (R286) to the KOH solution reduced the rate of dissolution and the solubility of ZnO. When aluminum oxide was added to a secondary zinc electrode the cycle life of a silver zinc battery was also increased although the capacity was low. However when the charging rate was less than 5 mA/cm$^2$ the loss of capacity relative to unmodified cells became small after 10 to 20 cycles. Several workers have claimed (R516, 602, 657) that addition of calcium oxide to a secondary zinc electrode reduced the zincate concentration and increased the cycle life of nickel-zinc cells but later studies (R254) showed that the addition of this material resulted in a 25% loss of capacity of the zinc electrodes at the end of 20 cycles. Further work is required to determine the effect of charge rate control on the electrode capacity. However, the volume increase resulting from the inclusion of Ca(OH)$_2$ in the zinc electrode adversely affects cell volume, energy and power density (R254).

Electrolyte composition is quite important since factors which promote the rapid transport of zincate ions, such as high zincate concentration or a low electrolyte viscosity, suppress the tendency toward dendritic growth. Thus (R474) dendrites were formed in NaOH solution more readily than in KOH solution. Both rate of zinc oxide dissolution and the solubility of zinc oxide are important factors; both of these parameters are larger in NaOH solution than in KOH solution of the same concentration (R512).

*Cell and Negative Plate Redesign.* One design concept, which reportedly (R217) overcomes many of the problems associated with the secondary zinc electrode, is the fluidized bed electrode. This consists of small inert glass spheres on which zinc is deposited. The particles are fluidized by agitation of the electrolyte and can be charged or discharged at 1 A/cm$^2$ when they are brought into contact with a permanent conductor. The high Reynolds numbers achievable in such systems facilitates rapid transfer of zincate to the surface of the particles resulting in the formation of smooth adherent deposits. The use of small particles provides areas similar to conventional porous negative plates but when fluidized they are not subjected to limiting zincate concentration gradients.

## FUTURE RESEARCH

It has been shown both fundamentally and in silver/zinc cells that dendrite growth, deposit adherency and gross zinc redistribution are all capable of causing cell failure. It has also been demonstrated that these effects can be individually controlled by modification of the operating and design parameters. However, since there are several possible sources of cell failure, all must be closely controlled for it is pointless (and expensive) to overcontrol one failure mode and not to control the others adequately.

At present it is not possible to correlate failure modes of silver/zinc cells statistically to design or operational parameters. More effort is required to identify the rate of occurrence of the various failure modes so that applied research can be directed to eliminating or lowering the rate of occurrence of the most common mode. It is in this way that steady progress in increasing battery cycle life can be achieved.

Another desirable approach, which could be conducted parallel to the preceding one, is the optimization and combination of operating and design parameters. Design factors to be considered are electrode and electrolyte composition and electrode dimensions, porosity, uniformity, and tolerances. These must be combined with the optimum values of the operating parameters such as temperature, charging current and potential, and the use of intermittent charging. A serious attempt should be made to identify the major interactions between the design and operating parameters before a tradeoff is attempted.

This will, of course, be somewhat limited by our ability to interpret correctly the fundamental phenomena reviewed in this paper.

Lastly, in the area of cell technology, there is a serious lack of published information on porous electrodes. It is vital that changes in the physical nature of the porous zinc electrode be monitored and recorded during cycling for, at present, our knowledge is limited to secondary phenomena such as change in dimensions, dendrite formation, dischargeable capacity, and $E/i$ characteristics. More basic physical measurements characterizing the electrode structure are badly needed. New electrode structures can probably be developed if the critical parameters of electrode design can be established.

From a fundamental viewpoint we still lack an understanding of many phenomena. In particular we feel that the mechanism of the formation of mossy deposits, the role of hydrogen evolution in determining the deposit morphology, the effect of lead on the deposit morphology, and the mechanism of zinc oxide formation and its electroreduction should be elucidated.

This review demonstrates that the fundamental studies have helped greatly to understand the limitations of the cycle life of silver/zinc batteries. Continuation of these studies is necessary to aid the optimization of the parameters affecting operation of the secondary zinc electrode. Many of the findings will be equally applicable to nickel/zinc and zinc/air cells.

# 7.

# Improvement of Performance of Zinc Electrodes

### George A. Dalin

Zinc is selected for the negative electrode in aqueous cells because of its high half-cell voltage, low polarization, and high limiting current density. In addition, its equivalent weight is fairly low. Although it is thermodynamically unstable in aqueous alkali, light amalgamation brings the corrosion rate within tolerable limits. As a result of these properties, it has proved useful in combination with silver oxide in both primaries and secondaries.

## ZINC IN PRIMARIES

A suitable method of preparation of zinc electrodes in the charged state is by electrodeposition on copper sheet. The product, after washing, is amalgamated either in mercuric chloride or acetate, with the latter giving somewhat better results. At the 2% amalgamation level, the corrosion rate in 31% KOH is cut by an order of magnitude. The use of potassium hydroxide containing 80 g/liter of zinc oxide also cuts the corrosion rate; the effect is stronger at lower concentrations of KOH (R166).

For use in primaries, the zinc is protected against shorting to the positive by a porous sheet, which may be cellulose, asbestos, or synthetic fiber. A difficulty that appears after some hours of activated wet stand is shorting of the cell due to the formation of zinc growths. In studies of this problem in our laboratory, it has been shown that zinc grows only after silver has migrated to the negative. The silver migrates as soluble argentous ion and is reduced on contact with the zinc. Consistent with this finding, zinc does not grow on wet stand when the counterelectrode is a charged nickel hydroxide positive.

The mechanism of growth may involve accelerated corrosion at points to which silver has migrated. It has been shown (R251) that zinc corrodes more rapidly in contact with a grid of copper or silver, which supports the above hypothesis. The accelerated corrosion at local regions would lead to concentration cell effects and formation of zinc growths. A study (R534) of the effect of various ions on the corrosion rate showed that ions containing Cu, Fe, Sb, As, and Sn increased the zinc corrosion rate while ions with Cd, Al, and Pb decreased the rate.

At low temperature the limiting current density drops off sharply as does the solubility of Zn(OH)$_2$. That the surface film is Zn(OH)$_2$ is consistent with a four-step mechanism proposed for a zinc electrode discharge (R199). In this mechanism the last step is the dissolution of the zinc hydroxide to form the doubly charged ion. It has been shown (R354) that Zn(OH)$_2$ forms zincate ion rapidly on discharge when the mean activity coefficient of hydroxyl ion in the KOH solution exceeds one. This is the case for the high KOH concentrations used in silver-zinc cells. When the activity coefficient is less than about 0.9, the rate at which this hydroxide goes into solution is low and the orthorhombic hydroxide is present as the solid phase. The conversion of the zincate ion (R551) into the negatively charged colloid with the stoichiometry, Zn$_2$(OH)$_6^{2-}$ would account for the fact that zinc can be driven into solution electrochemically in quantity which greatly exceeds the equilibrium solubility for potassium zincate.

The peak current density for zinc discharge was shown to increase with stirring by a potential sweep method (R175). The time to passivation at a given current density was found (R573) to drop as the temperature was lowered. These findings point to the formation of a zinc hydroxide film during discharge which must be removed by dissolution for the reaction to continue. At sufficiently high discharge rates, the rate of dissolution is inadequate to maintain the surface free of hydroxide film.

Such a film can also form in the normal method of preparing the zinc electrode (R142). The film was removed by washing in succession with concentrated ammonium chloride, dilute ammonium chloride, and, finally, water.

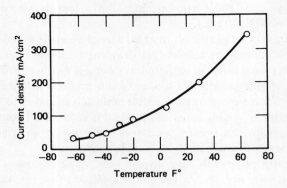

**Figure 34**    Limiting current density as a function of temperature; zinc electrode, 2% Hg, acetate method.

The limiting current density for electrodes amalgamated to 2% mercury and treated by the ammonium chloride procedure is shown in Figure 34 (R166). The utilization of similarly treated zinc, in zinc limiting cells, is shown in Figure 35 from the same work. Utilization as presented in Figure 35 is based on a theoretical value of 1.22 g of zinc/Ah. From the data in Figure 35, the appropriate quantity of zinc for design of primary batteries can be calculated.

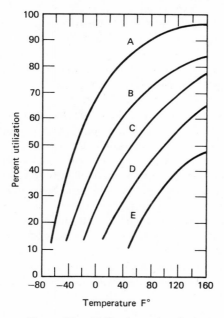

**Figure 35**  Utilization, zinc limited cells. Current density (mA/cm$^2$): (A) 10, (B) 30, (C) 60, (D) 150, and (E) 300.

## ZINC IN SECONDARIES

Since zinc is highly reversible, it is suitable for use in secondaries also. Electrodes can be made by mold pressing or pasting zinc oxide with mercuric oxide and other additives or by electrodeposition. This latter method is preferred where the cell is to be supplied in the charged state.

Under the usual charging conditions, the zinc grows out from the surface and if unchecked would soon short out the cell. However, André demonstrated (R15) that the use of one or more layers of semipermeable membrane as separator could delay shorting for many cycles.

Since discharged zinc goes into solution as zincate, the zinc is free to migrate, mostly along the face of the plate, but also out into the edge spaces of the cell and over the top of the separator. On recharge, the zinc can be deposited either as fine, whiskery moss or as dendrites. The form (R471, 472) depends on the overpotential during deposition. At low overpotentials (below 75 mV) moss is formed; above 100 mV dendritic growth occurs. Moreover, it (R571) has been shown that penetration of separators by zinc occurs by growth through the separator and not by mechanical puncture. The growth consists of deposition of zinc within the membrane, and this growth is always dendritic in form. Separator penetration (R440) can be avoided by limiting the overpotential of the zinc electrode during charge. So long as the overpotential is held below a critical voltage, overcharge can be continued indefinitely. The critical voltage varies with the separator; for cellophane, it is about 75 mV, and it is higher by 25 to 50 mV for separators with higher resistance.

The mechanism of zinc dendrite formation has received much attention. Powers' (R500A) particularly fine photomicrographs of zinc deposition show deposition on basal planes. Also, the initial formation of blunt pyramids and subsequent growth of dendrites can be seen. One approach (R145) has postulated that growth takes place preferentially on those points of the microrough surface protruding into the diffusion layer and that the growth at time $t$ is proportional to exponential $t/\tau$ where $\tau$ is a time constant which depends on the overvoltage, the limiting current density, and the thickness of the diffusion layer. The type of deposit then depends on whether spherical diffusion to the tip takes over.

A study (R357) on preventing dendritic penetration of separators by the use of a surfactant led to the proposal that such a compound be incorporated in the separator. This material would be adsorbed on the tips of dendrites approaching the separator and would prevent growth in that direction.

A series of reports (R344) describes studies of various surfactants incorporated in the zinc electrode to prevent penetration. Of these, the most effective was Emulphogene BC610 (GAF Corp.; Reference 173). In one test, the BC610 was used both with and without 2% of PVA as binder. The cells were made with three layers of regenerated viscose sausage casing to ensure that failure would occur by capacity loss and not by zinc penetration. The results for cycling on a 40% depth of discharge regime are given in Table 12. It is evident that 2% PVA in combination with BC610 is harmful. In other tests it was shown that concentrations above 1% BC610 give results which are less favorable than those obtained at 1% and 0.5%. Peculiarly, results obtained at 0° and 38° are inferior to those obtained at room temperature.

When the thickness of separator is increased to the point where failure by shorting is postponed beyond 40-50 cycles, loss of capacity becomes a problem. Examination of the negatives at faces of transparent cell cases shows that large areas of the plates have been denuded of zinc. This loss of zinc has been variously termed "slumping, washing, shape change," etc. It first becomes evident as loss of zinc from the upper corners of the grid. After 15 to 25 deep cycles the edges of the grid are bare and zinc will have thickened in a region about $\frac{2}{3}$ of the way down the plate; there will also be sponge zinc in the lower part of the edge space of the cell.

TABLE 12.   EFFECT OF BC610
AND PVA ON CYCLE LIFE

| BC610 Concentration, % | No PVA | 2% PVA |
|---|---|---|
| 1.0 | $559 \pm 35$ | $423 \pm 23$ |
| 0.75 | $509 \pm 23$ | $385 \pm 16$ |
| 0.5 | $613 \pm 65$ | $335 \pm 46$ |

Examination of dissected cells which have displayed loss of active zinc area always shows that the positive immediately opposite the denuded region is black and therefore undischarged, regardless of the state of charge of the cell. Obviously, the loss of active zinc area results in a proportionate loss in capacity. In addition, the loss of area also leads to penetration of the separator by the following mechanism. Silver-zinc cells are usually charged at constant current, the current being unchanged as the cell ages. As the active area decreases, the current density based on active zinc area then increases, and eventually the critical overvoltage for zinc dendrite growth and penetration of the separator is exceeded.

In studies of the mechanism of zinc deposition and of the morphology of the deposit, experimental cells without separator have been used. In these studies it was found that adhesion of the zinc to the substrate was weak. This, of course, could contribute to slumping. For plating (R21) under diffusion control (30 mA/cm$^2$), adherence is best on Sn, next best on Pb, and poorer on Cu, Ag, Cd, Zn, and amalgamated Zn. Since most cells use a negative grid of copper or of silver-plated copper, this poor adherence of the plated zinc might seem to be a factor in shape change; gravity would therefore be involved. Also, the net movement of the zinc is toward the bottom of the cell. Since zincate solution is denser than KOH, this again would indicate that gravity is a factor. Cells cycled (R439) in a variety of positions showed that the distribution of zinc on the plate is independent of the cell orientation, so gravity cannot be the principal factor involved. By holding the electrolyte level constant, it was shown that movement of the electrolyte between positive and negative compartments was not responsible for the shape change. In cycling a 2-plate cell in which the silver and the zinc plates were deliberately placed out of register, the zinc during 15 cycles slumped from the plate edge overlapped by the silver; at the other side of the cell, where the zinc extended beyond the silver, the zinc was unaffected. In Figure 36, the cell on the right shows shape change of the end negative when the plates are aligned. The left-hand cell shows the effect of misalignment. The zinc overlaps the silver at the left edge; the converse is true at the right edge.

The local current density must obviously be high on the zinc edge overlapped by the silver plate and low on the zinc edge which extends beyond the silver. The conclusion is that erosion takes place where the current density is high. In cells as usually built, all plates are fairly well aligned, but current density must

**Figure 36**     Shape change as a function of alignment.

be higher at the edges because of the extra current path in the solution in the edge space. It was also found that mechanical removal of a small quantity of zinc from a point in the center of the plate during construction of the cell would generate erosion around the blemish during cycling. The same is true of unintentional flaws such as holes or cracks in the plate.

The mechanism of the transfer of zinc must involve diffusion since the rate of migration is so low. For diffusion to take place, the concentration of zincate must vary over the face of a plate. McBreen (R439) cut a plate in half horizontally, connected the halves in parallel to a voltage source, and then measured the current to the two half-plates during discharge. At one stage the bottom plate charged while the top was discharging. Evidently the combination of the two half-plates formed a concentration cell, implying a substantial difference in zincate concentration between the top and bottom of the cell.

As a solution to this problem, cells are constructed with negatives slightly larger than the positives. They are then positioned in the cell pack so that the negative overlaps the positive at all edges. The extent of overlap is not critical, about 3 mm being adequate.

The addition (R581) of a small quantity of Teflon emulsion to the zinc oxide mix can decrease zinc mobility and increase capacity retention. Quantities from 1 to 5% are effective. The cell voltage is not adversely affected by the presence of

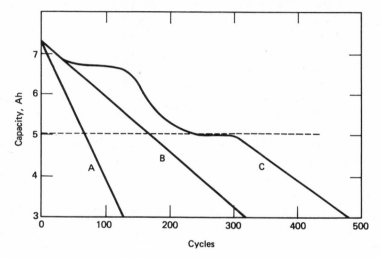

FIGURE 37    Capacity as a function of cycle life for dished electrode cells: (A) control; (B) extended edges plus Teflon; (C) extended edges plus Teflon plus dished profile.

Teflon, even at current densities as high as 0.22 A/cm$^2$. Figure 37 (R256) shows the effect of Teflon in combination with oversize negatives.

Since capacity loss is associated with loss of zinc from portions of the negative, it should be possible to delay the onset by starting with a greater mass of zinc per nominal ampere-hour (R371). Additional data (R438) show that a zinc-to-silver ratio of 1.5 was superior to 1.2, which in turn was superior to 1.0. Unfortunately, high ratios of zinc to silver cannot be used where cell weight is of primary importance. Where long cycle life is the principal requirement, this design modification is useful.

A further modification (R256) is based on the migration of zinc from the edge of the plate toward the center. Plates constructed with additional zinc at the edges and decreased zinc at the center were assembled into cells whose performance may be compared by the capacity-cycle curves in Figure 37.

The incorporation of insoluble hydrous metal oxides (R43) in negative plates and their use in cells with ceramic separators for shallow cycling would appear to be particularly suitable for elevated temperature operation.

The fact the the nature of the plated zinc is affected by the overvoltage was mentioned above. Some control can also be effected by the use of pulse plating, periodic reversal of current, and asymmetric alternating current. The latter at frequencies from 50 to 200 Hz was claimed to improve cycle life (R514-517). Simultaneously, the ends of the plates were wrapped with polyethylene film to discourage growth of zinc at the edges. Also, the combination of intermittent discharge (periodic reversal) and edge wrapping was recommended. (R472-473). Apparently, the tips of the advancing dendrites deplate on reversal more readily than do the edges, so that the resultant growth is thinner and denser. With respect to pulse plating, the benefit is thought to derive from decrease in the thickness of the diffusion layer during the OFF period.

It should be noted that these techniques are inapplicable when electrical efficiency is important. Periodic reverse and asymmetric a-c obviously transform electrical energy to heat. Interruption of the current requires that oversize components be used to handle the pulses.

Work (R560) using circulating electrolyte in a zinc-air battery warrants mention, since the circulation is a means of controlling zinc deposition which may prove applicable to the Zn/Ag system. The procedure involves continuous circulation with the following actions:

1. During discharge the concentrated zincate is transferred to a tank where the hydroxide (or oxide) settles out. The solution returns to the cells to collect more zinc.

2. During charge, zinc is deposited from the solution on the negatives. The depleted solution is returned to the tank, where it dissolves more zinc hydroxide.

The phenomenon on which the process is based is the deposition of zinc in relatively smooth, nondendritic form from flowing electrolyte. The rapidly moving electrolyte cuts the thickness of the diffusion layer so that a dense deposit forms.

As can be seen, the problem of zinc growth has been treated in a surprising variety of ways. Some of these ways, such as incorporation of Teflon and other additives, extension of the negative edges, variation in thickness over the plate and increase in the zinc-to-silver ratio, can be used in essentially standard cell constructions. Others, such as the use of ceramic separators and circulating electrolyte, require radical redesign of the cell. The early limitation of the zinc-silver oxide cell to a few cycles has been lifted to the point where as many as 500 cycles at 60% depth of discharge and at the 2.5-hour discharge rate have been achieved.

Several of the above techniques are still under investigation. One objective is to find the optimum combination of the electrode modifications. The search for better additives will undoubtedly continue. Such additives may also prove useful in increasing wet shelf life of primaries.

# PART III

## SILVER ELECTRODE, FUNDAMENTAL CHEMISTRY AND ELECTROCHEMISTRY

# 8.

## Chemistry of the Silver–Silver Oxide Electrode

T. P. Dirkse

Silver is one of the second-series transition metals. As such it is likely to have several oxidation states. In the periodic table it is a member of the subgroup which also includes copper and gold. As a member of this group it could have oxidation states of +1, +2, and +3. The corresponding oxides are $Ag_2O$, $AgO$, and $Ag_2O_3$.

Metallic silver is stable to ordinary environments. It is also stable with respect to an alkaline environment such as aqueous KOH. Silver is well known for its high electrical conductivity. Its specific resistance is $1.59 \times 10^{-6}$ ohm cm (R387A). Silver is permeable to oxygen. The ability of silver to allow oxygen to diffuse through it (R387) has been explained as being caused by the formation of $Ag_2O$ (R564).

The crystal structure of silver has been the subject of much investigation. It appears that there is but one crystal form, a face-centered cubic arrangement. The edge of the unit cube has a length of 4.0856 Å, and the distance between the centers of the nearest silver atoms is 2.888 Å (R35).

Although the crystal lattice seems fairly well established, there is some question about the electronic configuration of the silver atoms. The K, L, and M shells of the silver atoms are complete, as are the 4s and 4p orbitals. This leaves 11 electrons to be allocated. One possibility is to assign 10 of these to the 4d orbitals, leaving one for the 5s orbital. This is in agreement with the very common monovalent state of silver. However, it has also been suggested that there are 3 unpaired electrons (R336).

In order to decide on the allocation of the electrons, one must also confront the fact that silver is diamagnetic (R237). This is surprising because the silver atom has an odd number of electrons so that at least one would be unpaired. Such an unpaired electron gives rise to paramagnetic behavior. Thus the diamagnetism of metallic silver suggests that there is electron pairing between the silver atoms.

## SILVER(I) OXIDE

This deep brown oxide is a well-defined one so far as its structure is concerned. It gives a definite X-ray pattern, has but one crystalline modification and is isomorphous with $Cu_2O$. It can be prepared by treating a solution of silver ions with KOH or NaOH if no strong oxidizing agents are present. A white precipitate, likely AgOH, may form at first but it almost immediately changes to the $Ag_2O$. However, AgOH has not yet been isolated. For its structure, $Ag_2O$ has a face-centered cubic arrangement of silver atoms (or ions) interpenetrated by a cubic arrangement of oxygen atoms (R122) so that each oxygen atom is in the center of a tetrahedron of silver atoms. The formation of $Ag_2O$ may be considered as the spreading out of the basic silver structure by increasing the length of the cube edge. Two oxygen ions are introduced into the unit silver cell. The distance between centers of the nearest silver atoms is 3.336 Å and that between silver and oxygen is 2.043 Å, suggestive of covalent bonding.

A significant characteristic of $Ag_2O$ for its use as an active material is its high electrical resistance, the specific resistance being about $10^8$ ohm-cm (R392). This high resistance and the diamagnetic property of $Ag_2O$ indicate the absence of mobile or unpaired electrons in the $Ag_2O$ lattice.

Another characteristic which has a special importance is the solubility of $Ag_2O$ in alkaline solutions. The solubility increases with hydroxide ion concentration, suggesting amphoteric behavior (R389). The solubility product is of the order of $2 \times 10^{-8}$ at 25° (R324, 389, 457) and increases with increasing temperature (R317). The soluble silver and its transport to the zinc electrode have been considered the root source of certain problems in the behavior of zinc-silver oxide cells.

Solubility measurements made at KOH concentrations ordinarily used in batteries (up to 45%) show that the solubility of $Ag_2O$ reaches a maximum ($4.7 \times 10^{-4}N$) at about $7N$ KOH (R12), Figure 38. No satisfactory explanation has been given for this maximum. It may not be precisely reproducible. Kovba and Balashova (R355) found that the solubility increases with KOH concentration up to $10N$ KOH, in which solution it is $6 \times 10^{-4}N$. When solubility values are determined by adding solid $Ag_2O$ to KOH solutions with stirring, the amount of dissolved $Ag_2O$ as determined by analysis goes through a maximum after about an hour and then decreases to a steady, or equilibrium, value in another couple of hours.

The solubility values of $Ag_2O$ in the higher KOH concentrations are not very reproducible. The values one obtains appear to depend on the time required for filtering the solution and on the method of shaking or agitation (R168). Thus it is difficult to state precisely what the solubility values are in the more concentrated KOH solutions.

This erratic behavior may be caused by the tendency of silver(I) oxide

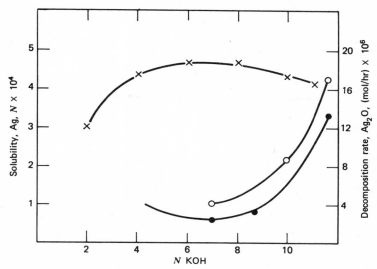

**Figure 38**   Properties of $Ag_2O$: (x) solubility in KOH solutions (R12); (○) and (●) decomposition rate of dissolved $Ag_2O$ at room temperature; (○) show clear flasks; (●) show opaque flasks.

dissolved in KOH solutions to undergo a decomposition. This was noted by Laue (R389) and has been studied extensively more recently (R163, 167). Solutions of $Ag_2O$ dissolved in KOH slowly form a finely divided black deposit of metallic silver. This reaction is hastened when reducing agents such as cellulose are present, but it still occurs to some extent in scrupulously clean solutions. Laue (R389) suggested that this reaction was caused by traces of oxidizable impurities. However, the reaction is thermodynamically spontaneous in the absence of such impurities (R58).

The rate of decomposition increases with increasing concentrations of dissolved $Ag_2O$ (R168). The effect of KOH concentration is more complex (R167). There is a reversal in the rate of decomposition with increasing KOH concentration (Figure 38), which also shows that there is a photodecomposition of the dissolved silver oxide. However, the variation of rate of decomposition with KOH concentration is the same whether or not the samples are exposed to ordinary daylight.

The rate of decomposition increases with increasing temperature and it also increases when solid zinc oxide is in contact with the solution (R168).

There has been some disagreement as to the nature of the dissolved $Ag_2O$. Pleskov and Kabanov (R492) suggested that it is a polynuclear complex, $Ag_3O(OH)_2^-$. Others, however, consider it to be $Ag(OH)_2^-$ or perhaps $AgO^-$ (R18, 157, 324). For $Ag(OH)_2^-$, $\Delta G°_{298}$ has been calculated to be $-57.07$ kcal/mol.

Silver(I) oxide does not decompose when dry at room temperature. The equilibrium oxygen pressure at $25°$ is 0.26 mm and it does not reach 1 atm until the temperature is $180°$ (R248). A thorough study was made of the thermal decomposition of $Ag_2O$ (R6) by heating $Ag_2O$ in a vacuum; the mechanism of the decomposition depended on the temperature range. Less than 10% of the total oxygen was evolved at temperatures below $300°$.

## SILVER(II) OXIDE

Silver(II) oxide has been known for some time and is commercially available as a grayish black solid. It has been designated as $Ag_2O_2$, AgO, and $Ag(I)Ag(III)O_2$. The first formula probably was suggested when this substance was thought to be a peroxide. It is sometimes called silver peroxide but there is no positive evidence that it is a peroxide.

It is possible to prepare AgO in a variety of ways. Silver or silver salts can be oxidized with ozone, persulfate, or permanganate. It can be prepared from acid as well as from alkaline solutions. When silver or $Ag_2O$ is electrolytically oxidized in strongly alkaline solutions, AgO is produced. However, all these preparations do not give exactly the same X-ray diffraction patterns (R157, 326, 553). Samples prepared from an alkaline solution have slightly different $d$ values than have the samples prepared from acid solutions.

So far AgO has not been prepared in a pure form. Ordinarily the samples that have been studied have had a purity of about 98%. This analysis has generally been based on a measurement of oxidizing power. The fact that the preparations of AgO always contain some (about 2%) impurities probably accounts for many of the differences that have been noted and reported for the characteristics of this substance.

The specific resistance of AgO has been reported as 0.012 ohm-cm (R326), 15 ohm-cm (R460), 10 ohm-cm (R392), and more recently as 59.3 ohm-cm at 2.1 kilobars (R597). Some of this discrepancy may be explained by the presence of impurities in the AgO and also by the fact that AgO is a semiconductor which has an excess or deficient oxygen lattice (R446). Very likely AgO is a nonstoichiometric oxide.

### STRUCTURE

X-ray and neutron diffraction studies (R260, 445, 543) have been interpreted to indicate a monoclinic structure with two Ag–O distances, namely 2.18 Å and 2.03 Å, corresponding to Ag(I)–O and Ag(III)–O distances. X-ray studies (R444) of a tetragonal form of AgO have been interpreted as having two Ag–O distances but, in this case, Ag(II)–O and Ag(I)–O distances.

This raises the question concerning the oxidation state of silver in AgO. Electrochemically, AgO behaves as if all the silver has an oxidation number of +2

(R141). This would also follow if half the silver were +1 and half +3. The arguments for the latter position are the two Ag–O distances and the fact that AgO is diamagnetic (R446, 460). Both Ag(I) and Ag(III) have an even number of electrons. These could all be paired and the resulting product would then have no paramagnetic moment. In Ag(II), on the other hand, there is at least one unpaired electron which should give rise to some paramagnetism. However, metallic silver also has an odd number of electrons and shows diamagnetic characteristics, as noted earlier. Some (R557) have suggested that the magnetic characteristics are a function of the lattice arrangement and not of the oxidation state of silver. Furthermore, the argument from the number of electrons is based on an ionic form of silver in these oxides, whereas the bonding may be covalent.

## SOLUBILITY

Because of the higher oxidation state of silver in AgO one would expect it to be more soluble than $Ag_2O$ in alkaline solutions. However, there is some doubt as to whether AgO will dissolve in KOH solutions. Solubility measurements have been made and the values obtained are substantially the same as for $Ag_2O$ (R157). As with $Ag_2O$, the solubility of AgO in KOH solutions passes through a peak value within an hour or so and then reaches a lower steady-state value in a few days. Some who tried to make solubility measurements were unable to obtain reproducible values and they attributed this to the strong oxidizing power of AgO (R355).

A study of the diffusion currents in solutions prepared by dissolving AgO and also $Ag_2O$ in aqueous KOH gave the same values for both types of solution (R12). From this it was concluded that AgO does not dissolve as such in KOH solutions but is decomposed and produces merely a solution of $Ag_2O$ in KOH. However, evidence (R104) has been marshaled for the existence of $Ag(OH)_4^-$ in KOH solutions. These solutions were not prepared by adding AgO to aqueous KOH but by anodizing silver in KOH solutions. The solutions had a yellow color. The solubility of the $Ag(OH)_4^-$ increased with increasing KOH concentration up to $12N$ and was considerably greater than that of $Ag_2O$. The solubility of $Ag(OH)_4^-$ in $12N$ KOH was $3.2 \times 10^{-3}N$ while that of $Ag_2O$ in the same KOH solution was about $4 \times 10^{-4}N$. McMillan (R448) made a magnetic-moment study of solutions of AgO in KOH and concluded that the solute species was $Ag(OH)^-_4$.

In summary, when AgO is added to KOH solutions the equilibrium solubility values obtained are the same as when $Ag_2O$ is used. The uncertainty has to do with the question whether the solute species contain silver in a higher oxidation state than +1.

Using the phase rule (R212) it was shown that, as long as some $Ag_2O$ is present with the AgO, there is only one solubility value possible. Because AgO does decompose slightly to $Ag_2O$, or may even contain some $Ag_2O$, both oxides are likely to be present when solubility values of AgO are determined. Thus, while the solubility of AgO may appear to be the same as that of $Ag_2O$, this does not, in itself, say anything about the oxidation state of the dissolved species or about the extent of solubility of pure AgO.

### STABILITY

While AgO is stable at room conditions it is more sensitive to thermal decomposition than is $Ag_2O$. It has been estimated that dry AgO should decompose at room temperature in about 5 to 10 years (R469). As the temperature increases, the rate of decomposition also increases, and above $100°$ decomposition is complete in a matter of hours or less. The rate appears to depend on the method of preparation. It is likely that there is also an induction period (R7). This thermal decomposition (R480) has an activation energy of 30 kcal/mol. The rate of decomposition appears to be governed by the progression of the interface towards the center of the crystal.

The following solid-state reaction has often been mentioned

$$AgO + Ag \rightarrow Ag_2O$$

but there are no good data available on the rate of this reaction.

The stability of AgO in contact with aqueous KOH has also received attention. The AgO dissociates to $Ag_2O$ and the rate of this dissociation increases with increasing temperature and increasing KOH concentration (R469). It has been estimated that the decomposition rate of AgO in 40% KOH is such that it would take about 1500 days to be complete at $25°$ (R72). The presence of ZnO reduces this rate of decomposition, apparently because it has an inhibiting effect on the further reduction of $Ag_2O$ to Ag (R58). However, it should be noted in this connection that solid ZnO increased the rate of decomposition of dissolved $Ag_2O$ (R168).

### SILVER(III) OXIDE

The evidence for the existence of $Ag_2O_3$ is indirect. It is extremely unlikely that a pure sample has been isolated. Jirsa (R320) claims that $Ag_2O_3$ can be prepared but cannot exist alone. Furthermore, it is readily decomposed by heating. The substance is said to have a cubic lattice (R577). Other attempts to determine the structure by the use of X-rays gave inconclusive results (R59). A cubic structure was suggested but the length of the cube edge depended on the method of preparation.

Two types of evidence have been presented for the existence of $Ag_2O_3$. The first is that of electrode potentials. Several investigators have obtained values about 150 to 200 mV above that associated with the $AgO/Ag_2O$ couple, that is, about 0.75 to 0.80 V on the hydrogen scale. Such values (R401) were found for a substance produced by oxidizing silver nitrate or sulfate anodically. It appeared to be a mixture of AgO and $Ag_2O_3$. It could not be produced by the anodic treatment of AgO in alkaline solution. Later (R24) it was suggested that these products were $Ag_3O_4$ or $Ag_4O_5$. However, these products also contained $AgNO_3$ or $Ag_2SO_4$.

In an attempt (R89) to study $Ag_2O_3$, the temperature $-40°$ was selected to slow down the decomposition of any $Ag_2O_3$ that would be produced. Silver was oxidized anodically in KOH solutions and the voltage-time curves were observed under a variety of conditions. Above a critical current density of 15 mA/cm$^2$ these curves change shape and completely bypass the $AgO/Ag_2O$ voltage level. Instead, there is a rise to the voltage level associated with oxygen evolution. On open circuit, the voltage then decays to a level well above that for $AgO/Ag_2O$. The authors suggest, but do not claim, that this may be due to the formation of $Ag_2O_3$. From other measurements the authors estimate the activation energy for the formation of $Ag_2O_3$ to be about 10 kcal/mol. This seems to be a rather low value for such a process. But, if any $Ag_2O_3$ was formed, it was transient and was not available for handling.

A second type of evidence is analytical. Some of this is a measure of the oxidizing ability of the substance (R322). However, this measures the total oxidizing ability of the compound and not necessarily just that of the silver. This is an important consideration when impurities are present, as they always are. Some of these may be pernitrates and other like substances which also have oxidizing ability. One such preparation had 9% impurities present (R484).

Other analytical evidence is based on total silver and oxygen content. This is insufficient to establish the oxidation state of silver. It has been shown (R158) that AgO can absorb considerable amounts of oxygen without altering significantly the electrode potential value or the X-ray diffraction pattern. In fact, on the basis of simple analysis, compounds such as $Ag_2O_3$ have been prepared. Yet on electrochemical reduction the quantity of electricity obtained corresponded to the existence of Ag(II) and not Ag(III).

Much of the evidence for tripositive silver has been obtained from products prepared in an acid solution or by the direct treatment with ozone. Compounds such as $Ag_7NO_{11}$, $Ag_7SO_{12}$, and $Ag_7FO_8$, obtained by electrolyzing solutions of $AgNO_3$, $Ag_2SO_4$, or AgF between platinum electrodes, have frequently been the source of so-called trivalent silver oxide. Such products may also contain pernitrates, peroxysulfates, or other per-compounds (R624).

Thus there seems to be some indirect evidence for the existence of $Ag_2O_3$. (There is considerably more evidence for the existence of tripositive silver in complex or coordination compounds). Some of the evidence may be interpreted in other ways but, whatever the interpretation, this $Ag_2O_3$ is of no significance for use in alkaline batteries. Its ampere-hour capacity and its voltage curve are the same as those for AgO.

# 9.

## Crystal Structures of the Silver Oxides

### Alvin J. Salkind

The silver oxides of most interest to manufacturers and users of silver-zinc cells are: $Ag_2O$ and $AgO$. Interest in the electrochemical properties of these materials has led, in recent years, to several investigations of their crystal structures. However, early efforts to understand their crystal structures were made over half a century ago.

Vogel (R616), in 1852, reported the preparation of $Ag_2O$ crystals which appeared as 4- or 6-rayed stars. He associated this with the cubic system when he obtained "microscopic isotropic crystals which could be grown in small octahedra." The earliest X-ray studies in 1922 (R122, 462, 643) reported the structure as cubic. The arrangement consisted of a body-centered cube of oxygen atoms and a face-centered cube of silver atoms. These two cubes interpenetrated as in $Cu_2O$ and had unit lengths of 4.69 and 4.26 Å, respectively. The position of the oxygen atoms (R643) in $Ag_2O$ is in the center of a tetrahedron of Ag atoms (Figure 39).

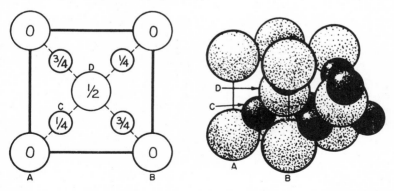

**Figure 39**    Left, a projection on the cube face of atoms in the unit of $Cu_2O$. The small circles represent the copper atoms. Right, a perspective drawing showing the packing of the atoms in $Cu_2O$ if they are given their ionic sizes. Letters identify corresponding atoms in left-hand and right-hand drawings (R644).

Since these early reports, subsequent investigations listed in Table 13 have shown only refinement of the lattice parameter. However, in his recent book, Wyckoff lists the space group of the $Cu_2O$ family as $T_h^2$ (Pn3). This is similar to Pn3m but lacks the mirror plane of symmetry. Recently, a new $Ag_2O$ phase was obtained at high temperatures and pressures (R332) but this hexagonal type structure is not of importance in the operation of silver-zinc batteries.

The structure of the higher valence oxide, AgO, has not been as clearly understood as that of $Ag_2O$, and the exact positions of the atoms were long in doubt. Reports on the preparation and reactions of silver monoxide date back to the middle 1800s, and among the list of investigators was Alessandro Volta. The

TABLE 13.   STUDIES OF STRUCTURE OF SILVER(I) OXIDE, $Ag_2O$

| Investigator | Results |
|---|---|
| 1922   Ralph W. G. Wyckoff (R643) | $Ag_2O$ structure considered similar to that of cuprous oxide: cubic, $a_0 = 4.768$ Å Space Group $O_h^4 - Pn3m$ (No. 224) with two molecules $Ag_2O$ per unit cell |
| 1922   W. P. Davey (R122) | $Ag_2O$ consists of a body-centered cube of $O^{2-}$ and a face-centered cube of the metal ion. These cubes interpenetrate like $Cu_2O$. The position of $O^{2-}$ in $Ag_2O$ is in the center of a tetrahedron $Ag^+$ identical with $I^-$ in AgI. |
| 1922   P. Niggli (R462) | Cubic, $a_0 = 4.728$ Å, sp. gr. 7.798. |
| 1926   V. M. Goldschmidt (R250) | Cubic, $a_0 = 4.73$ Å. |
| 1938   J. D. Hanawalt, H. W. Rinn, and L. K. Frevel (R275) | Identifying pattern reported (without indexing). |
| 1944   R. Faivre (R196) | Cubic, $a_0 = 4.734$ Å; reported that lattice constants vary between 4.697 and 4.736 Å, depending on heating |
| 1961   H. E. Swanson, M. C. Morris, R. P. Stinchfield, and E. H. Evans (R584) | $a_0 = 4.736$ Å at 25°. |
| 1963   Ralph E. G. Wyckoff, *Crystal Structures* (R644) | $a_0 = 4.72$ Å; The general classification of the cuprite family ($Cu_2O$) is listed as $T_h^2$ (Pn3) (No. 201). This is more general (without mirror symmetry) than $O_h^4 - Pn3m$. |
| 1963   S. S. Kabalkina, S. V. Popova, N. R. Serebryanaya, and L. F. Vereshchagin (R332) | A new $Ag_2O$ phase obtained at high temperatures and pressures; classified as having $CdI_2$ type structure (hexagonal) with $a = 3.072$ Å; $c = 4.941$ Å. |

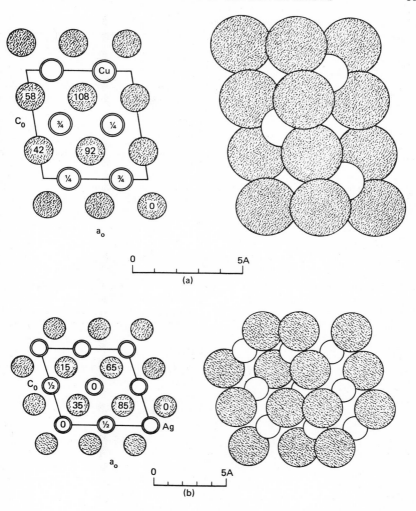

**Figure 40** Comparison of AgO arrangement with CuO. (a) The monoclinic structure of CuO projected along its $b_0$ axis.

$$\text{Cu:} \quad (4c) \pm (\tfrac{1}{4}\ \tfrac{1}{4}\ 0;\ \tfrac{3}{4}\ \tfrac{1}{4}\ \tfrac{1}{4})$$
$$\text{O:} \quad (4e) \pm (0\ u\ \tfrac{1}{4};\ \tfrac{1}{2}\ u + \tfrac{1}{2},\ \tfrac{1}{4})$$

(b) The monoclinic AgO arrangement projected along its $b_0$ axis.

$$\text{Ag(I):} \quad (2a)\ 0\ 0\ 0;\ 0\ \tfrac{1}{2}\ \tfrac{1}{2}$$
$$\text{Ag(III):} \quad (2d)\ \tfrac{1}{2}\ 0\ \tfrac{1}{2};\ \tfrac{1}{2}\ \tfrac{1}{2}\ 0$$
$$\text{O:} \quad (4e) \pm (xyz;\ x,\ \tfrac{1}{2} - y,\ z + \tfrac{1}{2})$$

## TABLE 14. STRUCTURE STUDIES OF SILVER MONOXIDE, AgO

| Investigator | Results |
|---|---|
| 1924 G. R. Levi and A. Quilico (R397) | Early X-ray; verified that the powder is a real compound with a low crystal symmetry. |
| 1954 J. A. McMillan (R445, 447) | X-ray powder pattern indexed on the assumption that AgO is isomorphous with CuO (monoclinic) with a single valence type silver; proposed C2/c structure type with 4 AgO groups per unit cell of a = 5.79 Å, b = 3.50 Å, c = 5.51 Å, $\beta$ = 1.07°. |
| 1954 P. Jones and H. R. Thirsk (R326) | X-ray powder pattern indexed on the basis of f.c.c. with $a_0$ = 4.96 Å; single valence type silver. |
| 1955 G. M. Schwab and G. Hartmann (R553) | X-ray powder patterns obtained which did not fit any high order of symmetry (e.g., hexagonal, cubic, tetragonal). |
| 1957 V. Scatturin, P. Bellon, and R. Zannetti (R541-2) | From X-ray powder diagrams, pattern indexed as monoclinic, b.c., similar to tenorite, CuO. The authors discussed the presence in the structure of two bond distances between silver and oxygen, 2.04 and 2.35 Å and suggested that the bonds are partly atomic and partly ionic. |
| 1958 W. S. Graff and H. H. Stadelmaier (R260) | X-ray powder pattern indexed as monoclinic with tenorite structure. |
| 1958 B. Stehlik and P. Weidenthaler (R576, 578) | X-ray powder pattern indexed as ZnS type structure (f.c.c.) |
| 1960 J. A. McMillan (R446) | Discussed magnetic properties of AgO and showed that previous structures failed to account for lack of paramagnetism; concluded that AgO should be a Ag(I) and Ag(III) covalent oxide; the latter coordinated with four oxygens ($dsp^2$), the former with two (sp). |
| 1960 V. Scatturin, P. Bellon, and A. J. Salkind (R543-4) | The existence of two kinds of silver confirmed by neutron diffraction; accurate positions for oxygen provided; lattice is monoclinic; space group $P2_1/c$; four formula weights of AgO in unit cell; Ag–O distances are 2.18 Å corresponding to colinear Ag(I)–O bonds, and 2.03 Å corresponding to square planar Ag(III)–O bonds. |

material was generally thought to be a peroxide until the clarifying work of Barbieri (R29–30). In retrospect, there were many indications that AgO was a true oxide and not a peroxide. For example: (1) the compound does not yield hydrogen peroxide when acidified; (2) the compound in concentrated nitric acid solution does not reduce lead dioxide, manganese dioxide, or potassium permanganate; and (3) the difference in specific volumes of AgO and $Ag_2O$ is less than would be expected if AgO were a peroxide (R460).

Structural studies are summarized in Table 14. The earliest X-ray work (R397) in 1924 verified that the powder is a real compound with a low crystal symmetry. Then McMillan (R445) assumed that AgO is isomorphous with CuO (Figure 40) and contained a singly valent silver; he proposed a monoclinic structure similar to that of tenorite (CuO). In additional studies by others, a cubic indexed pattern was used for identification (R326) and then it became evident that the X-ray powder patterns did not fit any high order of symmetry (R553). The early work was probably influenced by Klemm's (R352) report in 1931 of a low order of paramagnetic susceptibility. The single valence type of silver as the model did not conflict with this magnetic property. The structure was refined by additional work listed in Table 14. However, by this time, the compound was shown to be diamagnetic (R460). This characteristic could not be satisfactorily accounted for by a structure with a single valence type, Ag(II), which would contain a $4d^9$ unpaired electron.

On the basis of the magnetic and semiconductor properties, McMillan (R446) proposed that AgO is not a true bivalent silver oxide but an oxide of Ag(I) and Ag(III). Trivalent silver ions are coordinated with oxygens filling one d orbital, being, therefore, diamagnetic. Monovalent silver ions are sp coordinated with two oxygens. In the previous discussion of the bonding (R460), diamagnetism was explained on the basis of either a metallic bonding of silvers or a covalent bonding between two Ag(II), pairing the unpaired electrons, and ionic bonding with oxygens. However, this metallic bonding should lead to much higher values of conductivity in the compound (R446).

A reinvestigation of the structure was started using neutron diffraction. The amplitudes of nuclear scattering of Ag and O (Ag = $0.61 \times 10^{-12}$ cm and O = $0.58 \times 10^{-12}$ cm) make the two atomic species almost equal in scattering, whereas with X-rays the atomic scattering factors of silver and oxygen ions are roughly in the ratio of 5:1 (in the forward direction). Moreover, with neutron diffraction, the magnetic structure of crystalline solids can be easily determined. The work was carried out with a Corliss and Hastings spectrometer, and was first reported in 1960 (R543). The diamagnetic behavior (R460) was confirmed to the liquid helium temperature. Scatturin and associates reported the lattice to be monoclinic, space group $P2_1/C$, with four formula weights of AgO in the unit cell. The silver atoms are not equivalent and in the structure there are two Ag–O distances, 2.18 Å and approximately 2.03 Å. These correspond to the colinear Ag(I)—O bonds and the square planar Ag(III)—O bonds (R446).

In the structure of AgO, the oxygen is not $sp^3$ hybridized as in $Ag_2O$ and utilizes p orbitals in the formation of the bonds with silver: two of these are used to link Ag(III) and one to link Ag(I). This explains the presence of oxygen in octahedral positions of the silver lattice and its displacement from the center of the octahedron towards a face.

This point of view is in agreement with the crystal field theory (R445, 466A): the configuration $4d^8$ characteristic of $Ag^{3+}$ permits a regular octahedral grouping around the metal only if it is the type $d_\varepsilon^6 d_{z^2}^1 d_{(x^2-y^2)}^1$ with two unpaired electrons (where $d_\varepsilon$ is the notation for the lower energy orbitals sometimes represented as $d2_g$). For the high valence and atomic number of silver, its $d_\gamma$ orbitals are easily utilized to form covalent bonds, forcing the electrons to be paired in one of the two orbitals. The Jahn-Teller (R466A) effect is therefore operating in a manner similar to that present in the square planar complexes of Ni or Pd; there is distortion of the octahedra of coordination habitually with four bonds closer and two farther away from the central atom and approaching square planar coordination as a limit.

This is apparently what is happening in AgO; the $d_{z^2}$ orbital becomes stabilized and receives a pair of electrons. The $d_{(x^2-y^2)}$ orbital, destabilized, hybridizes with 5s and $5p^2$ orbitals, to give the square planar hybrid, which makes covalent bonding with the oxygen orbitals.

Similarly, the oxygen interacts with Ag(I), which can have a regular octahedron around itself if the bonds are ionic. The metallic atom is 5s 5p hybridized, and forms two colinear bonds with the p orbitals of the oxygen. The sixfold coordination becomes distorted, and indeed near the Ag(I) there are four coplanar oxygens at the distance of approximately 2.8 Å and two oxygens at 2.18 Å.

These facts seem to explain the structure of AgO, which can be considered a distortion of NaCl type lattice, caused by the formation of covalent bonds between the $d_\gamma sp^2$ orbitals of Ag(III) and sp orbitals of Ag(I) with the p orbitals of the oxygen.

Further confirmation of this structure was supplied in 1964, in studies of the valence of the species in solution (R101).

The deep blackness of AgO can be attributed to the bonding, following the rule (R490) that the color of a compound formed from colorless ions is related to the amount of covalent character in the bond between these ions. This relationship is illustrated in Table 15.

Over the years, the structures of other silver oxides have been reported; some of these are listed in Table 16. In almost all cases, these materials were prepared under nonbattery conditions or, if in batteries, were identified with surface layers. It has been shown that nitrogen, sulfur, and some halogens stabilize the

TABLE 15. COLOR OF SILVER COMPOUNDS AND
RELATIONSHIP TO BOND COVALENCY [a]

| Compound | Ag–O Å | Radius Sums | | Color |
|---|---|---|---|---|
| | | Ionic | Covalent | |
| $Ag_2SO_4$ | 2.50 | 2.46 | 2.19 | Colorless |
| $Ag_3PO_4$ | 2.34 | 2.46 | 2.19 | Yellow |
| $Ag_3AsO_4$ | 2.34 | 2.46 | 2.19 | Deep red |
| $Ag_2O$ | 2.06 | —— | 2.12 | Brown-black |
| AgO(I) ⎱ | 2.18 | —— | 2.19 | Black |
| (III) ⎰ | 2.03 | —— | 1.97 | —— |

[a] Source: in part from Reference 280.


TABLE 16.  STUDIES OF CRYSTAL STRUCTURE OF OTHER SILVER OXIDES

| Investigator | Results |
|---|---|
| 1935  H. Braekken (R59) | X-ray pattern of $Ag_2O_3$; exact structure could not be stated. |
| 1958  P. Weidenthaler and J. Vlach (R632) | $Ag_2O_3$ indexed as cubic with a lattice constant between 4.904 and 4.963 Å and symmetry group $O_h^4$–Pn3m. |
| 1962  A. S. McKie and D. Clark (R444) | Identified tetragonal AgO as an intermediate in the ozonation of silver to monoclinic AgO<br>  a = 4.816 Å; c = 4.548 Å; c/a = 0.944<br>Authors stated that tetragonal has *not* been found in silver electrodes, and also cast doubt on the validity of previous work reporting cubic AgO. |
| 1963  I. Naray-Szabo and K. Popp (R455) | Authors studied $Ag_2O_3$ stabilized with foreign ions (i.e., N, F, Cl); indexed as f.c.c.<br>  $a_0 \cong$ 9.8-9.9 Å<br>Both Ag(III) and Ag(I) bonds believed present. |
| 1965  I. Naray-Szabo, G. Argay, and P. Szabo (R456) | $Ag_7NO_{11}$ structure discussed on basis of X-ray and neutron diffraction data; $Ag^+$ and $Ag^{3+}$ ions identified in structure |
| 1965  E. J. Casey and W. J. Moroz (R89) | $Ag_2O_3$ identified electrochemically after charging silver electrodes at low temperatures |
| 1967  M. Feller-Kniepmeier, H. G. Feller, and E. Titzenthaler (R 208) | $Ag_2O_3$ identified on silver catalysts used for the oxidation of ethylene; electrodiffraction and X-ray patterns described which make questionable the previously accepted f.c.c. structure |

$Ag_2O_3$ structure, and these peroxy compounds have also been shown to contain valence(I) and valence(III) silver ions. The anodization (R89) of silver electrodes at low temperatures produced a new phase which may be true $Ag_2O_3$ or $Ag_2O \cdot [O]$ where the [O] is trapped oxygen. In any case, the material exists only as a thin layer. Tetragonal AgO (R444) was prepared by ozonization of hot silver powder. This material also contains four formula units of AgO and, as in the monoclinic structure, two crystallographically different groups of silver atoms are equally populated in the structure.

The physical and structural properties of silver and the silver oxides are compared in Table 17. In the oxidation of silver to oxides, the silver ions retain

TABLE 17. COMPARISON OF PHYSICAL AND STRUCTURAL
PROPERTIES OF SILVER AND SILVER OXIDES

| | Silver | $Ag_2O$ | $Ag_2O_2 (AgO)$ |
|---|---|---|---|
| Crystal structure | Faced-centered cubic $a_0 = 4.0862$ Å | Cubic $a_0 = 4.736$ Å | Monoclinic $a_0 = 5.852$ Å $b_0 = 3.478$ Å $c_0 = 5.495$ Å $\beta = 107° 30'$ |
| Atomic positions | $O_h^5$ (Fm3m) | $T_h^2$ (Pn3) or $O_h^4$ (Pn3m) | $C_{2h}^5$ (P2$_1$/c) |
| Unit cell formula | 4 Ag | $Ag_4O_2$ | $Ag_4O_4$ or 2 Ag (I)−Ag(III)0 Ag(I): 000; 0 ½ ½ Ag(III): ½ 0 ½; ½ ½ 0 0: ±(xyz; x, ½ − y, z + ½) |
| X-ray density, g/cc | 10.500 @ 25° | 7.243 @ 25° | 7.713 @ 25° |
| Heat of formation, kcal/mol | 0 | −7.2 @ 25° | −6.2 @ 25° |
| Free energy of formation, kcal/mol | 0 | −2.5 @ 25° | +6.6 @ 25° |
| Free energy of formation, kcal/mol | 0 | +0.6 @ 225° | +15.1 @ 225° |
| Magnetic susceptibility | Diamagnetic | Diamagnetic | Diamagnetic |
| Electromagnetic cgs units/g | $\chi = 0.18 \times 10^{-6}$ | $\chi = 0.58 \times 10^{-6}$ | $\chi = 0.155 \times 10^{-6}$ |
| Electrical conductivity, mho/cm | | | |
| bulk value | $6 \times 10^5$ | − | − |
| pressed powder sample | | $1.4 \times 10^{-9}$ | 0.0002 or 0.07 |

essentially the same cubic position. The unit cell always contains four silver ions; in the change from metallic silver to $Ag_2O$, the unit cell grows in volume by 55% (R70) and changes little thereafter in the further oxidation to AgO.

The conductivity of AgO is considerably better than that of $Ag_2O$. This could be associated with an excess or defect oxygen lattice as in ZnO. The material used in batteries was shown (R541-542) to have a stoichiometry of $Ag_2O_{1.95}$ and recent results (R448) show that oxygen-defective AgO has paramagnetic centers that can be attributed to Ag(III) coordinated with only three oxygens having an $sp^2$ hybridization. Another possibility is that the conductivity is enhanced by the dual valency coordination of AgO, similar to that proposed for other metallic oxides (R608) for other metallic oxides.

# I O.

## Recent Studies on the Nature and Stability of Silver Oxides

### Paul Ruetschi

The stability of $Ag_2O$ electrodes is limited by the thermal decomposition of $Ag_2O$ at low oxygen partial pressures, by dissolution of the oxide by the electrolyte with subsequent precipitation of Ag by Zn electrodes and by cellulosic separator material, and by reduction by hydrogen.

The decomposition of AgO to $Ag_2O$ is governed by the anodic $O_2$ evolution reaction, and the rate varies with $a_{OH^-}^{1-\alpha}$ and $a_{H_2O}^{\alpha}$, where $\alpha$ is the transfer coefficient. Dissolution of AgO appears to be accompanied by decomposition to $Ag_2O$. Reduction of AgO, as well as of $Ag_2O$, by hydrogen is extremely fast for unwetted electrodes but inhibited in the presence of electrolyte films. No conclusive evidence is available to date for the existence of a higher oxide phase, such as $Ag_2O_3$, acting as intermediate in the formation of AgO or $O_2$.

### MONOVALENT OXIDE ($Ag_2O$)

Thermodynamically, the univalent oxide is stable in aqueous alkaline solutions saturated with $Ag_2O$ and oxygen at 760 torr, its free energy of formation being $-2.70$ kcal/mol (R268). The electrode potential of the $Ag/Ag_2O$ couple at $25°$ is given by

$$E_{Ag/Ag_2O} = 0.342 - 0.059 \log a_{OH^-} + 0.030 \log a_{H_2O} \tag{1}$$

For the present discussion, it is to be noted that the reversible oxygen electrode depends on activities of hydroxyl ions and water, $a_{OH^-}$ and $a_{H_2O}$, in the same manner, namely,

$$E_{OH^-/O_2} = 0.401 - 0.059 \log a_{OH^-} + 0.030 \log a_{H_2O} + 0.015 \log p_{O_2} \tag{2}$$

such that at constant partial oxygen pressure of 760 torr, the potential difference between the two couples has a constant value of 0.059 V, independent of KOH concentration (Figure 41).

THERMAL DECOMPOSITION

In discussing the stability of $Ag_2O$ electrodes in silver-zinc cells one must take into account that the partial oxygen pressure in such cells can decrease to very low values because of the oxygen-gettering action of the zinc electrodes. The reversible potential of the oxygen electrode may under these conditions attain values below those of the reversible $Ag/Ag_2O$ electrode. This is in fact the case when the partial oxygen pressures in the cell are below the dissociation oxygen equilibrium pressure of $Ag_2O$. The latter has been determined over wide ranges of temperature (R40, 236, 347, 468,) and may be represented approximately by the equation

$$\log p_{O_2} = (-2859/T) + 6.258 \tag{3}$$

where the pressure is expressed in atmospheres. Figure 42 demonstrates that the equilibrium dissociation pressure increases rapidly with temperature and amounts, for example, at $60°$ to 3.5 torr. If the partial oxygen pressure in a silver-zinc cell becomes smaller than the equilibrium dissociation pressure, $Ag_2O$ decomposes according to the overall equation

$$Ag_2O = 2Ag + \tfrac{1}{2}O_2 \qquad (p_{O_2} < p_{O_2(eq)}) \tag{4}$$

The resulting oxygen is being consumed by the zinc electrode in a reaction which can generally be described by

$$Zn + \tfrac{1}{2}O_2 = ZnO \tag{5}$$

Theoretical models developed for partly wetted porous fuel cell electrodes (R416, 640) and known data on oxygen solubility and diffusion coefficients (R124) may be employed to calculate the limiting rate of $O_2$ diffusion to a sparsely wetted porous Zn electrode in an Ag/Zn cell (R587). Thus, assuming a reactive film-meniscus surface area of $0.01$ $m^2/g$ (corresponding to about 2% of the BET area of $0.5$ $m^2/g$) and assuming an effective electrolyte film thickness of 1 $\mu$ (R437), the maximum $O_2$ consumption rate in $5N$ KOH at $60°$ and at 3.5 torr partial $O_2$ pressure becomes $0.74$ $cm^3$ STP $O_2$ per g of Zn and per hour. This would correspond to a loss of positive electrode capacity of 3.5 mAh per g of Zn and per hour. This high figure represents the limiting diffusion rate. Actual $O_2$ loss would be smaller when controlled by the $O_2$ evolution step on the $Ag_2O$ electrode rather than by diffusion alone. However, this calculation shows that thermal decomposition could contribute significantly to the formation of

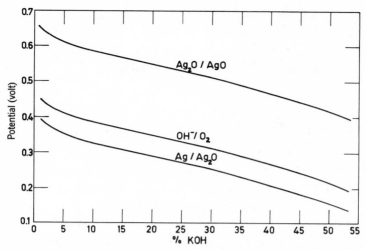

**Figure 41** Electrode potential of the electrode couples $Ag_2O/AgO$, $OH^-/O_2$, and $Ag/Ag_2O$ as a function of KOH concentration at $25°$ and 760 torr $O_2$. Data calculated with Equations 1, 2, and 7 using water activities from vapor pressures and $OH^-$ activities, from Reference 5.

metallic silver trees and short circuits during hot activated stand of $Ag_2O/Zn$ cells, as is in fact observed experimentally (R197, 451)

### SOLUBILITY

Because of the appreciable solubility of $Ag_2O$ in alkaline electrolytes (R12), the positive electrode suffers a loss of active material. Dissolved $Ag_2O$ is reduced

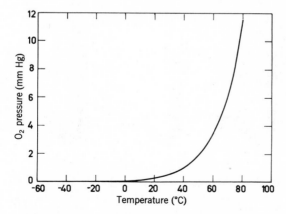

**Figure 42** Oxygen dissociation vapor pressure $(O_2)$ as a function of temperature.

to metallic silver by the zinc electrode or by the cellulosic separator material. However, stand loss of activated cells because of this process is relatively small. For instance, even at an average diffusion path length between positive and negative electrodes of only 0.1 mm, the loss by diffusion of $Ag_2O$ at $25°$ and in 5 $N$ KOH would amount to about 0.2 mg of $Ag_2O$ per hour and per $cm^2$ of positive electrode geometric plate surface. This is 0.046 mAh per hour and per $cm^2$, a significant figure, but still considerably less than the loss caused by thermal degradation as estimated above.

### REDUCTION BY HYDROGEN

Under certain conditions hydrogen evolved from the Zn electrode may react with $Ag_2O$ according to

$$Ag_2O + H_2 = 2Ag + H_2O \qquad (6)$$

The device shown in Figure 43 was used to study this reaction. Sample pellets of 6.8 mm diameter were pressed at 1000 kg from 0.5 g of a mixture of 20% Ag powder (325 mesh) and 80% $Ag_2O$ (Johnson Matthey, London, min 98.9% $Ag_2O$). The Ag powder was added to facilitate pelletizing and to provide Ag nuclei for the reaction. These samples were introduced into the carefully dried glass cells. Then electrolyte was added to the indicated level and all the air in the

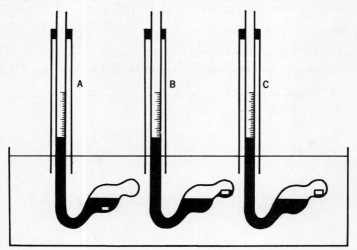

**Figure 43**    Glass cells for volumetric measurements. The calibrated sections had an inner diameter of 7.5 mm to allow introduction of the sample pellet. They were shielded thermally with glass tubes. The precision of the readings was 0.02 $cm^3$. Case A, sample submerged; case B, sample wetted; case C, sample dry.

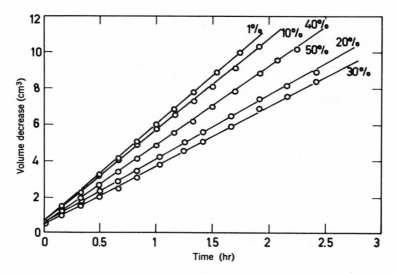

**Figure 44**    Hydrogen consumption of a sample of 0.5 g (80% Ag$_2$O, 20% Ag) Ag$_2$O + H$_2$ → 2Ag + H$_2$O. Volume decrease in cm$^3$ at 60° and 725 torr as measured without correction for H$_2$O vapor pressure.

cells was replaced by H$_2$ introduced through thin flexible tubing and by flushing with an amount of H$_2$ corresponding to 200 times the volume of the cell. In case A, the sample was completely submerged in the electrolyte such that it was covered by a layer of KOH solution of 5 mm depth. For condition B, the sample was wetted with KOH solution by tilting the cell for a few moments and by draining the nonadhering KOH to the bulk of the solution. For case C, the sample remained completely dry. The cell was thermostated at 60° and the hydrogen consumption rate measured by the decrease of the liquid level in the calibrated tubular section.

Figure 44, pertaining to data obtained for case C of Figure 43, shows that the rate was zero order with respect to the Ag$_2$O present. The ordinate represents volume of H$_2$ at 60° and 725 torr consumed by 0.5 g of sample. These data indicate that the reaction rate is extremely fast under these experimental conditions, in the order of 10 cm$^3$ per hour or gram. For wetted samples, case B, the hydrogen consumption rate was about 50 times smaller, for example, 0.2 cm$^3$ per hour. Finally, for case A, the consumption rate was only in the order of 0.01 cm$^3$ per hour, that is, 1000 times smaller than for case C. Evidently, hydrogen diffusion is greatly impaired in KOH. In the experiments corresponding to case B, the pores of the sample pellet were only partially filled with electrolyte. Here, a sharp break was sometimes observed in the H$_2$ consumption curve. This break was followed by a low rate, corresponding to case A. This phenomenon was probably caused by complete drowning or by deactivation of the surface as a result of adsorbed KOH electrolyte.

Since the reaction is exothermic, local energy dissipation effects might play a role. In presence of electrolyte, heat dissipation might be enhanced such that water could condense and drown the reaction site.

Conditions similar to C could possibly be realized in cells by partly waterproofing the electrodes, thus facilitating operation in a sealed condition.

Figure 45 demonstrates that the reaction rate does not significantly depend on KOH concentration.

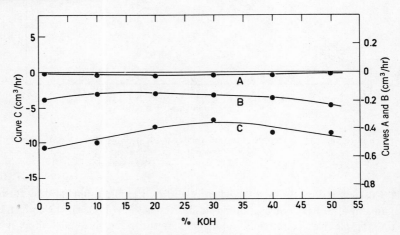

**Figure 45**    Hydrogen consumption rates per gram of sample and per hour, $Ag_2O + H_2 \rightarrow Ag + H_2O$. Volumes as read, at $60°$ and 725 torr, uncorrected for $H_2O$ vapor pressure.

## BIVALENT OXIDE AgO

The potential of the $Ag_2O/AgO$ electrode (R56, 162) depends on $OH^-$ ion activity and $H_2O$ activity in the same way as the oxygen electrode, and one has at $25°$

$$E_{Ag_2O/AgO} = 0.601^a - 0.059 \log a_{OH^-} + 0.029 \log a_{H_2O} \qquad (7)$$

As is apparent from Equation 2 and 7 and as illustrated in Figure 41, the $Ag_2O/AgO$ electode is at any KOH concentration 0.200 V above the $O_2$ electrode. It follows that AgO is thermodynamically unstable at any practically occurring $O_2$ pressure in an Ag-Zn cell. Experimentally, slow oxygen evolution is

[a]   The standard value of 0.601 differs from that given by Bode, Chapter 2, whose value is the same as given by DeBethune (R126-127).

observed when AgO is placed in contact with alkaline electrolyte. The decomposition of AgO

$$2AgO = Ag_2O + \tfrac{1}{2}O_2 \tag{8}$$

has been thought to be the result of two concurrent electrochemical component reactions (R12), namely

$$2AgO + H_2O + 2e^- = Ag_2O + 2OH^- \tag{9}$$

$$2OH^- = H_2O + \tfrac{1}{2}O_2 + 2e^- \tag{10}$$

with the overall rate being determined by the anodic oxygen evolution step, Equation 10, the latter reaction being driven by an overvoltage of 0.200 V. If indeed the decomposition process is as indicated, it will depend in a characteristic manner on KOH concentration. This relation is derived in the following manner.

Anodic oxygen evolution involves the discharge of $OH^-$ ions, leading to some activated complex, here not further defined. In the terminology of the absolute reaction rate theory (R244) the velocity of the process (Equation 10), under proper consideration of activity coefficients, is given by

$$v = k_f \cdot c_{OH^-} \cdot (y_{OH^-}/y_+) \exp(-\Delta G^{\ddagger}/RT) \tag{11}$$

where $k_f$ is the standard rate constant, $c_{OH^-}$ the molar concentrations of $OH^-$ ions, the $y$s are the molar activity coefficients of the species indicated by the corresponding subscripts and $\Delta G^{\ddagger}$ is the free energy of formation of the activated complex. For electrochemical reactions the free energy of activation $\Delta G^{\ddagger}$ depends on the electrode potential across the interface.

$$\Delta G^{\ddagger} = \Delta G_0^{\ddagger} + \alpha\, Z_i\, F E \tag{12}$$

where $Z_i$ is the valence of the reacting species, for $OH^-$ equal to $-1$, and where $\alpha$ is the symmetry factor (transfer coefficient). The electrode potential $E$ may be written as the sum of the equilibrium potential $E_e$ and the overvoltage $\eta$

$$E = E_e + \eta \tag{13}$$

For the oxygen electrode (Equation 10), the equilibrium potential $E_e$ is given by

$$E_e = E_0 + (RT/2F)\ln(a_{O_2}^{1/2} \cdot a_{H_2O} \cdot a_{OH^-}^{-2}) \tag{14}$$

Combining Equations 11 through 14 and considering that

$$c_{OH^-} \cdot y_{OH^-} = a_{OH^-}$$

one deduces

$$v = (k_f/y_{\mp}) \cdot \exp(-A/RT) \cdot a_{O_2}^{\alpha/4} \cdot a_{H_2O}^{\alpha/2} \cdot a_{OH^-}^{(1-\alpha)} \tag{15}$$

where $A = \Delta G_0^{\neq} + \alpha Z_i F E_0 + \alpha Z_i F\eta$ and $Z_i = -1$.

At constant overvoltage $\eta$, as realized on the $Ag_2O/AgO$ electrode (Figure 41) at constant temperature and constant partial oxygen pressure, the relation between rate $v$ and KOH concentration can be expressed by

$$\log a_{OH^-} - \log v = \alpha (\log a_{OH^-} - \tfrac{1}{2} \log a_{H_2O}) + \text{const} \tag{16}$$

A plot of the term $\log a_{OH^-} - \log v$ against $(\log a_{OH^-} - \tfrac{1}{2} \log a_{H_2O})$ should then yield a straight line with the slope of $\alpha$.

Considering the temperature dependence, one has

$$(1 - \alpha) \log a_{OH^-} + (\alpha/2) \log a_{H_2O} - \log v = (A/2.303RT) + \text{const} \tag{17}$$

Equations 16 and 17 were tested experimentally by using the apparatus shown in Figure 43, case A. The sample consisted of 0.5 g of pure AgO (Johnson-Matthey, London, min AgO content 98.9%) compressed at 1000 kg to a pellet of 6.8 mm diameter. The cell was carefully thermostated to ±0.1° and oxygen evolution measured by reading the liquid level in the calibrated section. Results are illustrated in Figures 46 and 47. The experimental volume changes were subsequently corrected for water vapor pressure of the solution, atmospheric pressure, hydrostatic head and temperature and the decomposition rates expressed in $cm^3$ STPO$_2$ per gram of AgO and per hour. Figure 48 demonstrates that Equation 16 is holding over the entire KOH concentration range between 1 and 50% KOH, the transfer coefficients being 0.36, 0.40, and 0.45 for 45°, 60° and 75°, respectively. The $H_2O$ activities were computed from the vapor pressures, $a_{H_2O} = p_s/p_{H_2O}$, and the hydroxyl ion activities from the data of Akerlof and Bender (R5) using the relation (R511)

$$a_{OH^-} = c_{OH^-} \cdot y_{\pm} = m \cdot \gamma_{\pm} \cdot d_0$$

Pertinent figures are listed in Table 18.

The plot of $[(1 - \alpha) \log a_{OH^-} + \tfrac{1}{2} \log a_{H_2O} - \log v]$ against $(1/T)$, including the data for all concentrations and temperatures yields an activation energy of $A = 28.4$ kcal/mol (Figure 49). This result is close to values obtained by other investigators for the "dry" decomposition of AgO, namely, 31.7 kcal and 25.4

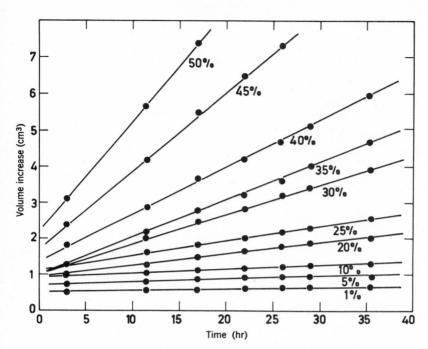

**Figure 46** Oxygen produced by a sample of 0.5 g AgO at 60° and 725 torr.

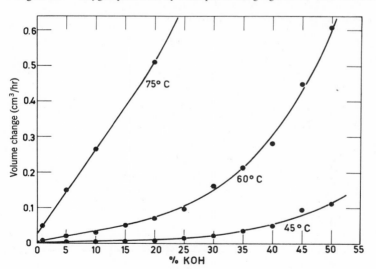

**Figure 47** Decomposition rate of AgO in cm³ per g of AgO as a function of KOH concentration at 45°, 60°, and 75°. Volumes at 60° and 725 torr not corrected for $H_2O$ vapor pressure.

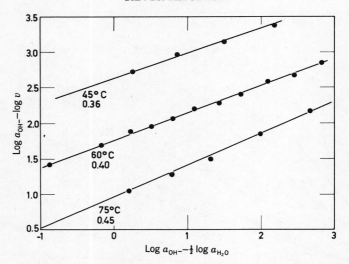

**Figure 48**     Plot of $\log a_{OH^-} - \log v$ against $\log a_{OH^-} - \frac{1}{2} \log a_{H_2O}$. The rate $v$ is expressed in terms of $cm^3$ of $O_2$ at STP per gram and per hour.

kcal (R451, 197), but lower than values obtained for the "dry" decomposition of $Ag_2O$, namely, 35 kcal (R40) and 43 kcal (R236).

The value of $A$ = 28.4 kcal found here pertains to a constant overvoltage $\eta$ of 0.203 V. With a mean value of $\alpha$ = 0.4 one has

$$\alpha Z_i F \eta = -1.9 \text{ kcal}$$

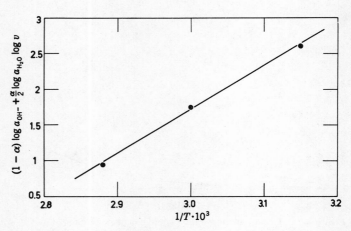

**Figure 49**     Activation energy ($A$ = 28.4 kcal) plot for the decomposition reaction, $2AgO \rightarrow Ag_2O + \frac{1}{2}O_2$.

TABLE 18.   ACTIVITIES OF OH⁻ IONS AND $H_2O$ IN CONCENTRATED KOH
SOLUTIONS AND GASSING RATE OF AgO

| Temperature °C | KOH Concentration % | $Log\ a_{OH^-}$ | $Log\ a_{H_2O}$ | $Log\ v$ ($v$ in cm³ STP $O_2$ per hr per g AgO) |
|---|---|---|---|---|
| 45° | 10 | 0.218 | −0.079 | −2.500 |
|  | 20 | 0.794 | −0.141 | −2.165 |
|  | 30 | 1.380 | −0.244 | −1.770 |
|  | 40 | 1.973 | −0.426 | −1.396 |
|  | 50 | 2.645 | −0.664 | −1.047 |
| 60° | 10 | 0.194 | −0.064 | −1.692 |
|  | 20 | 0.740 | −0.121 | −1.329 |
|  | 30 | 1.325 | −0.219 | −0.952 |
|  | 40 | 1.888 | −0.404 | −0.697 |
|  | 50 | 2.510 | −0.630 | −0.346 |
| 75° | 10 | 0.165 | −0.074 | −0.880 |
|  | 20 | 0.715 | −0.140 | −0.566 |
|  | 30 | 1.196 | −0.246 | −0.296 |
|  | 40 | 1.785 | −0.424 | −0.064 |
|  | 50 | 2.336 | −0.665 | 0.156 |

The activation energy for the $O_2$ evolution reaction on silver electrodes at the *reversible* oxygen equilibrium potential then becomes

$$A - \alpha Z_i F\eta = \Delta G_0^{\ddagger} - \alpha F E_0 = 30.3 \text{ kcal}$$

For the anodic oxygen evolution process on Pt electrodes a value of 22.5 kcal at the $O_2$ equilibrium potential was reported (R118).

The data presented here appear to provide ample proof that the decomposition of AgO proceeds via anodic $O_2$ evolution. This fact is further underlined by the observation that the decomposition is quite slow in the absence of KOH, as illustrated in Figure 50, where the curves A, B, and C relate to corresponding conditions of Figure 43.

The decomposition of AgO is strongly dependent on impurities present, as well as on particle shape and surface area (R12). The data presented here hold for the particular powder utilized. The gassing rate of AgO can be decreased by the addition of lead oxides (R76) or certain organic materials to the electrode (527).

Just as found for $Ag_2O$, the "bivalent" oxide AgO may under certain conditions react very rapidly with hydrogen according to $AgO + H_2 = Ag + H_2O$. Figure 51 shows $H_2$ consumption rates expressed in cm³ per gram of AgO and per hour. Curve A relates to conditions where the sample was submerged in KOH,

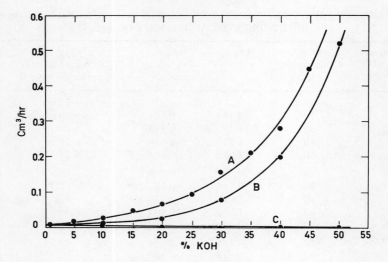

**Figure 50** Gassing of AgO as a function of KOH concentration for the three cases: A, sample submerged; B, sample wetted; C, sample dry. Volumes at 60° and 725 torr, uncorrected, per gram and per hour, $2AgO \rightarrow Ag_2O + \frac{1}{2}O_2$.

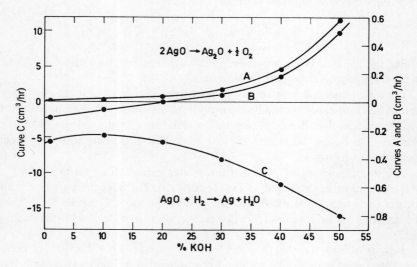

**Figure 51** Hydrogen consumption of AgO. The volume is expressed in $cm^3$ at 60° and 725 torr per gram and per hour, uncorrected for water vapor pressure.

curve B to the case where the sample was wetted but out of the electrolyte and curve C to a completely dry sample, that is, exposed to the water vapor pressure only. In the latter case AgO reacts vigorously with $H_2$, a fact which hitherto had not been realized (R372).

From curve B it can be seen that the presence of a thin KOH film strongly quenches the reaction.

For the interpretation of Figure 51, one must take account of the fact that the decomposition process (Equation 8) increases rapidly with increasing KOH concentration (Figures 50 and 51) and can mask the hydrogen consumption under conditions A and B, such that actually a volume increase is observed. Comparison between Figures 50 and 51 indicates that oxygen evolution and hydrogen consumption may occur simultaneously.

While the stability of AgO in activated silver-zinc cells is decreased because of thermal decomposition and hydrogen reduction, as described above, loss of AgO by solubility may be of less concern. A polarographic study had revealed (R12) that KOH solutions, thoroughly equilibrated with AgO, did not contain any dissolved silver species, other than monovalent ones. The latter are formed by the reaction in Equation 8 and subsequent dissolution. Although, crystallographically, AgO appears to contain an array of monovalent and trivalent silver ions (R544), the latter do not dissolve since they are stabilized in the crystal field, and a disproportionation of AgO to free monovalent and free trivalent silver is energetically unlikely unless coupled with immediate decomposition of $Ag_2O_3$ to $Ag_2O$ with evolution of $O_2$.

HIGHER OXIDES, $Ag_2O_3$

Proof for the transient existence of a defined higher oxide phase, such as $Ag_2O_3$, is still on weak grounds. It is based principally on the observation of an ill-defined arrest (Figure 52) during decay of oxygen overvoltage of silver oxide electrodes on open circuit (R11, 75, 247, 327, 449, 466, 609, 626). The length of this plateau depends on temperature, KOH concentration, and anodization rate preceding open circuit decay (Table 19). However, to date, crystallographic evidence for the presence of a higher oxide is lacking (R629). Therefore the possibility cannot be ruled out that the observed voltage plateau might be caused by adsorbed or dissolved oxygen or peroxide species (R626).

Recently, the old theory of Hickling and Taylor (R285), supposing $Ag_2O_3$ \to be an intermediate in the formation of AgO, has been revived (R442) in order to explain the additional current peak observed during cathodic voltage sweep of silver electrodes (R586).

It is pointed out here that a similar peak is present with $PbO_2$ electrodes in alkaline solution (R94) apparently caused by adsorbed oxygen or peroxidic compounds (Figure 53). In this connection it is also interesting to note that the

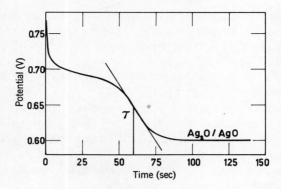

**Figure 52**   Decay of $O_2$ overvoltage of porous silver electrodes; illustration for Table 19.

reaction between Ag and $H_2O_2$ and $O_3$ leads to different types of silver oxides (R258, 402, 444).

The voltage peak occurring at the transition from the $Ag/Ag_2O$ level to the $Ag_2O/AgO$ level during galvanostatic oxidation has in the past also been used as argument for the transient existence of $Ag_2O_3$ but is now explained more satisfactorily in terms of resistance and nucleation overvoltage.

Unequivocal proof for the presence of trivalent silver during anodization has been claimed recently through polarographic analysis of solutions in which silver electrodes had been anodized (R102-3). It would be desirable to pursue these

TABLE 19.   OXYGEN OVERVOLTAGE DECAY OF POROUS SILVER OXIDE
ELECTRODES

| KOH Concentration | Temperature °C | Anodization Current mA/g Ag | Anodization Time hr | $\tau$ sec |
|---|---|---|---|---|
| 5 | 40 | 4.9 | 100 | 126 |
| 10 | 40 | 4.9 | 100 | 91 |
| 20 | 40 | 4.9 | 100 | 72 |
| 30 | 40 | 4.9 | 100 | 26 |
| 20 | 40 | 98 | 5 | 36 |
| 20 | 40 | 49 | 10 | 46 |
| 20 | 40 | 24.5 | 20 | 85 |
| 20 | 40 | 4.9 | 100 | 93 |
| 20 | 0 | 24.6 | 20 | 168 |
| 20 | 0 | 4.9 | 100 | 300 |
| 20 | 50 | 24.6 | 20 | 29 |
| 20 | 50 | 4.9 | 100 | 55 |

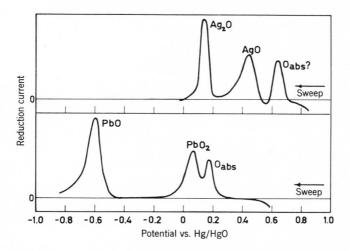

**Figure 53**    Cathodic potential sweep of silver and lead electrodes in KOH solution.

findings by work with the rotating disk-ring technique (R231) using a silver disk
and a platinum ring. By anodizing the silver disk at fixed potentials, above the
$Ag_2O/AgO$ level, silver species dissolving into the electrolyte could be detected
rapidly in form of a reduction current on the surrounding platinum ring when
keeping the latter at potentials where such species should be reduced as, for
example, at or slightly above the $Ag_2O/AgO$ level. Since at these potentials no
oxygen reduction could take place, presence of a reduction current would
indicate the existence in solution of some very electropositive species. In
performing the same experiment but with the silver disk being replaced by a
nickel disk, one could establish that these species are indeed specifically formed
on silver only. In fact, the complete reduction polarogram of transient species
can be obtained on the ring electrode (R561, 562).

In studies of this nature, one has to make sure also that no errors are
introduced by the possible presence of colloidal suspensions of AgO particles
(R12).

# I I.

## Electrochemical Kinetics of Silver Oxide Electrodes

### G. D. Nagy and E. J. Casey

Much work has been devoted to studying the anodic reactions that occur at the silver electrode. While the overall reactions are fairly well established, the mechanisms are not known. However, the picture is gradually becoming clearer. A smaller amount of work has been done on the reduction processes. It suffices to say at this time that the discharge reactions have a very high coulombic efficiency and are basically the reverse of the oxidation processes. Factors that affect the efficiency of the processes and the stability of the products will be discussed briefly.

Most of the work reviewed in this paper has been carried out under galvanostatic or potentiostatic conditions in alkaline solutions, either NaOH or KOH, over a wide range of electrolyte concentrations. The common type of electrodes used were silver-plated platinum, pure silver, and battery plates. Some work has also been done in acid solutions and other electrolytes in further attempts to elucidate the mechanisms of oxide formation and reduction.

By kinetics in this paper we mean measurements of rate of reaction per unit area, made as a fraction of the possible experimental variables with a view to interpretation of the results in terms of molecular and ionic mechanisms which are consistent with the structures, compositions, and changes accompanying the reactions of the electrode. As it turns out, not always are the mechanisms unequivocal. Since the questions: What is there? In what form is it? and How does it change as reaction proceeds? all have to be answered before a proposed mechanism can be accepted with any credibility, in this paper there are sections on these topics preceding the sections in which the kinetic information is reviewed. In the sections on kinetics the formal development of models is not presented, unless they can be used for interpretation and prediction. Rather, the emphasis is placed on the strength of the evidence supporting particular proposed mechanisms.

* DREO report number 591.

## ANODIC OXIDATION OF SILVER AND ITS OXIDES

### PHYSICAL AND COMPOSITIONAL COMPLEXITIES OF
### MULTIPLE OXIDATION PROCESSES

A typical galvanostatic anodic polarization curve (R75) is shown in Figure 54 for silver sheet in 30% KOH at 28° with a current density of 3 mA/cm$^2$. Slight variations from this curve are observed under different conditions, but this curve was chosen because it also shows variations in the resistance and capacitance as the oxidation proceeds. At the plateau A—B, silver is oxidized to Ag$_2$O. In the region D—E, the Ag$_2$O is converted to AgO. Beyond F, oxygen evolution occurs. There is some evidence that a higher oxide, possibly Ag$_2$O$_3$ is formed at the end of the AgO plateau. Curves for silver in H$_2$SO$_4$ solution are very similar, except that Ag$_2$SO$_4$ instead of Ag$_2$O is formed at the first plateau (R326).

The potential during Ag$_2$O formation for a smooth uncycled sheet electrode was usually higher than for subsequent cycles (R75, 148). In some cases a maximum potential was observed at point A of Figure 54 and for cycled electrodes a slight hump at B (R148, 628). This behavior was more pronounced at low current densities and low temperatures. These effects may be caused by the nucleation of Ag$_2$O sites at A and oxidation of virgin metal at B. The short

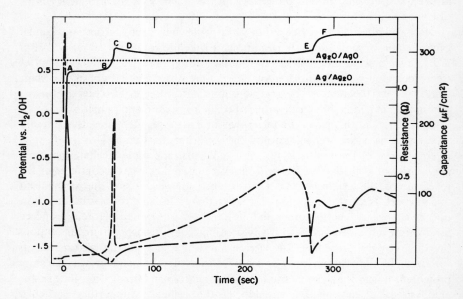

**Figure 54**    Electrode potential (solid line), ohmic resistance (dashed line), and double-layer capacitance (dotted-dashed line) of silver sheet, 1 cm$^{-2}$ in area during anodic oxidation at 3 mA/cm$^2$ in 30% KOH at 28° as a function of time; from Reference 75.

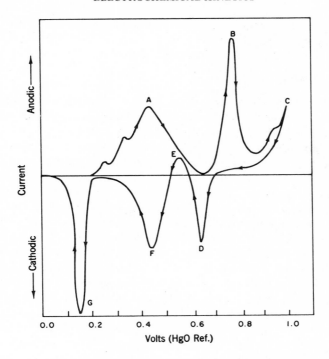

**Figure 55**    Potentiostatic sweep curve for silver in KOH (from R442). The current peaks are not to scale.

arrest in potential at approximately $-0.75$ V on Figure 54 has been observed by others (R449, 489) and is said to be the result of the oxidation of dissolved hydrogen.

Some work has been done using potential sweep methods. A curve (R442) is shown in Figure 55. Under the conditions used, during the anodic part only peaks associated with $Ag_2O$ and $Ag_2O_3$ formation (peaks A and B) were observed. No AgO was present at the start in this case. However, if some was present a peak was observed at a potential near peak E. Also AgO formation from $Ag_2O_3$ was observed (peak D) during the cathode sweep. At peak C, $O_2$ evolution occurred, and at peaks F and G the reduction of AgO and $Ag_2O$ occurred. The two small peaks on A were described as changes in either the mechanism or in the solid state process occurring within the $Ag_2O$ layer. In another study (R155) in which silver wire was used, three peaks on the anodic sweep were ascribed to the formation of an intermediate AgOH, $Ag_2O$, and AgO. Both sets of investigators observed that the potentials of the peaks and magnitude of the currents at the peaks were dependent upon the sweep rate, temperature, and alkali concentration. Larger electrodes (R110) yielded correspondingly larger currents; the anodic oxidation current steadily increased with increasing potential, with no distinctive peaks.

## IDENTIFICATION OF REACTION PRODUCTS

The existence of $Ag_2O$ and AgO at the first two plateaus has been confirmed by X-ray and electron diffraction work for polarizations in alkali solutions (R69-70, 327, 624, 628, 651) while $Ag_2SO_4$ and AgO have been shown to form in acid solutions (R61, 158, 185, 326, 624). The change in intensity of the diffraction lines for $Ag_2O$ and AgO was followed on smooth silver electrodes in 20-50% KOH at room temperature (R69-70, 624, 628). During the first plateau (Figure 54, A-B), $Ag_2O$ lines appeared first and then intensified while the silver lines present decreased in intensity. During the second plateau (Figure 54, D-E), AgO lines appeared and intensified; the intensity of the $Ag_2O$ lines diminished while the intensity of the silver lines remained constant (R651).

Although X-ray diffraction patterns for $Ag_2O_3$ in acid have been reported (R484, 577, 601), no such evidence has been obtained in alkali solutions (R69, 327, 626, 651). This may be caused by the instability of the oxide in alkali solutions (R89, 579) so that little is left at the time of measurement (R89, 257). The X-ray patterns usually show $Ag_2O$ and AgO lines (R321, 583, 624). The amount of oxygen contained in electrodes that had been oxidized to oxygen evolution (R75) and in charged battery plates (R19) has been measured. In both cases, extra oxygen was present. In alkali solutions (R159, 161), there were obtained compounds that had the compositions $AgO_{1.2}$ and $AgO_{1.6}$, the latter showing only X-ray diffraction lines for AgO. The existence of a surface peroxy compound with the formula $AgO_2$ on silver catalysts has been reported (R617-618). In this paper, the formula $Ag_2O_3$ refers to material of a higher oxidation state than that in AgO, whatever may be the nature of the higher silver oxide.

## PHYSICAL CHARACTERISTICS OF ANODIC OXIDE FILMS

The charge passed during the anodic formation of $Ag_2O$ is usually less than for the subsequent formation of AgO (R141, 285, 401, 450, 504, 625). On the first cycle the first plateau is always shorter. In general the thickness of the film at oxygen evolution is determined by the amount of $Ag_2O$ formed, which depends upon the current density, electrolyte concentration, and physical state of that electrode. The amount of $Ag_2O$ increases as the current density decreases or the temperature increases (R165, 174, 221, 348, 628). The maximum quantities of oxides were produced in 30-35% KOH (R69, 174, 626). Low-temperature cycling appeared to have little effect on subsequent behavior at room temperature (R165). In porous plates, small isolated pockets of $Ag_2O$ and silver remain when oxygen evolution begins. These can then be partially converted to AgO during oxygen evolution (R111, 141, 169, 513, 626, 631).

Measurements of the thickness of the $Ag_2O$ film formed have given values ranging up to 30,000 Å (R70, 75, 148). This film is porous, very rough, slightly amorphous and usually brown or black. Glassy films up to 7,000 Å thick have been formed in 0.7N NaOH at low current densities with stirring (R148). The thickness of the film and hence the capacity, especially that caused by $Ag_2O$ formation, increases with the number of cycles for both sheet and porous electrodes (R148, 625-626), the initial conditions determining the amount of material produced. The effect is more pronounced on sheet electrodes. Periodic increases by a factor of 4 to 20 in the current (R627), asymmetric current (R339), and pulsed current with open circuit stands between pulses (R625, 630) all increase the capacity, the last technique being the most effective (up to 40% increase). These effects were also more pronounced on sheet electrodes than on battery plates. The maximum effect was obtained in 35% KOH.

The crystal size of $Ag_2O$, 0.1 to 10 $\mu$ (R19, 70, 628, 631, 651) varied inversely with the current density. At high current densities the film was very tight and compact. Growth occurred by the formation of distinct crystals. During oxygen evolution the AgO crystals became more perfect (R624). Coulogravimetric studies on sintered electrodes (R384) suggest that on the first cycle the density of the $Ag_2O$ layer is slightly greater than normal. On subsequent cycles the density agrees with reported values. These differences may account for the lower stability in $H_2O$ of the $Ag_2O$ formed on the first cycle compared to that formed during later cycles (R452). During (R651-652) the plateau D-E of Figure 54, the size of the AgO crystals at oxygen evolution were about five times the size of the $Ag_2O$ crystals from which they were formed. No preferred orientation (R494) of the crystals was found but in many cases single crystals of $Ag_2O$ were formed with the (111) plane parallel to the (111) and (100) planes of the silver substrate. On the (110) silver plane the $Ag_2O$ was always polycrystalline. Crystals of AgO were always polycrystalline with the ($\bar{1}13$) plane often orientated on the (110) plane of the silver. No other preferred orientations were observed.

The growth of $Ag_2O$ crystals is not uniform. This is apparent at very low current densities when one observes dark areas developing and growing (R69, 75, 156, 450, 631). This resembles the case in sulphuric acid where growth of $Ag_2SO_4$ appears to be parallel rather than perpendicular to the surface (R467). For an anode-to-cathode distance that is greater than 2 cm in 1.0N or less KOH at room temperature, the current was concentrated at the edges of the plate (R73). This supported a prior statement that oxidation occurs initially at the outer edges and proceeds toward the center (R69). As mentioned earlier, in some cases free Ag and $Ag_2O$ may be left in the film especially in porous battery plates. Analysis (R111, 513) of various segments of charged battery plates showed that at the end of charge most of the AgO was near the grid. The outer portion contained mostly $Ag_2O$ with some AgO and Ag. The surface layer is claimed to be oxygen rich (R651).

ELECTRICAL PROPERTIES OF THE ANODIC OXIDE FILMS

The changes in resistance shown in Figure 54 are in the direction one would expect from measurements of the specific resistance of Ag, $Ag_2O$, and AgO: $1.59 \times 10^{-6}$ (R387), $10^8$ (R392), and 10-15 (R392, 460) ohm-cm, respectively. Other investigators found that, while they obtained different values, the general trends of change were the same (R57, 156, 170, 174, 333, 348, 395, 609, 612, 645, 647). The reported ranges of capacitance and resistance for the various points of Figure 54 were: (A) 15 to 455 $\mu F/cm^2$, 0.03 to 25 ohm; (B) 0.5 to 50 $\mu F/cm^2$, 0.05 to 100 ohm; (C) 0.5 to 20 $\mu F/cm^2$, 0.8 to 10,000 ohm; (D) 25 to 32 $\mu F/cm^2$, 0.1 to 99 ohm; (E-F) 10 to 40 $\mu F/cm^2$, 0.6 to 10 ohm. The divergence in the values appears to be quite large; however, the electrode surfaces themselves and the experimental conditions varied widely. As will be mentioned later these electrodes may have different true areas and types of films on the surface. The values (R57) obtained, especially the resistance, were very dependent upon the frequency used. The upper limits for the resistance ranges quoted above for points A, C, and D were obtained at 5 Hz; at $10^4$ Hz, 1.2, 9.2, and 1.8 ohm, respectively. Good agreement (R395) was obtained between the measured impedance and that of a model in which the system is treated as a semi-infinite RC line in parallel with the double layer capacitance.

## KINETICS OF $Ag_2O$ FORMATION

SELECTED INTERPRETABLE MEASUREMENTS

In this section we discuss $I/V$ data, measurements of charge as a function of $I$ and $V$, and potential decays on open circuit.

The overall reaction for the oxidation of silver to $Ag_2O$ is (see for example, R52, 70, 72, 110, 155, 158-159, 164, 174, 247, 415, 442, 504, 624, 651-652):

$$2\,Ag + 2OH^- = Ag_2O + H_2O + 2e \qquad (1)$$

Various relationships have been obtained for the dependence of overvoltage on the current density. Tafel behavior with slopes of the order of 20-30 mV/decade of current have been observed at low current densities and electrolyte concentration, approximately $1N$ (R148). At higher current densities and electrolyte concentrations (approximately 7-9$N$) a linear relationship between the overpotential and current has been reported (R148, 174). In the former case, exchange currents of the order of $10^{-8}$ to $10^{-6}$ A/cm$^2$ were obtained, while in the latter case they were of the order of 2 mA/cm$^2$. For oxidation during amperostatic sweeps in 0.1$N$ KOH, a linear change in voltage was obtained as. (R494) the applied current density was changed. An exponential relationship is claimed to exist (R247) between the current and field across the oxide film. For the same thickness of film this is equivalent to Tafel behavior. Measurements of the limiting thickness of the $Ag_2O$ layer as a function of current density seem to be conflicting at first. Below 0.25 mA/cm² the thickness varies inversely with the current density. For higher current densities the thickness follows more closely a linear dependence on the logarithm of current density (R148). The literature discloses surprisingly few systematic studies of effects of electrolyte composition and temperature, for instance, on $E/I$ behavior.

Open circuit potential decays, following oxidation at about 0.4 mA/cm², on rough (and presumably porous) electrodes gave rise to two distinct regions (R148). Of a total of 30 to 40 mV overpotential the first 10 to 15 mV had Tafel slopes of 0.012 to 0.023 V/decade and were dependent upon the fraction of the $Ag_2O$ film formed. The lowest slope was near the midpoint of the plateau A-B of Figure 54. The overpotential was also a minimum at this point. The remainder of the overpotential had apparent Tafel slopes of 0.03 to 0.009 V/decade. These latter values are too small to be accounted for in terms of conventional electrode kinetic arguments. The length of time required for the very slow decay indicates that this "overpotential" may have a thermodynamic rather than kinetic origin. In the light of this the exchange currents quoted above could very well be low by an order of magnitude.

TRANSPORT OF IONS AS SOLE RATE-DETERMINING STEP

Based on the results of potentiostatic measurements and the effect of electrolyte concentration on the rates of reaction (R155, 159), it was postulated that, when a field is applied, electrons are removed from the silver to form $Ag^+$ ions in a distorted silver lattice. At the same time hydroxyl ions from the solution react with the silver on the surface to form a layer of $Ag_2O$ and hydrogen ions. When the entire surface is covered with oxide, there is further distortion of the silver lattice and splitting of the hydroxyl ions to form $O^{2-}$ ions which diffuse through the oxide and thicken the oxide layer. The process stops when the $O^{2-}$ ions can no longer diffuse through the lattice. At this time a second electron is removed from the $Ag^+$ in the $Ag_2O$ layer and more oxygen as $O^{2-}$ diffuses in to form AgO. Evidence from coulogravimetric investigations (R384) and from X-ray diffraction studies (R70) indicated that expansion of the silver lattice takes place.

Some workers (R513, 626, 652) argue that oxygen ions ($O^{2-}$ or $O^-$) are the ionic species that are transported through the oxide layer. Others (R247, 395, 450, 613, 647) suggest that the diffusing species may be $Ag^+$ ions. To complete the total number of possibilities, others suggest that both may move (R505, 625, 610). The question of the moving species has not been resolved. It is difficult to distinguish between the possibilities because in both cases the $Ag_2O$ layer should be rich in oxygen at the surface. The matter might be settled by the use of radioactive tracer implantation (R123) in which the electrode is partially oxidized, then the surface labeled, and the oxidation resumed. Probably the charge transfer process can occur by movement of silver ions, of oxygen ions, or both, depending upon experimental conditions. Dry oxidation studies (R19, 422, 523) are relevant.

In one approach (R513) for porous battery plates, the silver near the grid is oxidized first. When the grid is completely covered with an $Ag_2O$ layer it is converted to AgO while the remainder of the silver is converted to $Ag_2O$. Although this is what the analysis indicates, it is difficult to see how the silver in the plate can be so different from the grid that ionic transport is easier than electronic transport. Also this would imply gradients of AgO and $Ag_2O$ concentration in directions opposite to the oxygen species that is moving through the electrode.

## PORE-FILLING MECHANISMS

The original theory of solid-state ion transfer discussed in the previous section has been modified (R156). The premises that the lattice was distorted and that $O^{2-}$ ions were the transported species, were retained, but it was proposed that the film might be porous. When the film reached a critical thickness, the pores began to fill causing an increase in the resistance as well as overpotential because of concentration of the current in the pores. This variation was proposed to account for some of the discrepancies in the earlier attempts to explain the peak at C of Figure 54. Others (R139, 285) inferred that the peak was caused by the formation of $Ag_2O_3$ which then decomposed to AgO; however, it was also claimed (R327) to be the result of difficulty in forming AgO nuclei from the $Ag_2O$ lattice. It was noted that these theories could explain the peak in the oxidation curve but not the dip between plateaus in the reduction curve. (See section on Cathodic Reduction.) Others do not agree with this resistance-pore model because the calculated ohmic drop is not large enough to account for the magnitude of the peak (R75, 333, 647) and also because the potential change according to this model would not be as rapid as is observed in some cases (R450). Others (R75) agreed that the surface was not completely covered with $Ag_2O$ until very near the end of the plateau. In another variation of the pore-filling theory it is held that the reaction is stifled not by the increased resistance but by isolation of the material from the electrolyte by the film (R221, 348, 612, 625). No quantitatively described proposals of mass-transfer-controlled polarization could be located for silver oxidation.

## DISSOLUTION AND PRECIPITATION PROCESSES

In recent work (R164, 174), a dependence was observed for the rate of diffusion of a negatively charged ion in solution on the rate of reaction. A dissolution-precipitation type of reaction involving the following reactions was proposed:

$$Ag_s = Ag^+_{aq} + e \tag{2}$$

$$Ag^+_{aq} + OH^- = \tfrac{1}{2}Ag_2O_s + \tfrac{1}{2}H_2O \tag{3}$$

This scheme can better explain the dependence of the thickness of the film on hydroxyl ion concentration and the fact that the maximum thickness is obtained for any particular current density in 35% KOH. At lower concentrations a larger potential gradient is needed to supply sufficient hydroxyl ions while at higher concentrations there is greater ion pairing with the cations and a higher potential is needed to discharge the hydroxyl ions. Possible intermediates or complexes which have been reported and which may be involved are: AgOH (R47, 155), $Ag(OH)_2^-$ (R155, 174), or $Ag_3O(OH)_2^-$ (R492).

This last type of mechanism is partially in agreement with conclusions (R 148) suggesting that at least two different mechanisms are involved in the anodic formation of $Ag_2O$. One involves a dissolution-precipitation reaction while the other probably involves a direct interfacial reaction leading to a compact film. At low current densities (less than $0.25$ mA/cm$^2$) the formation appears to be limited by diffusion of some species through the film formed by the dissolution-precipitation reaction. At higher current densities the formation appears to be limited by some diffusion or migration process, or both, in the compact film. This duplex film is similar to the model (R613) suggested for silver and earlier for cadmium (R369).

PASSIVATION: TERMINATION OF GROWTH

Most recent evidence indicates that the rise in potential at the peak C of Figure 54 may be caused by the formation of a passivating layer. This may be in the form of chemisorbed oxygen at the surface (R221, 610, 613, 647), hydroxyl ions on the surface (R170, 348, 450), or some barrier-impeding contact of the electrolyte with the electrode interface (R221, 348, 612, 625). Adsorbed oxygen (R610, 647) on the surface may increase the positive hole conduction in the layer resulting in a nonequilibrium concentration of holes and electrons in the layer. The effect of light on electrodes is interesting; it decreases the potential and allows a thicker $Ag_2O$ film to be produced on the electrode. In darkness, once the peak is reached and the current is shut off without forming AgO, very little additional $Ag_2O$ can be formed even at current densities much lower than that used to form the $Ag_2O$ layer (R148). A proposed complex equivalent circuit (R647) involves the capacitance and resistance of the oxide layer, the double layer, the adsorbed species, and the barrier for the transfer of charge. The adsorbed layer is estimated to be as much as ten monolayers thick. Most of the overpotential at the peak is in this layer and not in the double layer (R647). This is reasonable since it is difficult to see how the double layer potential can decay once AgO is formed. Open-circuit potential decays from this region B-C of the anodic polarization curve show three distinct regions, a fast decay at the high potentials followed by two slower decays similar to those mentioned above (R148). The fast decay may be associated with the passivating layer.

The presence of an oxygen-rich layer is not unreasonable. Evidence (R658) has been presented for the penetration of oxygen into the oxide layer at the potential of the peak; also $Ag_2O$ (R451) formed by the thermal decomposition of chemically prepared AgO may have as much as 10 to 15% of the oxygen from the AgO adsorbed in or on the surface. Finally the existence of an oxide intermediate between $Ag_2O$ and monoclinic AgO with a NaCl type lattice structure has been reported (R258).

RECENT VIEWS OF MECHANISMS

The theories mentioned above are not in accord with recent results (R148). If the mechanism is one in which the rate-controlling processes occur at the metal-electrolyte interface and the electrode reaction is stifled as the electrode coverage approaches completion, the slopes of the potential decays would be independent of the amount of oxide produced. Furthermore, a mechanism in which the sole rate-controlling step is the diffusion of some species through the oxide film would lead to an exact inverse variation of limiting thickness and the applied current. Finally a mechanism in which ion or electron migration through the film is the sole rate-controlling process would result in a galvanostatic curve in which the potential varied linearly with time and need not reach any limiting film thickness other than when the film breaks down dielectrically.

An expression relating the overpotential, current, and the thickness of the film has been derived (R450). It is of the form:

$$I = \text{const} \left[ \frac{y}{1 + \alpha y} - \frac{1}{y(1 + \alpha y)(1 + \alpha)} \right] \exp \left[ \frac{a}{x + a} \cdot \frac{e(\Delta\phi_e + \eta)}{kT} \right] \qquad (4)$$

where $y$ equals $\exp (e\eta/kT)$, $\alpha$ is the ratio of the area of the surface covered with chemisorbed hydroxyl ions to that not covered, $x$ is the thickness of the film, $e$ is the charge on the electron, $a$ is equal to the thickness of the double layer times the ratio of the dielectric constants of the oxide and double layer, and $\Delta\phi_e$ is the equilibrium potential difference between the electrode and the solution. Reasonable but not striking agreement was obtained with experimental results. However the disagreement is not surprising since the expression ideally is for a compact type of film in which the rate-controlling reaction occurs at the oxide-solution interface and the type of film to which it was applied was undoubtedly of the duplex type described above. The real test of the expression is yet to come, but some such quantitative description is sorely needed.

## KINETICS OF AgO FORMATION

CONSTANT-RATE PROCESS (from Ag and from $Ag_2O$)

The overall reaction for the conversion of $Ag_2O$ to AgO is:

$$Ag_2O + 2OH^- = 2AgO + H_2O + 2e \qquad (5)$$

The longer second plateau of Figure 54 and the comparable amounts of total charge passed for the oxidation and reduction of the electrode indicate that once the $Ag_2O$ is converted to AgO, the following reaction also takes place

$$Ag + 2OH^- = AgO + H_2O + 2e \qquad (6)$$

Amperostatic sweep measurements yielded a linear relationship between overvoltage and current (R494). At high values (R33-4) of the overpotential (greater than 30 mV) Tafel behavior was observed. The slope was $RT/F$ (59 mV/decade at $25°$) and was independent of the thickness of the film. Over the entire range of overpotential studied (up to 170 mV), the dependence of overpotential and current is of the form:

$$i = 2i_0 \sinh (\eta F/RT) \qquad (7)$$

This is what one would expect if the slopes for the anodic and cathodic reactions were equivalent. Analysis of the results showed that the stoichiometric number and symmetry factor were 2 and ½, respectively. On the basis of these results, the following mechanism was proposed involving the adsorption of hydroxyl ion on the surface sites ($S$) followed by any one or all of three mechanisms for the formation of $O^{2-}$ ion:

$$(OH^-)_{aq} = (OH)^-_S \qquad (8)$$

$$(OH^-)_S + (OH^-)_{aq} = (O^{2-})_S + H_2O \qquad (9)$$

$$2(OH^-)_S = (O^{2-})_S + H_2O \qquad (10)$$

$$(OH^-)_S = (O^{2-})_S + H^+_{aq} \qquad (11)$$

Oxygen ions ($O^{2-}$) migrate to the $Ag_2O$/AgO interface where the rate-determining step, the oxidation of monovalent silver ion to trivalent silver ion or the migration of the oxide ions across the interface, occurs. Many authors suggest that $O^{2-}$ is the species migrating through the solid AgO during the anodic formation of AgO (R155-156, 159, 625, 627, 651, 653).

## PULSED-OXIDATION PROCESS

The increase in the amount of oxides formed using pulsed or periodically varying current may be caused by breakdown of the AgO film followed by oxidation according to Equation 6 (R625, 627, 630). The increased capacity produced by pulsed current was smaller with porous plates than with silver sheet (R627); in the former, very little free silver is present when $Ag_2O$ formation ceases. Breakdown of the film is suggested by the fact that the potential is lower after the pulse than before, possibly because of a mixed potential of the $Ag_2O/AgO$ and $Ag/Ag_2O$ systems. Also the differences in specific volumes of Ag and AgO is sufficient to promote cracking (R70, 630). It is possible that $O^{2-}$ ions may be able to move more easily at the higher potential during the pulse. However, it has been indicated (R339) that the asymmetric current forms $Ag_2O_3$ which decomposes to AgO with a more open structure enabling easier access of oxygen ions.

## PASSIVATION: FORMATION OF $Ag_2O_3$ AND OXYGEN

The reasons for the passivation of anodic formation of AgO are not clear. Once all of the $Ag_2O$ is converted to AgO, the oxidation of silver directly to AgO may require an additional overpotential which may result in changes in the double layer that permit the formation of an adsorbed passivating layer or the formation of $Ag_2O_3$ and oxygen evolution. The possible role of $HO_{2ad}$ has been discussed (R89).

The overvoltage of oxygen evolution on AgO obeys a Tafel relationship with a slope of the order of 0.1 to 0.15 V/decade of current (R284, 327, 651-652). Exchange currents between $10^{-6}$ and $10^{-4}$ A/cm$^2$ have been reported (R524). Evidence has also been presented (R626, 628) that in alkali solution the oxygen is formed by a peroxy mechanism, probably by $OH^-$ being converted first to $O_2^-$, then to $HO_2^-$ and finally to $O_2$.

## KINETICS OF $Ag_2O_3$ FORMATION

Evidence for the existence of $Ag_2O_3$ as a distinct oxide phase is not incontrovertible, its formation being usually accompanied by oxygen evolution. Because of the instability of $Ag_2O_3$, either fast measurements or low temperatures must be used to study the reactions. In the presence of AgO on the surface, the reaction is said (R89, 257, 504, 628) to be:

$$2AgO + 2OH^- = Ag_2O_3 + H_2O + 2e \qquad (12)$$

It has been suggested (R493) that $Ag_2O_3$ may be formed by a precipitation reaction involving the complex $Ag_3O(OH)_2^-$ according to:

$$2Ag_3O(OH)_2^- + 10OH^- = 3Ag_2O_3 + 7H_2O + 12e \qquad (13)$$

a reaction which involves the oxidation of monovalent to trivalent silver. In the absence of AgO formation there is evidence (R89, 442) that the reaction is:

$$Ag_2O + 4OH^- = Ag_2O_3 + 2H_2O + 4e \qquad (14)$$

It was suggested that, under the conditions used, the excess oxygen that enters the film at the end of $Ag_2O$ formation expands the lattice. It is reported that $Ag_2O_3$ has a face-centered cubic lattice with a unit cube length about 5% greater than that of $Ag_2O$ (R484, 577, 624).

Although at low currents AgO has time to nucleate and grow, at high currents sufficient oxygen is forced directly into the $Ag_2O$ lattice to form $Ag_2O_3$ before AgO nucleation can begin. If both are present in quantity, AgO can be formed by the reaction of $Ag_2O_3$ with $Ag_2O$. The thickness of the anodically formed $Ag_2O_3$ layer is estimated to be between one and five monolayers (R89, 257, 493).

Tafel slopes between 0.05 and 0.25 V/decade have been observed (R89, 493). However, hysteresis in the V/log $I$ curves occurs between runs for increasing and decreasing currents. This hysteresis is taken as evidence for an "oxygen precursor," possibly $HO_{2ad}$, which "blankets the electrode" (R89) and protects the higher oxide. Open-circuit potential decay measurements gave Tafel slopes between 0.046 and 0.078 V/decade.

## CATHODIC REDUCTION OF SILVER OXIDES

### POTENTIAL AND COMPOSITIONAL COMPLEXITIES
### OF MULTIPLE REDUCTION PRODUCTS

A typical cathodic reduction curve is shown in Figure 56. This curve (R75), comparable to others (R645), includes resistance and capacitance measurements. In some cases a small voltage dip is observed between the two plateaus (R156, 158, 285, 327, 450, 624). Under conditions in which $Ag_2O_3$ is formed a short discharge plateau near the $AgO/Ag_2O_3$ potential is observed (R89, 339, 406).

X-ray diffraction studies confirm that during the first plateau (A-B) AgO is being converted to $Ag_2O$. During the second plateau (C-D) the conversion of AgO to $Ag_2O$ and $Ag_2O$ to Ag occurs simultaneously at the $Ag/Ag_2O$ potential (R69, 70, 169, 628, 631, 634, 651, 653). It has been shown that the portion oxidized last is the first to be reduced (R628). Interruption of the current during the first plateau permits the potential to rise to the $Ag_2O/AgO$ potential. For most of the second plateau a mixed potential between the $Ag_2O/AgO$ and $Ag/Ag_2O$ levels is observed. The potential for the last part (greater than 75% discharge) of the second plateau decays to the $Ag/Ag_2O$ level (R156, 450, 651).

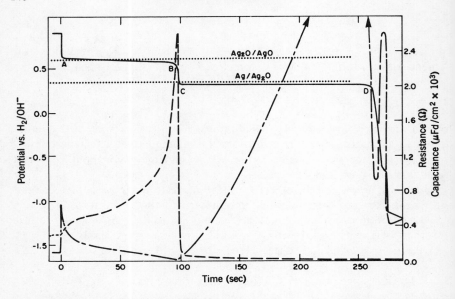

**Figure 56**   Electrode potential (solid line), ohmic resistance (dashed line), and double-layer capacitance (dotted-dashed line) of a silver oxide film produced by anodization shown in Figure 54 during discharge at 3 mA/cm², as a function of time (from R75).

Studies have been carried out at various temperatures and concentrations of electrolyte (R159, 164, 169, 221, 415, 625-626). Rather complex phenomenological behavior is observed, which so far has not been reduced to quantitative terms. For instance, the capacity on discharge is affected markedly by the charging conditions.

PHYSICAL PROPERTIES OF REDUCED ACTIVE MATERIAL

The cathodic reduction of silver oxides under ordinary conditions in which the oxides are relatively stable is highly efficient on a coulombic basis, although the coulombic efficiency (charge out/charge in) is usually less than 100% due to some oxygen evolution at the end of the prior oxidation or to dissolution or decomposition of the oxides (R450, 625-626, 651, 653). It can be seen from comparison of Figures 54 and 56 that the power efficiency is much lower. In most cases the first plateau (A-B) of Figure 56 is shorter than the second and is usually of the order of one-third of the total capacity (R75, 401, 327, 285, 450, 624). High discharge currents and low temperatures decrease the length of the first plateau because of the increased polarization. High temperature and very low current densities also decrease the plateau because of decomposition of AgO (R52, 221, 625-626, 653). At constant temperature and current density the length of the $Ag_2O/AgO$ plateau depends more upon the surface area than quantity of AgO (R69). The lengths of the two plateaus can be equalized in battery plates by putting a highly conductive matrix in the electrode (R645). Mechanical mixtures of $Ag_2O$ and AgO from battery plates gave a shorter first plateau than do the original plates (R513). The first plateau was due to AgO near the grid which when reduced to $Ag_2O$ assumed the $Ag_2O$ potential. It was also observed (R162) that for pressed pellets the potential of the electrode was that of the material near the grid regardless of the amount of the other oxide in the rest of the plate. Scraping layers of material from the surface of charged plates affected only the lower plateau (R513). Partial charges during the AgO plateau gave correspondingly shorter A-B (Figure 56) plateaus on reduction (R625).

During discharge the $Ag_2O$ and AgO crystals shrink (R651) leaving a heterogeneous surface even when the starting material before oxidation is very smooth (R69, 70, 148, 497). In the vicinity of silver particles which were unoxidized at the end of charge, the reduced silver tended to be granular in form, while the crystals were dendritic in regions where no silver was originally present (R631). The crystal size of the silver produced was found to increase with decrease in current density.

WHY THE SHORT $AgO/Ag_2O$ PLATEAU?

After a portion of the AgO, equivalent to the charge passed during the plateau A-B of Figure 56, has been converted to $Ag_2O$, the remaining AgO and $Ag_2O$ are reduced simultaneously. Recent work (R69-70) has produced evidence that the two reactions occur simultaneously with the two reaction interfaces separated by a layer of $Ag_2O$ of constant thickness. This thickness depends upon current density and the type of electrode but is roughly equivalent to the amount of charge passed at the upper plateau.

The shortened AgO plateau has been explained (R75, 160) as polarization caused by blocking off of the AgO sites. The resistance curve in Figure 56 is consistent with this idea. The decreased capacity at low temperatures is also attributed to polarization in the electrolyte, the smallest polarization being in 35% KOH (R160, 626).

The reason for the shortened $Ag_2O/AgO$ potential plateau under certain conditions of electrochemical reduction is more subtle than mere "polarization." Anions such as $Cl^-$ and zincate in the electrolyte have a dramatic effect. In the presence of $Cl^-$, the higher potential plateau is eliminated (R92). In the presence of zincate it is extended so that the two plateaus are of equal length (R197). Moreover, the total coulombic charge retrieved does not seem to be affected by addition of either. We assume that AgO must be present at the oxide/electrolyte interface if the higher potential is to continue to be observed because of the mixed potential of the two redox systems ($Ag_2O/AgO$ and $Ag/Ag_2O$).

Passivation of AgO by its reduction products, $Ag_2O$ and Ag, will occur unless electrolyte can penetrate to the remaining AgO. The lowest current density at which the AgO plateau begins to shorten, then, could be a measure of the limiting rate of electrolyte penetration through the "$Ag_2O$" product layer. Chloride, because of the attendant redox couples $Cl^-/ClO_2^-$ at 0.62 V and $ClO^-/ClO_3^-$; at 0.35 V may act as a mediator producing AgO via adsorbed $ClO_3^-$ at the lower end of the mixed-potential range. The zincate complex may have such a molecular structure that exchange of the $Ag^{2+}$ (or $Ag^{3+}$) and $Zn^{2+}$ ions occurs at the oxide/electrolyte interface, the net effect being to produce more effectively the higher oxide as a potential-determining species at the upper end of the mixed-potential range. Bromide ions and aluminate ions, among many others, remain to be studied from this viewpoint.

## KINETICS OF REDUCTION

Most mechanists who have studied the reduction reactions in detail offer reasons why the species which migrate through the solid oxide must be some form of oxygen, possibly $O^{2-}$ (R111, 160, 513, 625-626, 651). The early stages of the reduction of AgO to $Ag_2O$, and the reduction of $Ag_2O$ to Ag, are in principle the reverse of the reactions of Equations 5 and 1, respectively, and have been so considered (R72, 75, 174, 625, 634, 651). However, indications (625) that the reverse of the reaction of Equation 6 may occur to a small extent at the lower end of the mixed potential range. The reduction of the higher oxide $Ag_2O_3$ may be the reverse of the reactions of Equations 14 or 12, depending upon whether AgO is formed as a product (R442) or not (R89). Actually, as will become evident, the mechanisms of these cathodic reduction reactions are still poorly understood.

### CONCERNING REDUCTION MECHANISMS

*For AgO.* The cathodic reduction (R160) of AgO must involve the removal of $O^{2-}$ ions from AgO near the grid to form Ag, the $O^{2-}$ ions being then transported in the electric field through the oxide to the oxide-solution interface where they react with water molecules that are bound to cations ($K^+$ in this work). If the transport process is not fast enough, then the $O^{2-}$ ion concentration at the oxide-solution interface falls, the polarization is increased, $Ag_2O$ is formed, and the potential of the electrode falls to that of the $Ag/Ag_2O$ couple. Later work (R69-70) permits an elaboration of this basic idea assuming that the interface between the Ag and AgO is wider than previously suggested and that it is composed of a layer of $Ag_2O$ which remains fairly constant during the reduction process. A soluble species $Ag(II)(OH)_3^-$ may exist and also be an intermediate in the sequential steps of reduction.

*For $Ag_2O$.* For the cathodic reduction of electrodes containing only $Ag_2O$, the same type of mechanism (R160) applies but in this case reaction occurs only at the $Ag/Ag_2O$ interface and as a result the potential does not vary significantly with the extent of discharge. It is also proposed that the cathodic process for the reduction of $Ag_2O$ may proceed by a species in solution (R174) according to:

$$Ag_2O_s + 2OH^- + H_2O = 2Ag(OH)_{2aq}^- \qquad (15)$$

and

$$2Ag(OH)_{2\,(aq)}^- + 2e = 2Ag + 4OH^- \qquad (16)$$

For this case the overvoltage and any change in it would be due to the concentration of $Ag(OH)_2^-$ and would be governed by the availability of both hydroxyl ions and $H_2O$ for the dissolution of $Ag_2O$.

Hence, for both AgO and $Ag_2O$, there is the possibility that solid-state and liquid-state steps have to occur sequentially during the reduction. But exactly what they are and which one controls the rate under any specified conditions are not yet known.

## INFLUENTIAL CONCOMITANT REACTIONS

INSTABILITIES

During the course of the operation of a battery containing silver oxide electrodes, various nonelectrochemical reactions occur. One such reaction is that of silver with AgO (R75, 111, 160, 624, 626, 651):

$$Ag + AgO = Ag_2O \qquad (17)$$

Silver particles and the grid in battery plates are usually separated from AgO by a layer of $Ag_2O$ formed in this manner (R111, 624, 631). Because no net charge is lost, the reaction has little effect on the coulombic efficiency. However, the energy efficiency is lowered because the $Ag_2O$ is discharged at a lower potential than is AgO. In rechargeable cells the increase in capacity after long stands on open circuit may be due to reaction of AgO and Ag, either from the solution or in the electrode, to form $Ag_2O$, which is converted on recharging to AgO (R630). One would expect the resistance of the electrode near oxygen evolution to decrease, as shown in Figure 54, if the film was cracking. However, it has been suggested (R75) that the decrease could be due to the formation of an $Ag_2O_3$ having a lower resistivity.

For primary batteries, and to some extent secondary batteries, decomposition or dissolution of the oxides, or both, are the main causes of decreased capacities after prolonged stands. (See Ruetschi's paper, this volume, on decomposition studies of $Ag_2O$ and AgO).

The higher oxide phase, referred to generally in this paper as $Ag_2O_3$, formed anodically on Ag at $-40°$, decomposes rapidly, either dry or wet. The rate of decomposition is 2.5 $\mu A/cm^2$, and the product can be AgO or $Ag_2O$, depending upon whether or not AgO can nucleate (R89, 529, 601).

## ALLOYING OF SILVER ELECTRODES

Various modifications or additions have been made to the electrode in attempts to improve the capacity and the physical characteristics of battery electrodes. Although the referenced works are all phenomenological, the metal additions are noted here because they seem to offer a further approach to the study of kinetic influences, not yet been exploited.

The charge acceptance of Ag was increased up to 20% by the addition of Pd at 1-1.5% (R370, 404), by Au up to 10% (R405, 654) although Au accelerates the decomposition of AgO to $Ag_2O$ on stand, and by Sn and Pb at the 2% level (R654). No effect was noted for 2% Th, In, and Bi alloys (R654), for Pd at 5% (R197), and for Zn at 10% (R197) although Zn tends to corrode and affects the oxide potential. The charge acceptance was lowered by addition of Cd at 20% (R197), by addition of Pd at 10% (R404), and by combined Te and Sb addition at the 2% level (R654).

## STABILIZATION OF THE HIGHER OXIDES

Certain anions such as $F^-$, $NO_3^-$, $SO_4^{2-}$ and $ClO_4^-$ stabilize the higher oxides of silver (R447). The complex $Ag_7NO_{11}$ (variously referred to as $2Ag_2O_3 \cdot 2AgO \cdot AgNO_3$, as $2Ag(AgO_2)_2 \cdot AgNO_3$, or as $Ag(Ag_3O_4)_2NO_3$), which can be obtained by the anodic oxidation of silver nitrate or of silver salts with nitrate present, is stable in alkali hydroxides (R626). Reserve primary cells using this oxynitrate gave higher energy densities than comparable AgO cells (R106, 394).

Fairly stable periodate complexes can be formed in alkali solutions by the reaction of $KIO_4$ with $Ag_2O$ (R407) or $Ag_2SO_4$ (408) by reaction of $KIO_3$ with AgO (R100-101). These complexes have a bidentate form with the formula, $M_7^I Ag^{III}(IO_6)_2$, where $M$ can be $Na^+$, $K^+$, or $H^+$, separately or in combinations. Similar alkali argenti-tellurates ($M_9^I Ag^{III}(TeO_6)_2 \cdot nH_2O$) have been formed by action of alkali persulfate on $Ag_2O$ in potassium tellurate solution (R408). For these complexes the sodium salts are more stable than the corresponding potassium salts. Quadridentate silver (III) ethylenedibiguanidinium salts ($Ag^{III}C_6N_{10}H_{16})X_3$, where $X = NO_3^-$, $OH^-$, or $0.5SO_4^{2-}$ have been reported (R508). The hydroxide is obtained by treating the sulphate with KOH.

In $Na_2CO_3$ electrolyte (R285), AgO does not form; the reaction proceeds to oxygen evolution as if the carbonate ion displaced hydroxide ions from the surface sites. It is possible for $Ag_2O$ to react with $CO_2$ to form $Ag_2CO_3$ (R570), which has a greater solubility than has $Ag_2O$ (R276). Partial dissolution of the reaction products (R349) has been reported to occur during the anodic oxidation of Ag in $K_2CO_3$ solutions. The following carbonated silver oxides have been noted (R419) $2Ag_2CO_3 \cdot Ag_2O$, $3Ag_2O \cdot 2CO_2$ and $AgHCO_3$. We could locate no structural, stability, or solubility data for such complexes.

Of various additions directly to the electrolyte that have been studied, stannic, germanium (R529), lithium, and bivalent sulfur (R366) ions have been found to increase the charge acceptance. Zincate ions had little effect on the total capacity, but increased the lengths of the A-B plateaus of Figures 54 and 56 (R197) as was noted earlier.

## CONCLUSIONS

From the work reviewed on the kinetics and on factors which influence the kinetics, it can be seen that the reactions occurring at silver electrodes are not yet completely unraveled. The complexity, even during $Ag_2O$ formation, for example, arises from the fact that several processes may be occurring in parallel during oxidation.

Much good work has been done. The main difficulty in preparing this review was to find ways to correlate the results of work done in different laboratories. We suggest that a more complete physical description, such as particle size, true surface area, electrode geometry, and structural stoichiometry of the solid masses be included in the kinetic research on silver electrodes in the future. Knowledge of the migrating species, effects of electrolyte concentration, temperature, pressure in the case of reactions which may be accompanied by gas formation, potential control, and even of gravitational effects seems imperative if the mechanisms are ever to be fully described.

Proper quantitative expressions of the kinetics of oxidation and reduction of the silver oxide electrode will include terms for all of the named factors, probably more. Some general theoretical models now appearing in the literature contain at least some of these, but a more comprehensive theoretical attack could be a very useful guideline. Oversimplification would simply be misleading.

A final cautionary note about kinetic studies on this system: The probability seems very high that both anodic oxidation and cathodic reduction of silver oxide electrodes actually occur by different mechanisms under different experimental conditions.

# 12.

## Correlated Structural and Kinetic Studies in the Silver/Argentous Oxide/Argentic Oxide System

### H. R. Thirsk and D. Lax

A major interest in our laboratories over a number of years has been to advance both theory and experiment in the treatment of common electro-chemical reactions involving phase change and electron transfer. At different times both kinetic and structural investigations have been carried out with the silver oxide system and the present article is an attempt to summarize our work in the field. There are important gaps in the ground covered which seem to us to be of considerable significance and three points on which work is being pursued should be mentioned in this context: (1) the anodic dissolution of the metal prior to complete $Ag_2O$ coverage; (2) the reduction of AgO to $Ag_2O$; and (3) the reduction of $Ag_2O$ to Ag.

The work to be described falls naturally into three sections: (1) the texture, growth, and orientation of the oxides; (2) the phase change to $Ag_2O$; and (3) the phase change $Ag_2O/AgO$. Our concern is with the texture and thickness of the oxides.

The primary layer of argentous oxide has been regarded by previous workers as being coherent (R159), discontinuous (R75), or composite (R613). In addition, little information is available concerning the nucleation and growth of the silver oxide centers or factors influencing this growth.

It was concluded (R613) from the transients obtained under galvanostatic and potentiostatic conditions that the formation of a dense $Ag_2O$ layer, $10\,\mu$ thick, preceded the formation of a more porous modification. Galvanostatic experiments (R651) and associated electron microscopy indicated that spheri-cally shaped $Ag_2O$ centers formed on silver-plated platinum electrodes in $4.5N$ KOH solution and that these were approximately $1\,\mu$ in diameter at the end of the first stage of polarization and transformed into AgO crystals during the second stage. The presence of a primary compact $Ag_2O$ layer was not reported and, as the replicas used in the electron microscopy were not shadow cast, it was impossible to estimate the heights of the growth centers.

Examination (R624) of the working electrodes by X-ray diffraction during growth indicated that a thin layer of $Ag_2O$ was first formed, followed by a thicker layer of AgO. Both oxides had a random orientation, regardless of whether they were formed on randomly or highly orientated silver. It was also concluded that initially AgO was formed on the surface of the electrode next to the solution.

For the present work, electropolished polycrystalline and single crystal electrodes were used for anodic oxidations under constant current and potential conditions in potassium hydroxide solutions at $25°$. The electrochemically formed oxides were examined by electron microscopy, electron diffraction, and X-ray diffraction. Some specimens were also examined by a technique devised by one of our colleagues (R60).

## EXPERIMENTAL WORK

Smooth specimens suitable for electron microscopy and diffraction were prepared by electropolishing (R563).

The formation of the oxides was investigated in $0.1N$, $1N$, and $5N$ KOH solutions prepared from carbonate-free potassium hydroxide and triply distilled water and deoxygenated with purified nitrogen. In the constant potential experiments the electrodes were introduced into the cell at $-100$ mV with respect to the $Ag/Ag_2O/KOH_{aq}$ potential; all potentials are quoted on this scale. The cells were fitted with a second working electrode in parallel to the main electrode so that the specimens could be removed while being polarized. They were rinsed free from electrolyte with triply distilled water and placed in a desiccator, the whole operation taking about ten seconds.

Specimens were examined with an A.E.I. E.M.6. microscope, using both a single-stage positive and a two-stage negative replica technique (R266, 338). In the two-stage method for rod and wire electrodes, an impression in cellulose acetate was shadowed with gold-palladium (shadowing angle $cot^{-1}2$), and carbon was then deposited normally; single-stage positives were used for polycrystalline and single crystal block electrodes. A thin layer of gold palladium ($7Å$) was deposited from a point source at an angle of $cot^{-1}2$ or $cot^{-1}3$, followed by a layer of carbon ($130-180$ Å thick) deposited normally. The composite layers were stripped from the electrodes in the usual way.

For the electron diffraction an instrument having a camera length of approximately 50 cm was used at 50-55 kV.

X-ray diffraction patterns were taken with Co $K\alpha$ radiation with either a Philips powder camera (5.754 and 11.483 cm radius) or Unicam single-crystal goniometer with a 3.00 cm camera, the identity of the phases being established by the Debye-Scherrer powder method. Oxidized wire and rod electrodes were used directly in the cameras. These specimens were examined for preferred orientation by mounting vertically slightly off center to the beam. The relevant part of the surface was then replicated.

The following procedures were used for chemical analysis. Oxides were formed on abraded silver sheet at a current density of $0.1$ mA cm$^{-2}$; after washing, the deposits were dissolved in a dilute solution of ammonia and the silver was estimated volumetrically using Volhard's method. Gravimetric estimations with benzimidazole (R391) were also carried out and these were cross checked with Volhard's method after dissolving the deposit in $6N$ nitric acid. The agreement was $\pm0.2$ mg in 10-mg samples.

The results of chemical analyses are given in Figure 57, together with the relevant polarization curve. It is well known that at low current densities the second oxidation stage, corresponding to the step $Ag_2O \rightarrow AgO$, is much longer than the first, corresponding to $Ag \rightarrow Ag_2O$. As the whole of the argentic oxide formed cannot be reduced to $Ag_2O$ at the first arrest on cathodic polarization (R391), it was necessary to estimate total silver present chemically. Figure 57 shows that toward the end of the second plateau there is more silver present than would be expected by the conversion of the initial $Ag_2O$ layer to AgO. On the other hand, when oxygen evolution has started, the deposit only thickens slowly.

A large number of electron micrographs was taken in this study (R391), and the results described in this paper indicate the general changes which were observed. They have been described in more detail elsewhere (R63). After a primary formation of $Ag_2O$ there followed a growth of secondary oxide centers on both polycrystalline silver and single crystals of silver; the variation in size at

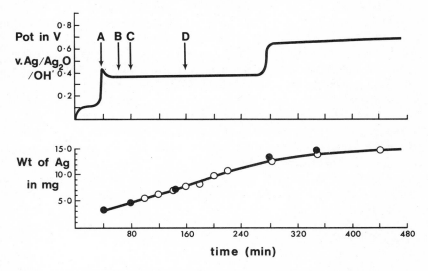

**Figure 57**    Polarization curve: The formation of silver oxides in $N$ KOH at $0.1$ mA cm$^{-2}$. A, B, C, D, typical points at which specimens were removed for X-ray examination. Chemical analysis: $\bigcirc$ = volumetric estimation; $\bullet$ = gravimetric estimation.

a particular stage of anodization is greatest for polycrystalline and (100) faces and least for (111) faces of the silver substrate. A magnification of 20,000 revealed a primary layer of argentous oxide between the growth centers while a magnification of 100,000 showed that the large centers are covered by tertiary hemispherical growth forms at the end of the first stage of polarization, the tertiary centers having a diameter of approximately 200 Å.

Abraded wire electrodes, in contrast to electropolished surfaces, showed a marked size distribution of the growth centers, and the surface was not completely covered at the potential maximum in the polarization curve. In the area between the centers, the texture had changes, and high magnification showed small crystallites. There is also an increase in size of secondary growths with increasing temperature. The variation in size of the growth centers is less in potentiostatic than in galvanostatic polarization. In addition, the size of the centers is smaller, consistent with a smaller charge under these conditions. Attempts were made to obtain replicas of argentic oxide centers at comparable stages of growth to that for $Ag_2O$; it would appear that the growth centers are much smaller than in the case of $Ag_2O$.

The material removed from the electrodes after constant current polarizations up to peak A in Figure 57 and at +100 and +300 mV at constant potential, was always found to be $Ag_2O$ by X-ray diffraction. At the reversible $Ag/Ag_2O$ potential, silver and argentous oxide alone were present, and there were no lines which could be attributed to a suboxide. An additional faint line corresponding to $d = 3.35$ Å was found in some glancing incidence patterns from thick deposits on wires. This line may be attributed to the (110) plane of $Ag_2O$ (R462) and has also been observed in electron diffraction measurements.

Simultaneous diffraction measurements and micrographs were made for thick deposits formed at low current densities; irrespective of the substrate pretreatment, the rings were always broken, indicative of the relatively large blocks of oxide on the surface. The size of the centers estimated from the micrographs lies between $10^{-5}$ and $10^{-3}$ cm and therefore cannot be compared with the diffraction patterns because this lies above the range of line broadening and below the value for which the spots may be measured; it has often been assumed that a thin layer only is present during this stage of oxide growth (R159, 285).

Under certain conditions the blocks of oxide adopted a preferred orientation. Electropolished substrates favoured the formation of orientated material while abraded surfaces in general did not give rise to orientated oxide, with the exception that sometimes specific areas of the surface did produce orientated blocks.

The results of chemical analyses at the points at which electrodes were removed for X-ray diffraction are also shown in Figure 57. It can be seen that the increase in the amount of silver is linear with time and only falls away during oxygen evolution. At point A, $Ag_2O$ alone was present, $Ag_2O$ together with

AgO at point B, while at point D the diffraction patterns corresponded solely to AgO. In contrast, experiments using oscillating electrodes gave evidence for the presence of AgO very shortly after the peak in the polarization curve; the greater intensity of the AgO lines in patterns obtained under these conditions suggested that AgO may form initially at the outside of the $Ag_2O$ layer.

The first major change which was found in the topography was the formation of a primary layer of argentous oxide irrespective of whether the electrodes were abraded, electropolished, or anodized at constant current or constant potential. Electron diffraction patterns consisted of rings of essentially uniform intensity, implying that the deposit at this stage must consist of randomly orientated small crystallites. The thickness of the layer, based on coulometric estimates, is of the order 50-100 Å. It is likely that the primary deposit retains the same form and thickness in the later stages because the appearance is unchanged. Observations at high magnification, however, suggest that there are local variations in thickness.

The second major stage is the formation of relatively large blocks of oxide on the primary layer; there is no essential difference between the deposit formed on abraded and electropolished surfaces but the texture, perhaps because of the effect of nucleation, is more uniform for polarizations at constant potential. However, there is no marked difference in the shape and size of the centers formed at different potentials, so that, in contrast to other oxides, the kinetics of growth of these centers cannot be controlled by a potential dependent step at the interface. At constant current, after 50 millicoulombs of charge had been passed, the centers on electropolished electrodes were hemispherical and about 1000 Å in diameter. Further charging to 100 millicoulombs led to an increase in number and size and with a charge greater than 100 millicoulombs the centers adopted a cubic shape. The electrode potential remains constant during these changes and finally rises to the maximum when the surface is almost completely covered by the secondary centers. High magnification shows that the secondary deposit does not completely seal off the primary layer from the solution at this stage. The secondary layer is itself in turn covered by "tertiary" hemispherical centers (~200 Å across) irrespective of whether abraded or electropolished substrates were used. These tertiary structures were absent at higher temperatures and occasionally at constant potential at low temperature. The complete coverage of the primary by the secondary layer is not a prerequisite initiating the oxidation to AgO but it is clear that the change in oxidation must be dictated by the primary layer which may, in fact, thicken in the later stages of polarization; such a change could not be detected by electron microscopy.

X-ray diffraction patterns showed strong evidence for a ⟨111⟩ fiber axis orientation; the overall geometry of the polycrystalline growth forms often conformed with the crystallographic symmetry of $Ag_2O$. Electron diffraction patterns support the same conclusion. In the later stages of growth the diffraction patterns for single crystal electrodes showed intensities corresponding to layer line reinforcement arising from a $\langle 111 \rangle$ fiber axis, showing the $\{111\}$ planes to be parallel to the (100) surface of the metal. This orientation is maintained for deposits reaching a thickness of several thousand angstroms. The marked orientation on the single crystal and polycrystalline electrodes is surprising in view of the formation of the random primary layer and might possibly be explained if only those nuclei in the primary layer having a preferred orientation with respect to the substrate continue to grow into large centers. The surface of these oxide centers appears to be polycrystalline but the evidence as to whether the secondary growths themselves are polycrystalline is not conclusive.

A further feature of the electron diffraction patterns is the intensity of the ring indexed as a reflection from the $\{112\}$ planes of $Ag_2O$. This line is forbidden and has been listed as a very weak reflection (R462). In the comparable case of $Cu_2O$, which has the same structure, the reflection from the (112) plane is listed for both X-ray and electron diffraction data (R285, 119). In the present study its appearance would be most simply explained if the deposit had an appreciable extent of $\langle 112 \rangle$ basal orientation, which would have a marked effect on the diffraction intensities. Other explanations might be advanced for the observations of the $\{112\}$ diffraction; thus, dynamic scattering of the electrons (R488) has been invoked to explain the forbidden reflections observed from large oxide polyhedra formed after prolonged oxidation of copper single crystals (R390). First-order twins would also have some $\{112\}$ planes parallel to the surface, as has been observed for electrochemically formed cuprous oxide (R98).

When the oxide layers on the single crystals are reduced, the silver deposits form as a highly twinned structure, showing a strong preferred three-dimensional orientation with the cube face and cube edge of the deposit parallel to the cube face and cube edge of the substrate.

The results reported here show that the argentous oxide layer has a composite structure as a result of the formation of the secondary deposit and agree in essence with the conclusions drawn (R613) from an examination of transients under potentiostatic and galvanostatic conditions and also amplify published observations (R651).

The conclusions which can be drawn from the investigation of the conversion of the argentous oxide layer to argentic oxide are not so definite. It has been noted that further silver oxide is formed at this stage and that the orientation of the $Ag_2O$ is destroyed at the same time as the centers of AgO grow. Orientated argentic oxide deposits were not observed under any conditions and this result therefore agrees partly with the observations (R629) that the $Ag_2O$ and AgO had random orientations irrespective of the nature of the substrate. The difficulty of replicating the surface found earlier (R159) was also experienced in this investigation. Nevertheless, it appears that AgO deposits formed at low current densities are also composed of centers, the diameter being in the range between $1\,\mu$ and $3\,\mu$. On the other hand, at high constant potentials the diameter appears to lie in the range between 0.1 and $0.2\,\mu$, which is in essential agreement with earlier observations (R629).

KINETIC STUDIES IN THE $Ag/Ag_2O$ SYSTEM

The anodic formation of argentous oxide in hydroxide ion solutions has been extensively investigated; much of the earlier work (R119, 327, 401) was confined to the interpretation of three potential plateaus seen under galvanostatic conditions. The peak before the AgO formation was attributed (R327) to the difficulty of forming centers of AgO in the lattice of $Ag_2O$. Others (R156) considered that the resistance of the electrolyte-solution interface increased due to the formation of a film of $Ag_2O$ and the potential maximum was attributed to ohmic resistance due to complete coverage of the electrode. From measurements of the double layer capacitance and of the resistance, it was concluded (R75) that the peak was not directly caused by ohmic resistance but was caused by concentration of the current into small localized areas where the conversion to AgO takes place; these areas were the last to be covered by $Ag_2O$ and the thinnest regions of the film.

Based on measurements of the a-c impedance of the electrodes (R333, 396, 613), it was concluded that, after a primary layer of $Ag_2O$ had formed, approximately 20 millicoulombs $cm^{-2}$ of charge had been passed, the anodic dissolution became limited by diffusion of silver ions through the layer. The increase in potential was attributed to concentration polarization caused by an increase in the oxygen concentration at the $Ag_2O$/solution interface leading to an enhanced passivation.

From an examination of current-time transients at constant potential, Croft (R110) concluded that the rate of oxidation increased monotonically with overpotential with solid state conduction through the reaction product.

*Experimental.* Abraded polycrystalline silver-block electrodes were polarized galvanostatically over a current density range 0.1-10 mA $cm^{-2}$ in an air thermostat maintained at $25 \pm 1°$. Potentiostatic experiments were carried out by holding the electropolished silver electrodes at a potential 100 mV negative to the reversible $Ag/Ag_2O/N$ KOH electrode and then oxidizing to $Ag_2O$ at a predetermined potential. A number of standard capacity $Ag/Ag_2O$ electrodes prepared by polarizing for five minutes at +150 mV, then a further six minutes at +250 mV, were reduced by switching the potential to the appropriate preset potential negative to the reversible $Ag/Ag_2O/N$ KOH potential.

*Results.* Typical current-time transients for the formation of $Ag_2O$ constant potential are shown in Figure 58. For oxidation potentials in the range +50 mV

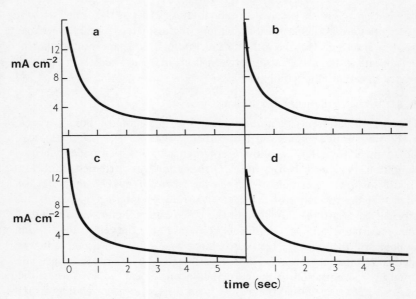

**Figure 58**    Current versus time curves for argentous oxide formation.
(a) 150 mV; (b) 200 mV; (c) 250 mV, $N$ KOH; (d) 150 mV, $5N$ KOH.

to +250 mV, that is, for potentials below the reversible $Ag_2O/AgO$ potential, and for oxidations in $0.1N$, $1N$, and $5N$ KOH (with and without nitrogen stirring), the same general shape of current-time transient is observed. There is a marked fall in the rate of $Ag_2O$ formation immediately after the positive potential is imposed on the electrode and a marked change in rate occurs after a small amount of charge has passed. Once the current has fallen to low values, a stepwise increase in the potential does not appreciably increase the capacity of the electrodes. This is consistent with data (R159) showing that when electrodes are charged at constant cell potential the current accepted by the electrode is practically zero above +100 mV.

Argentous oxide can be formed on silver electrodes at a potential +40 mV above the accepted $Ag_2O/AgO$ equilibrium potential. Thus when silver electrodes are charged at +308 mV the characteristic falling current-time curves of Figure 58 were observed and powder diffraction patterns of the deposit showed only the lines characteristic of $Ag_2O$; that is, the nucleation of AgO must be markedly inhibited. In addition $Ag_2O$ could be formed at a potential just less than the minimum potential required for oxygen evolution (0.06 V on the $Ag/Ag_2O$ scale); under these conditions the formation of oxygen is unlikely to occur, and it can be assumed that the current is due entirely to the formation of oxide at all stages of growth. The fact that the shape of the current-time curve remains constant over a 250 mV range of potential indicates that the mechanism for the formation of $Ag_2O$ remains unchanged.

The variation of the potential with time at constant current has been shown to be dependent on the current density and electrolyte concentration (R391). For freshly abraded or electropolished substrates it is found that decrease of current density and increase in electrolyte concentration favors the development of a potential arrest prior to the peak. In addition the quantity of electricity for this first stage increases with decrease of current density, Figure 59. The variation in thickness of the deposit with current density is confirmed by the results of the electron-optical investigation.

The potential-time transients at constant current and the current-time transients at constant potential were relatively insensitive to the method of preparation of the electrode and a primary layer of randomly oriented $Ag_2O$ is formed on the electrodes, followed by a secondary layer of discrete growth centers of $Ag_2O$ (R63). However, the orientation of this layer and the degree of coverage depend markedly on the nature and pretreatment of the substrate and

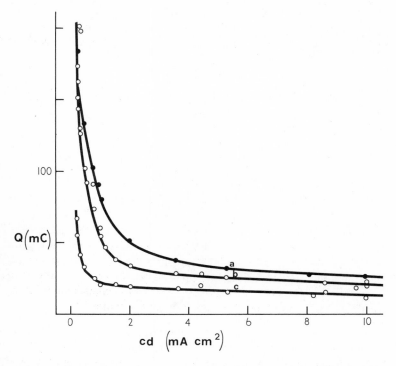

**Figure 59**    Argentous oxide formation. Plot of capacity of oxide deposit against current density. (a) $5N$ KOH; (b) $N$ KOH; (c) $0.1N$ KOH.

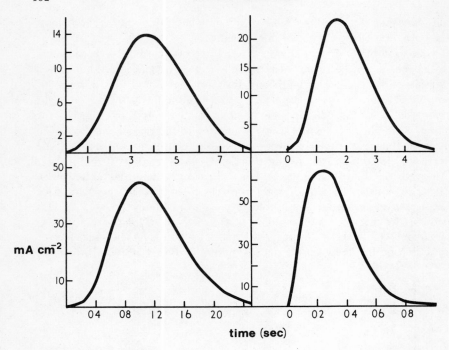

**Figure 60**   Current versus time curves for the reduction of "standard" capacity silver/argentous oxide electrodes in $N$ KOH.   (a) −50 mV; (b) −60 mV; (c) −80 mV; (d) −100 mV.

since the transients are independent of these factors, the rate-determining step must take place in the randomly oriented basal layer.

Current-time transients for the reduction of argentous oxide layers formed successively at +150 mV and +250 mV are shown in Figure 60. In general for reductions in the range −40 mV to −150 mV the transients display the same characteristics. The current has an initial low value, increases to a maximum (corresponding to the minimum in the potential-time curves at constant current), and then decreases to a final low value. As the overpotential becomes more negative the time taken to the maximum decreases implying an increasing rate of reduction with increasing negative overpotential.

*Discussion.* Electrode processes which involve the formation of a new phase at constant potential give currents which initially increase with time (R215) because of the nucleation of growth centers which expand with time, the current being proportional to the area of the expanding interfaces. In the present system this pattern has been observed for the oxidation of argentous oxide to argentic oxide, which will be described below, where growth centers of the new phase are nucleated and grow in three dimensions until they expand to the boundaries of the parent phase.

An analogous pattern would seem to be observed for the reduction of the argentous oxide layers where the electron microscopic evidence indicates the progressive nucleation and growth of two dimensional centers of silver in the parent phase. If this process is controlled by the electrode reaction at the interface having a rate constant $k$ (mol cm$^{-2}$ sec$^{-1}$), a current-time curve is predicted of the form (R215):

$$i = \frac{2F\pi Mhk^2At^2}{\rho} \exp\left(-\frac{\pi M^2k^2At^3}{3\rho^2}\right) \tag{1}$$

where $A$ (nuclei cm$^{-2}$ sec$^{-1}$) is the nucleation rate constant, $\rho$ is the density, $h$ the height of the growth centers assumed to be cylindrical in shape, $M$ is the molecular weight, and the other symbols have their usual significance. At short times where the growth centers do not interact with each other, Equation 1 reduces to

$$i = \frac{2F\pi Mhk^2At^2}{\rho} \tag{2}$$

The application of Equation 2 to the reductions at short times is illustrated in Figure 61A while an example of a whole transient treated according to Equation 1 is shown in Figure 61B.

In view of the fit of these equations to the data values of the composite rate constant $k^2A$ may be estimated from the maximum in the current-time curves

$$t_{max} = \left(\frac{2\rho^2}{\pi M^2k^2A}\right)^{1/3} \tag{3}$$

The plot of 3 log $t_{max}$ against electrode potential has the form of a Tafel line since the composite rate constant contains the term $k^2$ and this rate constant would be anticipated to have the form

$$k = k_0 \prod_j C_j^{\nu_j} \exp\left(\alpha z\phi F/RT\right) \tag{4}$$

A surprising feature of the data is that the slope is of the order 13 mV$^{-1}$ whereas the maximum value that would be predicted with $\alpha = 1$, $z = 1$ from Equation 4 would be 30 mV$^{-1}$. It is clear that the nucleation rate constant must have a

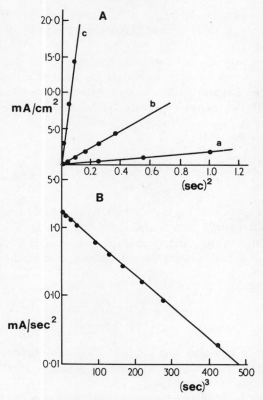

Figure 61 Reduction of argentous oxide deposits in $N$ KOH. (A) Test of Equation 2 at (a) −50 mV; (b) −60 mV; (c) −70 mV. (B) Test of Equation 1 at −50 mV.

marked potential dependence and in addition the mechanism of the formation of the silver lattice may be more complicated than the simple discharge

$$Ag^+ + e \to Ag \tag{5}$$

and may possibly involve a second order reaction.

In the phase change Ag → Ag$_2$O centers of the oxide grow three dimensionally on the electrode surface. It might therefore be expected that for the formation of this oxide, the current should vary initially with the square or the cube of the time depending on whether the centers are nucleated in the early stages or progressively with time. The current is, however, observed to fall with time. It is clear therefore that the rate-determining step cannot be at the expanding interfaces of the growth centers but rather in the randomly oriented thin basal layer which has been found to form the initial stages of the reaction. We assume that the surfaces of this basal layer are progressively covered by hemispherical centers and that the current is proportional to the area of uncovered surface. If the additional assumption is made that there is no overlap, theory and experiment do not agree (R214-215).

Unfortunately, an expression for the current-time variation has not been deduced to date if we allow for the overlap of centers but it is possible to derive a relation between the charge and the instantaneous current. If we assume the growth centers to be right circular cones of radius $r_0$ and height $mr_0$ then the charge is given by (R20):

$$q = \frac{\rho F}{M} \int_0^{mr_0} \left\{ 1 - \exp\left[-\pi N_0 r_0^2 \left(1 - \frac{x}{mr_0}\right)^2\right] dx \right\}$$

(6)

$$= \frac{\rho F mr_0}{M} - \frac{\rho F m}{2MN_0} \, \mathrm{erf}(\pi^{1/2} N_0^{1/2} r_0)$$

while the current, being proportional to the area of uncovered surface, is given by

$$i = \frac{dq}{dt} = i_0 \exp(-\pi N_0 r_0^2)$$

(7)

Substituting for $r_0$ from Equation 7 we obtain

$$\frac{q}{[\ln(i_0/i)]} = \frac{\rho F m}{M\pi^{1/2} N_0^{1/2}} - \frac{\rho F m}{2MN_0^{1/2}} \cdot \frac{1}{[\ln(i_0/i)]} \, \mathrm{erf}[\ln(i_0/i)]^{1/2}$$

(8)

A test of this equation is shown in Figure 62. Both the slope and intercept of this plot may be calculated if $m$ is known, taking the value of $N_0$ from the micrographs. With $m = 1$ and $N_0 = 9 \times 10^8$ a calculated slope of $16.5 \times 10^{-2}$ coulomb cm$^{-2}$ is obtained compared with the experimental value of $14.0 \times 10^{-2}$ while the intercept is $1.79 \times 10^{-2}$ coulomb cm$^{-2}$, the graph giving values in the range $1.6 - 1.9 \times 10^{-2}$.

It is concluded, therefore, that for the time range shown in Figure 58, the kinetic data are in good accord with the electron optical observations if we assume that the current is controlled by the area of the uncovered basal layer. At longer times the current deviates from the predicted Equation 8 and it is found that the relation is then parabolic (Figure 63); the slope of these plots cannot be explained by diffusion through the solution because the solubility of

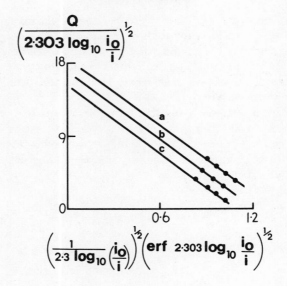

Figure 62    Argentous oxide formation. Plot of

$$\frac{Q}{[2.303 \log_{10} (i_0/i)]^{\frac{1}{2}}}$$

against

$$\frac{1}{[2.303 \log_{10}(i_0/i)]^{\frac{1}{2}}} \cdot \text{erf} \, [2.303 \log_{10}(i_0/i)]$$

Values of $q$ derived from integrated current-time curves
(Figure 58): (a) 150 mV; (b) 200 mV; (c) 250 mV.

$Ag_2O$ in $1N$ KOH is $1.5 \times 10^{-4}$ molar (R12), and this would give a slope at least two orders of magnitude smaller than that observed, particularly if it is borne in mind that only a part of the surface is uncovered. It follows that the parabolic relation is due to the thickening of the basal layer.

### THE $Ag_2O$/AgO PHASE CHANGE

Anodic charging curves for the oxidation of $Ag_2O$ deposits to AgO in KOH solutions have been characterized by various workers but no systematic potentiostatic studies have been reported (R156, 159, 215-16, 401, 629). Hitherto, kinetic studies of this phase change have relied on supposedly steady-state current/potential measurements (R33).

Monoclinic argentic oxide is the final oxidation product of silver in alkaline solution; the same material is formed in initial stages of the oxidation of silver in

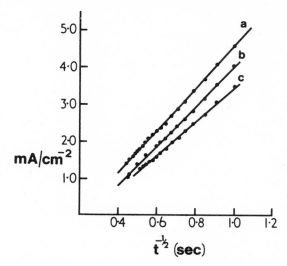

**Figure 63**   Argentous oxide formation. Variation of current with time $(t)^{-1/2}$. Curves identified as in Figure 62.

acid sulphate solution (R185). This work is of importance in connection with the present system. From the current-time transients obtained in various sulphuric acid solutions under potentiostatic conditions it was concluded that argentic oxide is formed by the cylindrical growth of centers instantaneously nucleated in the initial stages of the growing deposit. Microscopic evidence supports the view that the solid phase AgO is deposited at the $Ag_2O$/electrolyte interface but in contrast with the sulphate system the oxidation proceeds by the *three*-dimensional growth of centers which are progressively nucleated in the $Ag_2O$ layer covering the silver substrate. When due allowance is made for the difference between this form of crystal growth and that found for the oxidation of $Ag_2SO_4$, it is found that the kinetics of lattice formation in the two systems are identical.

*Experimental*. Electropolished silver electrodes were introduced into $1N$ KOH solution at $25°$ at a potential $100\,mV$ negative to the $Ag/Ag_2O/1N$ KOH potential and a deposit of standard thickness was formed by a polarization for five minutes at $+150\,mV$, followed by a further six minutes at $+250\,mV$. The current-time curves for the conversion of the $Ag_2O$ layer to AgO were then taken at a series of potentials by raising the potential to appropriate values. Current-time measurements were made either oscillographically or with a meter and stopwatch. For experiments in $0.1N$ and $5N$ KOH standard capacity electrodes were first formed in $1N$ KOH and then removed from the cell, rinsed in conductivity water, followed by potassium hydroxide solution having the different concentration and finally transferred to another cell containing the required electrolyte. The operation was carried out as quickly as possible in order to minimize capacity loss by dissolution or self-discharge. In all experiments the electrolyte was stirred vigorously with purified nitrogen.

*Results*. Typical current-time transients for the formation of argentic oxide from argentous oxide deposits, in potassium hydroxide solutions are shown in Figure 64. The phase change first occurs at an appreciable rate in $N$ KOH at an

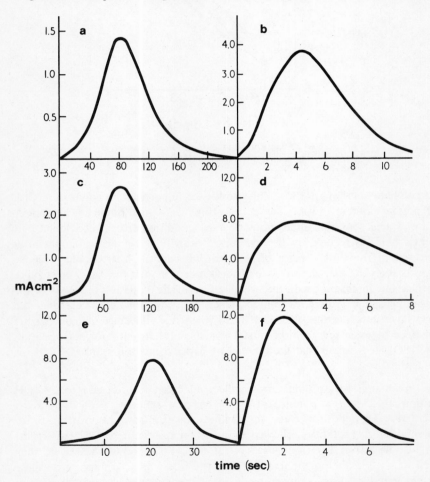

**Figure 64**    Current versus time curves for the oxidation of argentous oxide to argentic oxide in potassium hydroxide solutions. (a) 405 mV, 0.1$N$ KOH; (b) 500 mV, 0.1$N$ KOH; (c) 350 mV, $N$ KOH; (d) 450 mV, $N$ KOH; (e) 365 mV, 5$N$ KOH; (f) 460 mV, 5$N$ KOH.

overpotential of 80 mV; an overpotential of approximately 140 mV is required to give the same rate of oxidation in 0.1$N$ KOH. In general, the current has a low initial value which increases to a maximum and subsequently decays nonlinearly to a final low value; the time, $t_{max}$, taken to reach the maximum current decreases with increase of overpotential and with increase in electrolyte concentration. At potentials greater than +500 mV, oxygen is evolved and this additional reaction is reflected in the shape of the transients by an increase in the value of the final current.

The total quantity of electricity used for oxide formation was estimated by integration of the current-time curves. At low overpotentials more material formed during the oxidation to AgO while at high overpotentials only a part of the $Ag_2O$ was oxidized to AgO; thus, at high overpotentials the deposit probably has the form $Ag/Ag_2O/AgO$. The variation of capacity with overpotential shows that an accurate kinetic analysis of the current-time curves is only possible in a narrow range of potential where the capacity of the AgO layer is equal to that of the $Ag_2O$ layer.

The reactions taking place initially during the phase change to AgO occur at the outer surface of the electrode; the AgO centers nucleate on the $Ag_2O$ layer and progressively cover the surface with increasing time.

The general method of analysis has been described in detail elsewhere (R215). In order to obtain the current-time relation algebraically for the growth of the new phase, two assumptions are made: (1) there is a uniform probability with time of forming AgO nuclei at preferred sites (of total number $N_0$) giving a nucleation law

$$N = N_0 \left[ 1 - \exp(-At) \right]$$

where $A$, $sec^{-1}$, is the nucleation rate constant; and (2) the AgO centers grow in three dimensions and during the initial stages of formation the individual centers grow independently of each other without overlapping, giving

$$i = \frac{2zF\pi M^2 k^3 t^2}{\rho^2}$$

for a single hemispherical center, where $M$ is the molecular weight, $z$ the number of electrons transferred per molecule of reaction, and $\rho$ the density. The rate constant of crystal growth $k$ has the units moles $cm^{-2}$ $sec^{-1}$.

Two possible schemes for the oxidation of the $Ag_2O$ deposits may be envisaged; the AgO nuclei will either (1) form and grow at each discrete $Ag_2O$ center or (2) be distributed over a considerable volume of the overall $Ag_2O$ layer. It is not possible to distinguish between these schemes from the electron-optical evidence.

In Scheme 1, assuming hemispherical centers, the overall current time transient is

$$i = \frac{2zF\pi M^2 N_0 k^3}{A^2 \rho^2} [A^2 t^2 - 2At + 2 - 2\exp(-At)] \qquad (9)$$

which applies to short times. For very short times, when $At \ll 1$, Equation 9 may be simplified to

$$i = 2zF\pi M^2 N_0 k^3 A t^3 / 3\rho^2 \qquad (10)$$

If it is assumed that the growth centers have a different shape, then the coefficient in Equations 9 and 10 can be changed appropriately. Equation 9 applies to short times where there is no limitation on the size of the growth centers caused by the parent $Ag_2O$ crystals. For spherical growth centers Equations 9 and 10 must be multiplied by 2. In this case and for long times, when the AgO centers have grown to the limiting surface of the $Ag_2O$ crystals,

$$i = \frac{4zF\pi M^2 N_0 k^3}{A^2 \rho^2} [A^2 t_{max}^2 - 2At_{max} + 2 - 2\exp(-At_{max})]$$
$$\times \exp -A(t - t_{max}) \qquad (11)$$

In Scheme (2), assuming that the growth of spherical centers of AgO takes place in the overall $Ag_2O$ layer, the kinetics will be similar to a case which has been discussed previously. It has been shown that (R215) the growth of spherical centers of $\beta$-lead dioxide in large lead sulphate crystals follows the equation

$$i = \frac{4zF\pi M^2 N_0' V k^3}{A^2 \rho^2} [A^2 t^2 - 2At + 2 - \exp(-At)]$$
$$\times \exp \frac{-4\pi M^3 N_0' k^3}{3A^3 \rho^3} \cdot [A^3 t^3 - 3A^2 t^2 + 6At - 6 + 6\exp(-At)] \qquad (12)$$

in which $N_0'$ is the maximum density of nuclei per unit volume of the deposit and $V$ is the volume of substrate per unit area (thickness of the layer).

*Analysis of the experimental current-time curves.* Equations 9, 10, 11, and 12 may be tested in a variety of ways.

*The initial stage.* For $t \ll t_{max}$, it can be seen from Figure 65 that the current varies with the cube of time as predicted in Equation 10. This relationship is

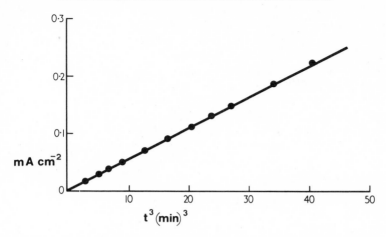

**Figure 65**    Current against (time)$^3$ plot; Ag$_2$O initially deposited in $N$ KOH. Oxidation to AgO at 400 mV, 0.1$N$ KOH.

observed at low and medium overpotentials ($100\ mV \leqslant \eta \leqslant 170\ mV$). At high overpotentials the current initially increases linearly with time. The overpotential region in which a $t^3$ dependence is observed corresponds to that in which the electrochemical capacities of the Ag/Ag$_2$O and Ag$_2$O/AgO electrodes are approximately the same; this overpotential range is further restricted by the requirement that $At \ll 1$ for Equation 10 to be valid so that the linear approximation for the exponential in Equation 9 applies. In addition, if Scheme (2) is correct there must be no appreciable overlap of growth centers in the time range considered.

The slopes of the $i/t^3$ graphs, which depend on the term $k^3 A N_0$, are plotted in Figure 66 as a function of the applied potential. Because of the difficulty of assessing the contribution which the dissolution of the oxide makes to the overall current-time transients for oxidations in 5$N$ KOH only the data for 0.1 $N$ and 1$N$ KOH have been treated in this way.

*The stage $t < t_{max}$.*    If Equation 9 is rewritten in logarithmic form (for spherical centers), the last term may be separately estimated and $k^3 N_0$ values deduced.

$$\ln i = \frac{\ln 4zF\pi M^2 k^3 N_0}{\rho^2 A^2} + \ln\left[A^2 t^2 - 2At + 2 - 2\exp(-At)\right] \qquad (13)$$

This equation will apply when $At \approx 1$ but when there is no interaction or overlap of growth centers so that deviations from the $t^3$ law are caused by the departure of the number of nuclei from the linear approximation for Equation

9. The value of the logarithmic term may be estimated (see below) and subtracted from $\ln i$. A plot of this difference against $t$ gives the value $k^3 N_0 / A^2$ as shown in Figure 67A (for an overpotential of 100 mV and A = 0.04 sec$^{-1}$); $k^3 N_0$ may therefore be obtained directly. It can be seen from Figure 67B that at higher overpotentials the growth constant cannot be separated out in this way. Hence, although the agreement of the predicted and actual current-time curves at low overpotentials substantiates the view that the AgO centers are being randomly nucleated and then grow three dimensionally, a different treatment is necessary to separate out $k^3 N_0$ values at high overpotentials.

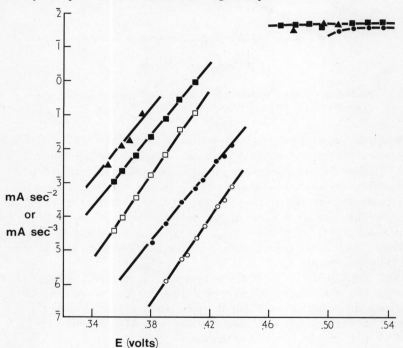

**Figure 66**     Variation of log mA sec$^{-2}$ and log mA sec$^{-3}$ with potential:

$$\text{Ordinate} = \frac{\log (4zF\pi M^2 k^3 A N_0 \times 10^3)}{3\rho^2}$$

where $\bigcirc$ = 0.1$N$ KOH and $\square$ = $N$ KOH and

$$\text{ordinate} = \frac{\log (4zF\pi M^2 k^3 N_0 \times 10^3)}{3\rho^2}$$

$\bullet$ = 0.1$N$ KOH, $\blacksquare$ = $N$ KOH, and $\blacktriangle$ = 5$N$ KOH. All values for oxidations in 5$N$ KOH and those in 0.1$N$ and $N$ KOH at high potential obtained from Figure 69 using approximate $t_{\max}$ values.

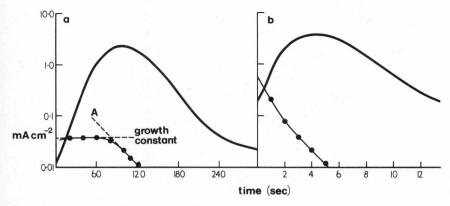

**Figure 67**    Argentic oxide formation. Variation of log current with time. (a) 360 mV, $N$ KOH; (b) 500 mV, $0.1N$ KOH. Construction showing the reduction of the current-time transient for the phase change to two sections giving the nucleation constant $A$ and the growth constant.

*The stage $t > t_{max}$.*    From Equation 11, when the current first becomes limited by restrictions on the size of the growing centers, the decrease after the maximum obeys the law of nucleation. Under these conditions Equation 4 reduces to the simple form

$$\ln i = At + \text{constant}$$

The variation of $\ln A$ with $\eta^{-2}$ is shown in Figure 68 according to the predicted potential dependence of this rate constant.

*The point $t = t_{max}$.*    A characteristic feature of the current-time curves is the dependence of $t_{max}$, the time at which the maximum current is reached, on overpotential and the solution concentration. Relations between the crystal growth constant and $t_{max}$ have been derived and tested in Figures 7 and 14. If the linear dimension of an AgO center increases by $dr$, then

$$dv = \left(\frac{Mk}{\rho}\right)dt \tag{14}$$

and $r = Mkt/\rho$. Two models can apply as before and their applicability must be established. First, if the centers grow in a crystal of $Ag_2O$ and become limited by interaction with the boundaries, then $k$ is determined by

$$k = r_0\rho/Mt_{max} \tag{15}$$

where $r_0$ is the linear extension of the $Ag_2O$ center. Because of the difficulty of

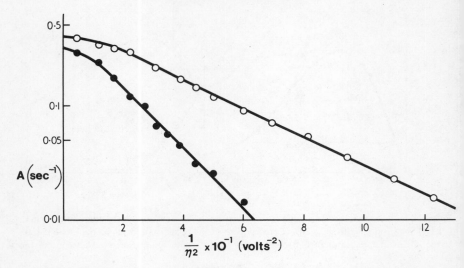

**Figure 68**    Variation of log nucleation rate constant, $A$, with $\eta^{-2}$, $\bullet = 0.1N$ KOH; ($\circ$) $= N$ KOH.

obtaining $N_0$, an alternative test of Equation 15 must be made. The volume enclosed by the $Ag_2O$ centers is given by

$$r_0{}^3 = \frac{Q}{zF} \cdot \frac{M}{N_0\rho} \tag{16}$$

and from Equation 15,

$$K^3 = \frac{r_0{}^3\rho^3}{M^3 t_{max}^3} \tag{17}$$

Hence by substituting the value of $r_0{}^3$ of Equation 16 in Equation 17,

$$K^3 N_0 = \frac{Q\rho^2}{zFM^2 t_{max}^3} \tag{18}$$

This equation may be tested by plotting $k^3 N_0$ against $1/t_{max}^3$. Values of $k^3 N_0$ at particular overpotentials may be obtained from Figure 66 by dividing by the appropriate value of $A$ obtained from the $\ln i/t$ plots. It can be seen in Figure 69 that there is a linear relationship between $\ln k^3 N_0$ and $\ln i/t_{max}^3$ over a large range of overpotential. In addition, the data from the oxidations carried out in $0.1N$ and $1N$ KOH lie on the same straight line.

Second, if the growth of centers takes place in the overall $Ag_2O$ layer the fall in current with time is due to the overlap of the AgO centers and Equation 12 applies. In order to obtain $t_{max}$, Equation 12 may be differentiated and $di/dt$ set equal to zero. Using the approximation

$$\exp(-At) = 1 - At + \frac{A^2 t^2}{2!} - \frac{A^3 t^3}{2!}$$

then

$$l = \frac{4\pi M^3 N_0' k^3 A t_{max}^4}{9\rho^3}$$

and hence,

$$k^3 N_0' A = \frac{9\rho^3}{4\pi M^3 t_{max}^4} \tag{19}$$

As $N_0$ is defined as the maximum density of nuclei per unit volume,

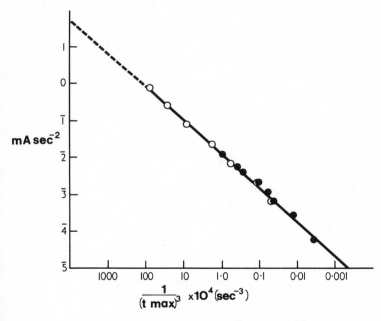

**Figure 69**  Variation of log mA sec$^{-2}$ with log $1/t^3_{max}$

$$\text{ordinate} = \log \frac{(4zF\pi M^2 k^3 N_0 \times 10^3)}{3\rho^2}$$

where $\bullet$ = $0.1N$ KOH, $\circ$ = $N$ KOH, and - - - - = extrapolation of plot to the higher potential region where $A$ is virtually constant (*cf.* Figure 62).

multiplication of the right-hand side of Equation 19 by a factor representing the thickness of the deposit gives,

$$k^3 N_0 A = \frac{9\rho^3}{4\pi M^3 t_{max}^4} \cdot \frac{QM}{zF\rho} = \frac{9Q\rho^2}{4zF\pi M^2 t_{max}^4} \qquad (20)$$

This equation has been tested in Figure 70, where $k^3 N_0 A$ values from Figure 60 have been plotted against $1/t_{max}^4$. However, the data do not fit the predicted current-time relationship as well as for the first case.

In view of the fit of the data to Equation 18, approximate values of $k^3 N_0$ may be estimated from this plot provided $t_{max}$ is known. In the region of the maximum it may be assumed that the bulk of the current is used to oxidize the $Ag_2O$ layer to $AgO$. A reasonable estimate of $k^3 N_0$ may therefore be made at overpotentials where Equations 10 and 12 cannot be fitted. These values are included in Figure 66.

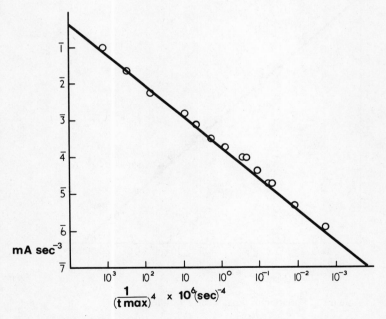

**Figure 70**    Variation of log mA sec$^{-3}$ with log $1/t_{max}^4$

$$\text{ordinate} = \log \frac{(4zF\pi M^2 k^3 A N_0 \times 10^3)}{3\rho^2}$$

where ● $= 0.1N$ KOH and ○ $= N$ KOH.

## DISCUSSION

It would seem established that discrete centers of $Ag_2O$ and $AgO$ nucleate and grow on anodized silver electrodes in alkaline solutions. X-ray diffraction examination of the working silver electrode by using an oscillating device (R60) has confirmed that in the potential regions investigated the initial phase change involved is always $Ag_2O$ to $AgO$ and evidence is given that this change takes place at the surface of the electrode next to the solution. What has not been stressed previously is the growth geometry involved in this phase change. The shapes and size of the $Ag_2O$ centers formed under constant current and potentiostatic conditions on silver electrodes have been characterized and some information presented on the texture of $AgO$ deposits. The growth of $AgO$ centers has been assumed to be determined to some extent by the environment of the $Ag_2O$ crystals in which they develop.

Constant potential measurements have shown that a critical overpotential of the order of 100 mV is necessary before the oxidation process $Ag_2O \rightarrow AgO$ will proceed at an appreciable rate in alkaline solution. This potential, which is concentration dependent, is taken to be that below which nucleation cannot be observed on a time scale of one hour, that is, $A \approx 10^{-4}$ $sec^{-1}$. This "nucleation overpotential" is comparable to those obtained for the formation of three-dimensional nuclei in other oxide systems (R607). A minimum overpotential of 40 mV is required to form $AgO$ from silver sulphate in $1M$ sulphuric acid solution whereas 100 mV is required to form $\beta$-$PbO_2$ from lead sulphate in $2N$ sulphuric acid (R185, 215, 607). The fact that the current-time curves characteristic of the phase change $Ag \rightarrow Ag_2O$ persist for oxidations carried out for potentials up to 300 mV, that is, 40 mV above the $Ag_2O/AgO$ reversible potential, also indicates the inhibition of the $AgO$ nucleation. In this connection it may also be noted that we have been unable to detect $AgO$ by means of powder diffraction patterns on electrodes charged at 310 mV in $1N$ KOH solution although it has been reported that this oxide does form on battery plates when these are charged at 308 mV in $8N$ KOH for long times (R629). Oxidations carried out at potentials between 310 and 340 mV gave very small currents, probably because of thickening of the $Ag_2O$ layer rather than the formation of $AgO$. Electrodes maintained at these potentials for one hour gave no indication of $AgO$ by X-ray diffraction. It is possible that the formation of small amounts of $AgO$ would be obscured by the simultaneous deposition of further $Ag_2O$ under these conditions.

The observation of a relatively large nucleation overpotential is consistent with the fact that a potential peak occurs during constant current charge of an $Ag/Ag_2O$ electrode to $AgO$. On the other hand (R156, 159), it has been suggested that this potential peak cannot be caused by the difficulty of nucleating $AgO$ centers in $Ag_2O$ because a similar peak is observed during the reduction of charged electrodes. We have shown above that the formation and growth of Ag centers on $Ag/Ag_2O$ electrodes takes place during reduction and it is probable that similar structural changes take place during the reduction of fully charged electrodes. The need to nucleate AgO centers has also not been recognized in other kinetic studies of this phase change (R33). Over an appreciable range of the Tafel plots which were reported it is likely that the main faradaic process was the further formation of $Ag_2O$. As has been pointed out, steady-state measurements of Tafel lines cannot be applied to this type of electrocrystallization process.

Thus, from this work, the overall rate constant of crystal growth, $k^3 N_0$, may be determined separately from the nucleation rate. As this rate constant contains the concentration- and potential-dependent terms, the mechanism of lattice formation may be examined. A precise determination of $k$ would demand a measurement of $N_0$, but for this particular phase change it is difficult to obtain an independent value of this parameter. If it is supposed that each $Ag_2O$ crystal is a site for AgO growth then the value for $N_0$ can be derived directly by counting the centers observable by electron microscopy; this could be equivalent to $\sim 10^9$ centers $cm^{-2}$. Electron and X-ray diffraction measurements suggest that the $Ag_2O$ centers are polycrystalline and not single crystals and consequently nucleation may take place in the subunits. The electron micrographs of the AgO deposits do not permit a decision as to which of these two possible situations actually applies. Development of more faithful replicating techniques for these thick AgO deposits may give an indication of the precise textural changes taking place and thus permit $N_0$ values to be obtained. However, for an examination of the kinetics, the constant $k^3 N_0$ may be used instead of $k$ since $N_0$ is a constant parameter determined by the conditions for forming the $Ag_2O$ layer.

In spite of the differences in the geometric factors controlling the formation of the higher oxide in alkaline solutions as compared with the acid solutions, the derived rate constant $k^3 N_0$ varies in the same way with potential (Figure 66) as for the formation of AgO in $Ag_2SO_4$ (R185; see Figure 9 there). This shows that the mechanism of lattice formation is similar in the two media although the formations take place at very different electrode potentials. The similarity in the kinetics of oxidation is not surprising in view of the fact that monoclinic argentic oxide is formed in both cases. In addition, the form of the plots of the rate constant against the electrode potential for the formation of silver oxide is related to that which has been observed for the electrodeposition of $\beta$-NiOOH and $\alpha$-MnO$_2$ (R62, 213). At low electrode potentials the Tafel slope is approximately $(60\ mV)^{-1}$ and the overall heterogeneous "rate constant" is proportional to the hydroxide ion concentration at constant electrode potential. The overall rate constant is always of the form

$$k = k_0 \prod_j C_j^{\nu_j} \quad \exp\left(\frac{\alpha z \phi F}{RT}\right) \tag{4}$$

where

$$\prod_j c_{-j}^{v_j}$$

represents the stoichiometry of the slow stage of the reaction together with any pre-equilibria. It is evident therefore that the reaction leading to the formation of the lattice is first order with respect to hydroxyl ion. At high electrode potentials the reaction becomes independent of potential and also independent of the composition of the solution.

These data are best interpreted by assuming a slow formation of the lattice by the reaction of $Ag^+$ ions with adsorbed OH radicals, these radicals being formed from $OH^-$ by an electrochemical pre-equilibrium.

$$OH^- = OH_{ad} + e$$

$$Ag^+ + OH_{ad} \xrightarrow{\text{slow}} AgO + H^+$$

At low electrode potentials the pre-equilibrium leads to a reaction of the first order with respect to $OH^-$ and a Tafel slope of 60 $mV^{-1}$ while at high electrode potentials the coverage of the sites where crystal growth takes place becomes complete so that the reaction becomes independent of the potential and concentration in agreement with the observed facts.

# PART IV
## MANUFACTURE OF ELECTRODES

# 13.

## Zinc Electrode Manufacture

### J. A. Keralla

Various types of zinc electrodes (R615) have been used over the years ranging from sheet stock (R181) through pasted (R36, 294, 546) and pressed (R16) zinc oxides on suitable grids (R64, 235) to plated or electrodeposited zinc plates. Requirements for both primary and secondary types have eliminated some electrode structures because of higher energy demands, excessive manufacturing costs, or lack of physical strength.

In general, today's battery requirements call for a zinc plate with high energy yields at high current densities. Zinc plates must possess good physical strength and integrity in order to withstand severe environmental stresses and in order to adequately retain their active material during repeated discharge-recharge cycles.

The purpose of this paper is to describe materials, processes, and control measures that have been successfully used to manufacture zinc electrodes by the electrodeposition and pressed-powder methods. Materials and procedures can, however, be expected to vary with individual manufacturers.

Both plate-making processes generally use a metallic grid or support structure, either silver or copper, in a woven screen or expanded metal form. The leads or plate terminals are either wire or flat stock of the same material as the grids. The choice of metal, grid, and terminal design is, of course, dependent on the battery application requirements. In this paper, expanded silver metal will be the base grid structure for both plate types with wire terminal leads for the pressed zinc oxide plate and flat-tab stock for the electrodeposited zinc plate. The quality control procedures are placed throughout the process steps, but methods of control are generally left to individual manufacturing practices.

## MATERIAL SPECIFICATIONS

Plate-making processes and plate performance can be adversely affected by changes in material properties and impurity levels. High purity materials are selected in order to achieve maximum reproducibility. Minimizing the level of impurities in basic materials also permits controlled use of addition agents to obtain reproducible beneficial effects.

The following descriptions cover all the basic constituent materials used in the manufacture of both types of zinc plates.

## ZINC OXIDE

Zinc oxide is the basic material used in both methods of plate manufacture. For best results, reagent grade zinc oxide is preferred. The chemical analysis should conform, as a minimum, to American Chemical Society (A.C.S.) requirements (R10A). On the physical side, the white powder should have an average particle size of $0.11\ \mu$ and a surface area of $10\ m^2 g^{-1}$.

## POTASSIUM HYDROXIDE

Potassium hydroxide is used for the preparation of solutions for both plating and plate formation. Mercury cell grade potassium hydroxide is preferred. The specification (R8) covers mercury cell grade KOH as shown in Table 20.

TABLE 20.   MERCURY CELL GRADE KOH SOLUTION

| Chemical Analysis | | Physical Properties | |
|---|---|---|---|
| Substituent | % by wt | | |
| KOH | 45-46 | Specific gravity at 27° | 1.45-1.46 |
| KCl | 0.005   max | Color | Colorless |
| $K_2CO_3$ | 0.06     max | Turbidity, max | 0.0005 |
| $KClO_3$ | None | Freezing point | $-32°$ |
| Fe | 0.0004 max | Viscosity at 15.6° | 7.3 centipoise |
| Hg | 0.0001 max | | |
| $Na_2O$ | 0.1       max | | |

## MERCURIC OXIDE

A small percentage of mercuric oxide is customarily blended with zinc oxide to be used for manufacture of the pressed plate for the secondary battery. Reagent grade mercuric oxide, red in color, is recommended. Its chemical analysis should conform to A.C.S. requirements (R10B).

## EXPANDED SILVER SHEET

The grid or supporting member for the plate active materials is usually a metal of low resistivity. Generally, pure silver or copper is used in the form of either woven screen or expanded metal. The latter is generally specified to be manufactured using commercial fine silver, 99.90% (R74), with optional certification of material purity.

Expanded silver sheet is fabricated usually by placing uniformly spaced slits in the sheet and expanding under tension. The resultant pattern must be uniform and conform to the dimensions and limitations specified on the part drawing. It is desired that the surface of the sheet be free of oil film or other contaminants. Dimensional tolerances for three mesh sizes are illustrated in Table 21.

TABLE 21. EXPANDED SILVER METAL SHEET DIMENSIONAL TOLERANCES

| Openings[a] | | | |
|---|---|---|---|
| no./linear in. | 23-25 | 15-17 | 12-14 |
| no./linear cm | 9-10 | 6-7 | 5-6 |
| Specified thickness | | | |
| inch | 0.0090 | 0.0140 | 0.0205 |
| ± | .001 | .001 | .0025 |
| mm | 0.229 | 0.356 | 0.521 |
| ± | .025 | .025 | .063 |
| Weight per unit area | | | |
| g/sq. in. | 0.220 | 0.300 | 0.270 |
| ± | .040 | .050 | .060 |
| g/sq dm | 3.41 | 4.65 | 4.19 |
| ± | .62 | .76 | .93 |

[a]Measured along the short dimension of diamond.

## SILVER TAB AND WIRE

Electrical connections to the individual plates are generally accomplished by welding to the grid either round wire or flat strips of the same metal. Commercial fine silver (R74) is specified for plates using silver expanded sheet; analysis of each lot is generally required. The wire diameter is generally 0.43-0.51 mm (0.017-0.020 in.) having an electrical resistance of 1.59 microhm-cm.

## WATER

Deionized or distilled water of high quality should be employed in the preparation of solutions and in the plate processing.

## MANUFACTURING PROCEDURES

### ELECTRODEPOSITED ZINC NEGATIVE PLATES

The basic equipment required for this procedure consists of tanks, frames to hold the grids and the dummy electrodes, an adequate power supply, and pressing fixtures. The tanks are constructed from Lucite plastic sheet, as are the grid holders and the frames. The size of the plates to be made will determine the size of the tanks and frames. Since only one grid can be plated per tank, it is common practice to design the strip of such size that several individual electrodes can be die cut from a strip. Figure 71 shows a grid strip before electroplating and the plated strip prior to die cutting into individual electrodes.

The anode, or positive electrode, is a nickel screen or a solid sheet, if preferred, enveloped in a frame identical to the cathode frame. The power

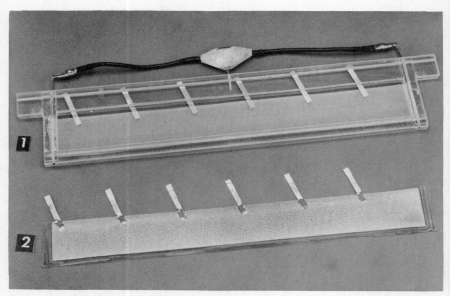

**Figure 71**     Part 1, unplated grid section in Lucite framework ready for plating; part 2, finished plated grid section ready for die-cutting to proper electrode size.

supply should be of a size sufficient to allow a current density of 120-160 mA/cm² (0.75-1.0 A/in.²) to be applied to the grid. Additional tanks can be added to the circuit depending on production requirements.

The plating operation should be done under a hood or a suitable exhaust system because considerable spray and gases are produced. The following description is an acceptable procedure for the manufacture of negative plates used in some primary zinc-silver oxide batteries:

*Preparation of the Plating Solution.*

1. Prepare a solution of 45% potassium hydroxide and deionized water. The specific gravity shall be 1.239-1.242 (20-30°) and the solution shall be inspected in accordance with the described control procedures.

2. Add 35.0 ± .5 g zinc oxide (ZnO) to each liter of KOH solution. After thoroughly stirring, the plating solution shall be inspected in accordance with the described control procedure.

3. The plating solution shall be identifiable with the raw material lots used therein and the plates made therefrom.

4. The plating solution shall be thoroughly stirred and stabilized at 24-35° before being added to the plating tank and beginning the plating operation. The temperature shall be maintained within these limits during the plating operation and shall be controlled in accordance with the described control procedure.

5. Thoroughly stir the plating solution before plating each grid section.

6. The plating solution in a tank shall be changed after a maximum number of sections have been plated, which is determined by the weight change of succeeding sections. This point is determined early in the process, and a safe margin is easily set.

*Preparation of the Silver Grid Section.*

1. Cut the silver grid sections from the appropriate expanded silver sheet material.

2. Cut the silver tabs and spot weld to the grid sections.

*Plating of the Silver Grid Section.*

1. Place the grid section in the Lucite frame and immerse the assembly in the tank of plating solution. The top of the grid section should be about 2.5 cm (1 in.) below the surface of the solution.

2. Position the anode frames containing nickel screen on either side of the silver grid. No exposed parts of the silver grid should be outside the dimensional limits of either nickel screen anode.

3. Connect both ends and a center tab of the silver grid to the negative terminal of the power supply.

4. Connect the two anodes in parallel to the positive terminal of the power supply.

5. When using more than one plating tank, the tanks shall be connected in series with the power supply.

6. Connect an ampere-time meter in the plating circuit.

7. Turn on power supply and plate zinc on the silver grid at a rate approximately 15.5 A/dm$^2$ (1 A/in.$^2$) of exposed grid area.

8. Continue the plating operation for the required number of ampere-hours to obtain the necessary weight of zinc from the relation, 1.22 g zinc equal to 1 ampere-hour (theoretical equivalent).

9. Turn off power supply and disconnect the plated zinc grid frame. Remove the section from the tank and immerse in deionized water to remove the KOH. The wash water temperature is not critical but should be maintained within a reasonable room temperature range. Wash until drops of water from the plated grid exhibit a pH of 7.

10. Using a spatula or similar tool, remove the excess zinc from the tab areas affected.

11. Remove the plated grid from the frame; place in a wet pressing fixture and press to the required plate thickness.

12. Place the pressed strip in an oven at 50-55° and allow to dry for a minimum time of 25 min.

13. Die cut the plated grid section into individual plates. Plates shall be stored in a temperature range of 15-30° and a maximum relative humidity of 10%.

### PRESSED POWDER ZINC PLATES

The basic equipment required for this procedure are cavity molds with covers, a mixer, a press capable of exerting 14 kg cm$^{-2}$ (200 psi) of plate area, formation tanks, and a suitable power supply. The cavity mold is constructed from steel plate about 13 mm (0.5 in.) thick and 50-75 mm oversize from the required electrode dimensions. A cavity is cut in the plate about 6 mm deep to electrode dimensions with a steel block to fit over the electrode to serve as a pressing fixture. Figure 72 shows such a cavity mold with an electrode in place during the course of construction. A piece of nonwoven material such as Viscon paper is

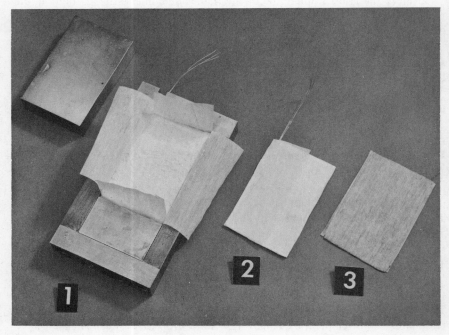

**Figure 72**    Part 1, cavity mold with grid, Viscon paper, and zinc oxide mix ready for pressing; part 2, finished unformed negative plate; part 3, formed, charged, and dry zinc plate.

wrapped around the plate as a strengthening measure, and it becomes an integral part of the electrode.

A dry mixer (Patterson-Kelly Vee type) is used to blend oxides and additives prior to plate assembly and is also suitable for wet mixes. In preparing a negative plate for secondary battery usage requiring, say, twenty cycles in 3-6 months activated life, it is customary to add 1-2% mercuric oxide (R648) by weight in the zinc oxide. For batteries requiring several hundred or more cycles, additives such as Teflon powders (R254) and Emulphogene (R376) solutions can be used in addition to the mercuric oxide, which are helpful in maintaining the zinc electrode cycle capability.

The following description is an acceptable procedure for the manufacture of negative plates used in some secondary silver-zinc batteries:

*Preparation of the Zinc Oxide and Mercuric Oxide Blend.*
1. Weigh out a required amount of zinc oxide powder.
2. Add 1-2% mercuric oxide of total batch weight of the zinc oxide.
3. Blend in the mixer for a minimum of 30 minutes. Remove sample and analyze for mercury content in accordance with the described control procedure.
4. At this point, additives or binders of the dry powder type can be added according to the amount required and the blending is then continued for another 30 minutes.
5. Remove blend from mixer. The mix is ready for application on the grid.

*Preparation of the Silver Grid.*
1. Cut the silver grid from the appropriate expanded silver sheet material.
2. Cut the silver wires and spot weld three to the grid. The number of wires depends on specific design requirements.

*Preparation of the Viscon Envelope.*
1. From stock material, cut to required size such that, when a grid complete with active material is placed on the paper center, the four sides of the plate can be securely lapped by the folds.
2. Envelope the grids prior to addition of ZnO mix.

*Preparation of a Plate.*
1. Place the grid with Viscon folds open in the cavity mold.
2. Weigh a given amount of zinc oxide mix and spread evenly over the grid with a spatula or similar tool.
3. Fold the Viscon flaps about the plate, and place a block over the plate in the cavity.
4. Place fixture in a suitable press and apply pressure to required plate thickness.
5. Remove from press; remove block cover from plate.

6. The addition of liquid additives is now made to the exposed surface of the plate in the cavity by pouring a measured quantity over the plate surface.

7. Place the block over the plate area and apply pressure of 0.07 kg cm$^{-2}$. This will disperse the liquid additive completely in the zinc oxide mix on the plate.

8. Remove the block and insert a spatula or similar tool between the plate and mold. Lift and place on paper. At this time plates can be stored indefinitely in sealed polyethylene bags or constructed into cell groups for jar formation or open-tank formation and dry charging.

### Preparation of Formation Electrolyte.

1. Prepare a solution of 5% potassium hydroxide and deionized water. The specific gravity of the prepared solution shall be 1.046-1.055 (21-27°), and the solution shall be inspected in accordance with the described control procedure.

2. This formation electrolyte shall be identifiable with the raw material lots used therein and the plates made therefrom.

### Assembly of plates in the formation tank.

1. Fold a piece of PUDO cellophane, oversize to the width and height of the electrode by 2.5 cm (1 in.).

2. Insert the electrode between the folds of the cellophane such that the bottom of the plate is encompassed but the sides and top are open.

3. Place the wrapped electrode between two pieces of inert separator material, such as microporous polyethylene or rubber, having the same dimensions as the cellophane wrap.

4. Place two nickel-screen dummy electrodes on either side of the material. This is the basic plate group. Additional negative plates are laid up in the same sequence of plate, cellophane, separator, and dummy electrode.

5. When a required number of plates has been assembled in the tank, the assembly is firmly blocked in place so that no movement of the electrodes is possible. The leads of the electrodes are soldered or clamped to a suitable metal bar such as copper 3.2 mm thick ($\frac{1}{8}$ in.) 12.7 mm wide (0.5 in.) and having a length as required. The leads of the dummy electrodes are similarly soldered or clamped to a copper bar.

### Formation of the Electrodes.

1. This procedure should be done under a hood or exhaust system because considerable spray and gassing are produced.

2. Pour a sufficient quantity of the prepared electrolyte into the tank containing the electrodes such that the level of the electrolyte is about 3.2 mm ($\frac{1}{8}$ in.) below the tops of the separators, and allow to soak for a minimum of one hour. If necessary adjust electrolyte level prior to charging.

3. Connect the positive terminal of the charger to the nickel dummy

terminal. Connect the negative terminal of the charger to the zinc oxide electrode terminal.

4. Turn on the power supply and commence charging at a rate of approximately 1.55 A/dm$^2$ (10 mA/in.$^2$).

5. Continue the charging for the necessary number of ampere-hours to reduce the available zinc oxide to zinc using the relation, 1.2 g Zn requires 1 Ah.

6. Turn off power supply; remove the zinc electrodes from the tank and immerse in deionized water to remove the KOH. Wash until drops of water from the zinc plate exhibit a pH of 7.

7. Place the wet zinc plates in a suitable pressing fixture and press to required plate thickness.

8. Place the wet plates in an oven at 50-55°. Allow to dry for a minimum period of 8 hours. Dried plates shall be stored in a temperature range of 15-30° and a maximum relative humidity of 10%.

PROPERTIES OF FINISHED PLATES

*The electrodeposited zinc electrode* produced by the methods herein described is rugged and easy to handle during processing and cell construction, yet it has a porosity in the range of 55-65%. These electrodes can have a practical thickness of 0.25 mm up. There is ample active surface area contained in the electrode to produce good voltage at the high current density discharges required in some primary battery applications.

For battery applications requiring long charged wet-stand times prior to discharge, the electrode must be amalgamated to raise the hydrogen overvoltage. Otherwise considerable gassing will ensue causing self-discharge and subsequent loss in capacity (R567).

*The pressed powder electrode* is not so rugged as the plated electrode, and care must be exercised in handling during processing and construction. These electrodes have a porosity of 60-70% and a minimum practical thickness of 0.635 mm up. It is not practical to attempt to produce this electrode at thickness levels of below 0.635 mm because considerable coining of the material takes place. The electrode contains an abundance of active surface area, and good voltages are produced over current density ranges required in secondary battery applications.

This electrode has good charged wet-stand characteristics and good cycle performance required by some secondary battery applications.

## QUALITY CONTROL PROCEDURES

If the negative plates made by either process are to be used in cells or batteries for applications requiring high reliability, then extensive quality control procedures are deemed essential. These quality control measures include chemical and physical analysis of raw materials, process inspections at control points during manufacture, and performance tests on finished plates.

The following set of procedures is an example of the minimum inspection requirements for the manufacture of electrodeposited plates used in primary silver-zinc batteries and pressed powder plates used in secondary batteries.

A. Chemical and Physical Analysis of Raw Materials.

    1. Raw materials shall be analyzed for compliance with specifications before release for use in the manufacturing processes.

B. Process Inspections at Control Points During Manufacture.

    1. The KOH solution for the plating tank shall have a specific gravity of 1.239-1.242 (20-27°).

    2. The plating solution shall not be released until the zinc oxide (ZnO) content has been measured as follows and found to be 2.5% or above (by weight). At no time during the plating operation shall the ZnO content drop below 1.25% (by weight).

    a. Pipette 10 ml plating solution into 100 ml flask.

    b. Add 25 ml 1:3 nitric acid ($HNO_3$) and stir.

    c. Dilute to 100 ml by adding deionized water; thoroughly mix.

    d. Pipette 10 ml aliquot into a 250 ml beaker.

    e. Add 50 ml deionized water and stir.

    f. Add ammonium hydroxide ($NH_4OH$) until pH of solution is 7.0-7.5. The pH shall be measured with a pH meter. Thoroughly stir solution while the ammonium hydroxide is being added.

    g. Add 0.5 ml saturated sodium acetate ($NaC_2H_3O_2$) and stir.

    h. Add 50% acetic acid ($HC_2H_3O_2$) until the pH of solution is 4.6-5.0. The pH is critical, and a pH meter shall be used. Thoroughly stir the solution while the acetic acid is being added.

    i. Add 2 drops cupric sulfate ($CuSO_4$)-EDTA solution and 4-5 drops PAN indicator solution.

    j. Heat solution to boiling and immediately titrate with 0.1$M$ EDTA solution. Thoroughly stir the solution as the EDTA is added drop by drop. The endpoint of the titration is a solution color change from purple to yellow.

    k. By weight,

$$\text{percent ZnO} = \frac{(\text{ml EDTA}) \times (\text{EDTA constant}) \times (100)}{1.2}$$

The EDTA constant must be determined for each new batch of prepared EDTA solution.

A record shall be kept of the plating solution batch number, raw material lot numbers used therein, plate group numbers made therefrom, date released, and inspector.

3. Determination of percent mercuric oxide in zinc oxide by X-ray fluorescence.
   a. Prepare concentrations of standards in the range from 1.50-2.50% mercuric oxide by weight to zinc oxide.
   b. Pack the concentration powders into slotted Bakelite slides which fit the sample holder.
   c. These standards are used to set up the analytical curve, combined with a fourth standard to check validity of the curve.
   d. Counts are taken over a 200 second interval and plotted versus the percent mercuric oxide.
   e. The equipment used in this procedure is a General Electric Company XRD-6 equipped with an emission spectrogoniometer. The tungsten target X-ray tube is used at 50 KVP and 50 mA. A S.P.G.#7 counter tube is used with P-10 gas and a LiF analyzing crystal dispersed the K beta-line of mercury at $30.19°$ Z theta.
4. The KOH solution for formation of pressed powder negative plates shall have a specific gravity of 1.046-1.055 (20-30°).
5. Preparation of testing solutions.
   a. 1:3 nitric acid ($HNO_3$): Add one part concentrated $HNO_3$ to three parts deionized water.
   b. Saturated sodium acetate ($NaC_2H_3O_2$): Prepare saturated solution of $NaC_2H_3O_2$ in deionized water.
   c. 50% acetic acid ($HC_2H_3O_2$): Add one part glacial $HC_2H_3O_2$ to one part deionized water (by volume).
   d. Cupric sulfate ($CuSO_4$)-EDTA solution: Dilute 25 g $CuSO_4$ -$5H_2O$ to one liter with deionized water. Determine equivalent factor of solution by titrating with $0.1M$ EDTA solution using Murexide as the indicator. The endpoint color change will be from dark blue to yellow. Then mix equivalent amounts of $CuSO_4$ and EDTA solutions as ascertained by the titration.
   e. PAN indicator solution: Dissolve 0.1 g PAN (1-2-pyridyl-AZO-2-naphthol) in 100 ml ethanol.
   f. $0.1M$ EDTA solution: Dilute 37.22 g EDTA (disodium ethylene-diaminetetracetate) to one liter with deionized water. Determine the EDTA constant by standardizing with spectrographic grade pure zinc oxide (ZnO). Use 0.2 g pure ZnO and perform the procedure as outlined in Paragraph B.
6. The temperature of the plating solution shall be measured at designated intervals.
7. The silver grid strip-silver tab weld and wire shall be inspected by performing a destructive test on welded parts that have been randomly sampled.

8.    The silver grid strips shall be weighed before and after plating at established intervals. The plated grid strips must be dried before weighing. The weight gain shall be as specified. The grids for use in pressed powder negatives shall be weighed before use.

9.    The thickness of the plated grid strips after pressing shall be measured at established intervals. The thickness of pressed powder negatives shall be measured at established intervals.

C.   Performance Tests on Finished Plates

The negative plates of both processes shall be sampled and tested as follows:

1.    Sample a sufficient number of negative plates during the interval between change of plating solution to construct a test cell.

2.    The negative plates constructed by pressed-powder methods shall be tested by sampling a sufficient number after each ZnO mix has been depleted.

3.    The positive plates used in the test cell shall be of known good electrical characteristics.

4.    Discharge the test cell at a nominal room temperature of 24° using the proper load profile specified.

5.    The test cell shall meet the requirements designated or the plates represented by the test cell shall be rejected.

## PRIMARY AND SECONDARY BATTERY ELECTRICAL CHARACTERISTICS

It is well known that Zn/Ag batteries, both primary and secondary, are used for missile and space applications, including any vehicle for propulsion requiring a high energy-to-weight performance over a wide range of extreme environmental conditions.

No attempt will be made here to touch on all of these applications but several characteristics of batteries common to most requirements are best illustrated by cells containing electrodes produced by both manufacturing processes. Again, it is stressed that the data presented are meant merely to compare the performance of voltage and capacity of the two electrodes in primary and secondary type cell construction at two common discharge rates at room temperature. It is to be understood that the cell parameters described can be varied many ways to meet desired battery specifications.

Figure 73 shows the initial capacity and voltage curves for two secondary cells of 25 Ah nominal capacity. Both cells contain two layers of a cellophane-type separation with seven positive plates, 0.38 mm (0.015 in.) thick, the one cell containing eight pressed-powder negatives, 1.4 mm (0.055 in.) thick and the other cell containing eight plated negatives, 1.14 mm (0.045 in.) thick. All negative plates contain the same amount of active material. All cells are activated

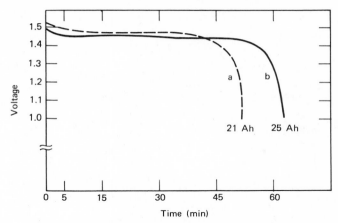

**Figure 73**     Discharge of 25 Ah secondary cell at C/1 rate and 25°; 45% KOH electrolyte; (a) pressed ZnO negative, (b) electrodeposited negative.

in 45% KOH electrolyte. The discharges are approximately at the one-hour rate, which is equivalent to a current density of 28 mA/cm$^2$ (0.18 A/in.$^2$). The cell containing the pressed powder negatives exhibited a slightly higher voltage plateau but delivered less capacity than the plated negative.

Figure 74 shows the voltage and capacity curves obtained by these same cells on the second cycle at a relatively high rate discharge of 100 amperes. This is equivalent to a current density of 100 mA/cm$^2$ (0.71 A/in.$^2$). Here again the cell

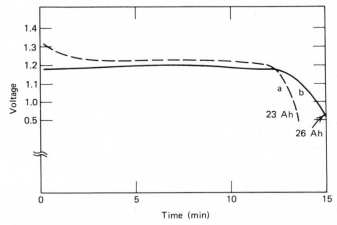

**Figure 74**     High-rate (4C) discharge of 25-Ah secondary battery at 25°; 100 A discharge; 45% KOH electrolyte; (a) pressed ZnO negative, (b) electrodeposited negative.

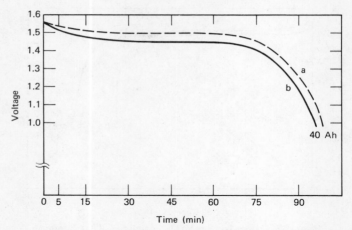

**Figure 75**    One-hour rate (C/1) discharge of primary cell at 25°;
25 A discharge; 45% KOH electrolyte; (a) pressed ZnO negative,
(b) electrodeposited negative.

containing the pressed-powder negatives exhibited a higher initial and plateau
voltage but delivered slightly less capacity than the cell containing the plated
negatives.

Figure 75 shows the voltage and capacity curves for two primary cells of
nominal 25 Ah capacity. Both of these cells, as well as those illustrated in Figure
76 contain the same number, thickness, and type of positive and negative plates

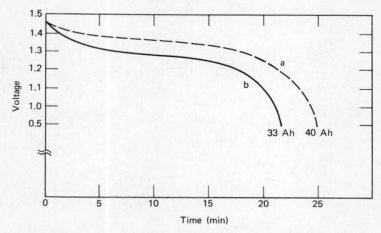

**Figure 76**    High-rate (4C) discharge of 25-Ah primary cell at 25°;
100 A discharge; 45% KOH electrolyte; (a) pressed ZnO negative, (b)
electrodeposited negative.

as described for the secondary cells, *except* the separator system contains two layers of a nonwoven paper material (Viscon) in place of the cellophane-type membranes.

The discharges shown are repeated at the same hourly rate (25 A) as the secondary cells. The voltage of the cells containing the pressed powder negatives is again slightly higher than the cells of the plated negatives, but both voltage and capacity levels are greater than those of the secondary cells.

Figure 76 shows the voltage and capacity curves of the last two primary cells. The voltage and capacity of the cells containing the pressed-powder negatives are higher than those of the cell containing the plated negatives. Both cells exhibited the same pattern of higher voltage and capacity than the secondary cells. This is to be expected since the secondary cellulosic separator is an electrolyte diffusion limiting system, a factor which is eliminated in the primary separator system. As for the higher voltage exhibited by cells containing the pressed powder electrodes, there is probably more surface area available per plate than is present in the plated negatives at the plate thickness used.

For Zn/Ag batteries requiring long cycle space applications the pressed zinc oxide powder negative has a good history of delivering many cycles of two-hour duration at 40% depth of discharge. There are unpublished data of cells containing plated negatives also delivering many cycles of two-hour duration at 25% and 40% depth of discharge. Additional development work is required on the plated negative to reduce the gassing and self-discharge properties, and this will involve the perfecting of techniques for amalgamation or suitable means of raising the hydrogen overvoltage and retention of same throughout cycle life.

The zinc negative plate is the main cause of secondary cell cycle failures due to material shedding, dendritic growths through and around the separators, and material densification prohibiting charge acceptance. Future research and development programs may evolve around procedures that could conceivably produce a zinc negative plate having improved properties leading to a more reliable and longer lived electrode than are presently available.

# 14.

## Sintered Silver Electrodes

### S. Uno Falk and A. Fleischer

The porous silver electrode patented by Jungner in 1899 (R329) was prepared by applying a mixture of finely divided silver and silver chloride onto a nickel grid, then pressing and heating to at least $450°$. After cathodic polarization in an alkaline solution, the electrode was ready for use. Perhaps unintentionally on the part of Jungner, the heat treatment accomplished a sintering of the silver particles. The use of silver chloride was adopted in the more recent pioneering development (R15, 17) of the zinc-silver oxide battery.

With the development of an industry and with experience gained in the manufacture of electrodes and their use in batteries, the technology moved to the use of silver and silver oxide powders as the base material. Various sintering techniques came into use in the middle nineteen fifties and made possible the manufacture of thin, mechanically strong plates. There are three principal methods based on particular combinations of raw materials and sintering procedure; electrodes produced by these methods will be referred to as, sintered silver oxide electrodes, sintered silver powder electrodes, and sintered resin-bonded electrodes.

### MANUFACTURE OF SINTERED SILVER OXIDE ELECTRODES

In the nineteen forties work was started (R634) under U.S. Navy contract to develop a technique based on applying a paste of univalent silver oxide and water onto a suitable grid and, after drying, thermally decomposing the oxide to a porous, cohesive silver deposit. In the present-day silver oxide method the steps are typically as follows:

A paste is produced by mixing finely divided, univalent silver oxide and water in a cylinder or some other simple mixer. The paste may contain 70-80% $Ag_2O$ by weight. A layer of this paste is applied by a spatula or by some pasting apparatus onto one or both sides of a grid of silver in the woven or expanded metal form or a grid of silver-plated copper or nickel gauze. The paste is adjusted with or without carboxymethyl cellulose to control plasticity in applying the paste (R210). The amount of silver oxide is checked by weighing after the pasting operating. The layer of paste is dried for instance at 70-85° and then thermally decomposed to metallic silver by exposure to temperatures of 400 to 600°. No protective atmosphere is necessary. The sintering time varies with the temperature. A typical value is 30 min at 450°, which is probably the optimum sintering temperature.

After the sintering the plates are often compressed. Typically, a pressure in the order of 90 kg/cm² (R435) develops an intimate contact between grid and porous silver without deteriorating the structure of the electrode. Pressures up to 200 kg/cm² have, however, been used with good results.

The silver is then electroformed to the divalent oxide ($Ag_2O_2$). This is generally done in dilute KOH solutions, approximately 5%, at low rates (C/8 or lower). The complete conversion usually takes from 16 to 22 hours. Low formation rates are necessary to achieve a high degree of conversion. The electrodes are washed in pure running water and dried in air at temperatures in the vicinity of 70°. If continuous strips are fabricated, the formed strips are now cut to proper electrode size.

A variation of this process (R292) utilizes a slurry of silver oxide which is applied to a permeable support such as filter paper while vacuum is drawn to drain water from the slurry. A conductive grid is placed on the silver oxide deposit and an additional quantity of silver oxide is deposited from the slurry. The structure is then compacted to reduce the water content and dried at a temperature below 100°. To avoid formation of cracks, the drying should not be too rapid. Suitable drying is normally accomplished in 4 hours at 65°. After drying, the permeable support is removed and the structure is heated to decompose the oxide and sinter the metallic silver, preferably in a muffle furnace at about 500°. After this an electrolytic formation as previously described may be carried out.

This variation is stated to give more uniform layers of silver oxide than the ordinary process. It is also considered to provide a less time-consuming pasting step.


## MANUFACTURE OF SINTERED SILVER POWDER ELECTRODES

Manufacturing methods based on the direct sintering of silver powder have found use to produce electrodes either in batch lots or continuous strip.

For sintered silver powder electrodes, expanded metal is often used for grids, the size of the diamond-shaped openings being selected on the basis of the processing conditions, powder characteristics, electrode thickness, performance requirements, and service life. Similar considerations apply in selecting other types of grids such as silver wire, silver foil, or silver wire cloth (R323, 565).

The silver powder electrodes are often fabricated in molds, made for instance by milling a suitable cavity in a graphite or stainless steel plate. A weighed portion of finely divided silver powder is filled into the mold to form an even layer either by manual scraping or by means of a mechanical powder filling device. The grid is then put on this layer and an additional layer of silver powder is filled into the mold. In a variation, all the silver powder is filled into the mold and the grid is then placed on the layer. The powder is sometimes compressed at this stage. The molds, one by one or in stacks, are placed in an electric furnace and are heated to 400-700° in air for a sufficient time to obtain sintering of the silver particles.

Sintered plates made in molds (R314) can be produced with densities ranging from 2.5-4.5 g/cm$^3$ corresponding to porosities of 57-76%.

When master plates are produced in molds they are cut to appropriate individual plate size after the sintering and coining operations for current leads are carried out if necessary. At this stage corrugation (R323) of the electrodes has also been tried as a way to provide additional space for electrolyte between the silver plate and the permeable separator.

Silver powder electrodes are also manufactured according to continuous methods. In a typical process (R149), silver powder is continuously deposited on a conveyor belt of paper web. A ribbon of expanded silver from a supply roll is imbedded in the silver layer. The composite sheet is compacted by passing between pressure rollers. The paper web is then wound upon a take-up roll and the silver sheet is passed through a furnace with a central zone heated to 650-800° where the silver particles are sintered together. Coming out of the furnace the sheet is divided to plates of suitable length by an automatic cutting knife. Leads are attached by coining and welding or by a clamping connection.

A plurality of modifications for the manufacture of sintered silver powder plates have been suggested in the literature (R120, 565), and some of these methods are probably used, for instance in the U.S.S.R.

Sintered silver powder plates to be used in secondary cells may or may not be electroformed in KOH solution in the manner previously discussed. In many cases the plates are assembled in the metallic state to form silver-zinc cells in the dry, unformed condition.

## MANUFACTURE OF SINTERED RESIN-BONDED ELECTRODES

The resin-bonding technique of making silver electrodes is relatively new. Basically, silver powder is mixed with a suitable resin and in some cases, a special pore-forming compound, to form a mass which can be rolled or extruded to a continuous sheet which is heated to burn away the resin and sinter the silver. Before or after the sintering the sheet is die-cut to appropriate plate sizes. Forerunners to this method appear in several patents (R27, 267, 434) dealing with powder metallurgical procedures.

Duddy was the first to develop a practicable process for silver electrodes based on resin-bonding techniques. Initially (R183) combustible materials such as filter paper or porous synthetic resins were impregnated with aqueous solutions of silver salts. After being heated to evaporate the water, the structures were ignited and the organic material was burned off to leave a sheet of porous silver. Continuing from this approach, thermoplastic resins were mixed with finely divided silver or silver compounds and in a patent (R184) the technique and the possibilities of continuous operation were discussed. Although there are some modifications of this so-called DP-method, the process is largely as described in the following paragraphs.

Finely divided polyethylene, such as "Alathon 14," flows from a supply bin to a mixing device, preferably two rolls (typical rubber rolls) operated at different speeds, for plasticizing the polyethylene. The rolls are heated to a temperature of about 120°. When plasticizing has been completed, which normally will take 2 to 3 minutes, silver is added in the form of silver powder. The proportions between polyethylene and silver may vary from 1:2 to 1:10 by weight. The components are mixed on the rolls for about 10 minutes to form a homogeneous mass. By means of a stripper blade the mass is withdrawn and the material is fed into a sheeting device such as calendering rolls operated at elevated temperature (about 110°). Here sheets of the desired thickness are produced and cut to suitable width as can be seen in Figure 77. The process from the sheeting device on may be continuous or intermittent.

One or two layers of the resin-bonded silver strip are provided with a grid, preferably of the expanded metal or screen type. This operation may be carried out by pressing between virgin Teflon sheets maintained at a temperature of about 120°. Here the strip is compressed at a minimum pressure of 14 kg cm$^{-2}$ (200 psi) to the desired thickness, which may vary from 0.2 to 1.5 mm. Pressing time depends on the assembly thickness. The assembly is fed into a furnace either as continuous strip or cut to electrodes of suitable shape. In the furnace the electrodes are ignited and the polyethylene is burnt away. This will take 5-10 min. The electrodes are then flattened by a rolling and then returned to the oven and sintered at about 550° for 20-25 min. Finally, the plates are pressed to the desired thickness. The subsequent treatment of the electrodes follows along the lines already indicated although the KOH concentration of the formation electrolite is usually higher than for the other types of electrodes.

## MANUFACTURE OF SINTERED NICKEL MATRIX ELECTRODES

Besides the three common methods just discussed there is an additional route that has been studied in detail for the manufacture of a sintered silver electrode. This is the sintered nickel matrix method. In a patent of 1954 Ameln (R9) mentioned the method of impregnating porous bodies of sintered nickel powder

**Figure 77**    Rolling and cutting of resin-bonded silver sheet (courtesy of J. C. Duddy, ESB, Incorporated).

with an active material comprising silver oxide. Fischbach (R209) in 1955 gave details of this method discussing the impregnation of a nickel matrix with a concentrated silver salt solution. This basic approach has been investigated in considerable detail (R93, 506) but the expected promise of a complete utilization of the silver oxide at the higher level of oxidation on repetitive cycling has not been achieved. This approach has apparently been abandoned at this time.

## SPECIFICATIONS ON BASIC RAW MATERIALS

The chemical and physical properties of the basic raw materials—silver oxide and silver powder—have a bearing on the processing steps in the production of plates and on the performance of the electrodes. Generally, each battery manufacturer adjusts the specification to that suitable for his particular variation of the manufacturing procedure. As a rule, the properties of major interest are the chemical composition, apparent density, and particle size.

Chemical specifications place limits on certain impurities; for example, iron has been indicated to have a deleterious effect (R120, 565). The moisture content may influence the flow properties of the powders. The range of values for the chemical composition of typical American silver oxides and silver powders is shown in Tables 22 and 23.

The apparent density plays a role in the preparation of paste and in the steps of assembling the green compact. It is the controlling factor in adjusting the porosity of the plates. High porosity is a usual requirement for electrodes that will be used at high discharge rates or in sealed cells whereas a lower range of porosities seems satisfactory for low rate and low temperature operation (R581).

Particle sizes must also be adjusted to suitable ranges in the achievement of suitable porosities and for the development of optimum surface area. On the basis of electrochemical studies (R75) the favorable particle size is about 1 micron for the optimum conversion to oxide during charging. The same size has been indicated (R582) as being advantageous for palladium additions to the silver in order to extend the charge acceptance on the silver(I) oxide level.

TABLE 22.   SPECIFICATIONS FOR
MONOVALENT SILVER OXIDES

|  | Goldsmith[a] | Ames[b] | Handy and[c] Harman |
|---|---|---|---|
| *Chemical composition:* | | | |
| $Ag_2O$        % min | 99.7 | 99.7 | 99.6 |
| $H_2O$ or loss on drying at | | | |
| 110°        % max | 0.25 | 0.25 | 0.1 |
| Cu        % max | 0.0005 | 0.002 | 0.003 |
| Fe        % max | 0.002 | 0.002 | 0.003 |
| Pb        % max | 0.0005 | 0.001 | – |
| Na        % max | – | 0.001 | – |
| Cl        % max | – | – | 0.001 |
| $H_2O$-soluble        % max | – | – | 0.15 |
| $NO_3$        % max | 0.01 | – | 0.05 |
| *Physical properties:* | | | |
| Apparent density by Scott volumeter | | | |
| $g/cm^3$ | 0.79 | – | 0.40-0.67 |
| $g/in.^3$ | 13 | – | 6.5-11 |
| Bulk density tapped | | | |
| $cm^3/g$ | 0.91 | 0.80-1.1 | 1.4-1.8 |
| $cm^3/av.$ oz | 26 | 23-31 | 40-50 |
| Screen analysis | | | |
| −325 mesh    % | – | 99-100 | 99-100 |
| Color | Brownish black to black | Brownish black | Brownish black |

[a] Goldsmith Brothers, Division of National Lead Company, Chicago, Illinois.
[b] The M. Ames Chemical Marks, Glens Falls, New York.
[c] Hardy and Harman, Fairfield, Connecticut.

TABLE 23. SPECIFICATIONS FOR SILVER POWDERS
MANUFACTURED BY HANDY & HARMAN, FAIRFIELD, CONNECTICUT

|  |  | Silpowder #120 | Silpowder #160 | Silpowder #150 |
|---|---|---|---|---|
| *Chemical composition:* |  |  |  |  |
| $Ag^a$ | % min | 99.9 | 99.9 | 99.9 |
| Cu | % max | 0.05 | 0.05 | 0.05 |
| Fe | % max | 0.05 | 0.05 | 0.05 |
| $H_2O$ | % max | 0.1 | 0.1 | 0.1 |
| *Physical properties:* |  |  |  |  |
| Apparent density |  |  |  |  |
| by Scott volumeter |  |  |  |  |
| $g/cm^3$ |  | 1.2-2.1 | 1.5-2.4 | 1.5-2.4 |
| $g/in.^3$ |  | 20-35 | 25-40 | 25-40 |
| Screen analysis[b] |  |  |  |  |
| +100 % |  | Trace | Trace | 0.5 max |
| −325 % |  | 85 min | 85 max | 30 max |
| Flow by rotary |  |  |  |  |
| orifice gauge |  | Good | Fair | Excellent |
| Particle size |  |  |  |  |
| range, micron |  | 4-8 | 5-25 | 8-12 |

[a] Determined by firing for 10 minutes at 540°.
[b] 100 g sample for ½ hour on Rotap sieve shaker.

Tables 22 and 23 show the physical data for silver oxide and silver powder. The values indicate, as for the chemical data, the ranges found for various specifications by American manufacturers. Although these specifications may not reflect the exact requirements that one would like to put on these powders to obtain optimum performance of the electrodes, they are typical values for powders which are today available and used.

## PROPERTIES OF SINTERED SILVER ELECTRODES

Quite a lot of information has been published on the properties of silver-zinc cells and batteries; however, the literature on the characteristics of silver electrodes is relatively meager. By this is meant that there is no possibility of achieving a complete correlation of the physical and mechanical properties with electrochemical performance and life.

### UTILIZATION OF ACTIVE MATERIAL

Sintered silver powder and sintered silver oxide electrodes show only fair utilization of the silver material. A typical value at room temperature and at moderate discharge rates is considered to be about 50% of the theoretical yield based on a two-electron change, 60-65% is achieved at best (R106). Higher values are, however, occasionally found in the literature. Resin-bonded electrodes seem to have about the same utilization as the other plates mentioned or somewhat higher (R105).

Generally speaking, the utilization of a silver electrode, especially at high rates, increases with increasing porosity and with decreasing plate thickness.

## DISCHARGE CHARACTERISTICS

There are no great differences between the general discharge characteristics of electrodes manufactured according to the three standard methods although the sintered silver powder plates are more uniform in electrical performance than are the pasted plates. After complete charging, the voltage-time discharge curves display the well-known two plateaus corresponding to two different reduction steps.

The bivalent plateau should, theoretically, be of the same length as the univalent but this is not the case in ordinary electrodes. At higher rates the bivalent step becomes less pronounced and usually completely disappears at rates higher than 1C. This is illustrated in Figure 78. It should be mentioned that a single-step discharge curve can be obtained under certain conditions (R92, 568). Where requirements for close voltage regulation on discharge are severe, the charge may be limited to the univalent stage. This restriction lowers the coefficient of utilization.

## CHARGE CHARACTERISTICS

No significant differences in charge characteristics have been reported between silver oxide, silver powder, and resin-bonded electrodes, providing that

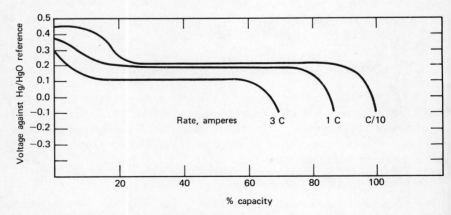

Figure 78    Discharge characteristics of silver electrodes at various constant rates, room temperature.

the plates have been properly formed. The tendency for early gassing of electrodes on charge is reported to be counteracted by the addition of lead to the silver electrodes (R76).

## CYCLE LIFE

With regard to cycling properties, the silver powder plates and the resin-bonded plates are to be preferred. Silver oxide plates seem to lose capacity more rapidly than the other kinds on cycling, and are mainly used for primary cells.

## MECHANICAL PROPERTIES

The sintered silver electrodes have, in general, satisfactory mechanical properties. The pasted plates are normally somewhat weaker than are the others. The tendency for cracking on sintering and forming of the pasted plates is one of the contributing factors to this. The silver powder and resin-bonded plates have very good mechanical strength.

## CONCLUDING REMARKS

Present-day silver electrodes based on sintering techniques have many advantageous features. The energy density is high, the discharge curve is, for the main part, very flat, and the high rate performance is excellent. Furthermore, the charge efficiency is outstanding, the self-discharge is very low, the cycle life is considerable, and the mechanical strength is satisfactory. Also, at least some of the paths of manufacture are of the continuous kind with low production cost as a result.

However, it is anticipated that a satisfactory solution of the life capabilities of the zinc electrode may impose more stringent demands on the silver electrode which at this time cannot be considered as limiting to performance or life.

# 15.

## Chemically Prepared Silver Oxide Plates

### Nicholas T. Wilburn

Although there is an extensive literature on zinc-silver oxide cells, the use of chemically prepared bivalent silver oxide (AgO) in the preparation of battery electrodes has not been reported extensively. The U.S. Army Electronics Command (R638, 639A) reported techniques and data for the preparation of dry process electrodes by pressing or rolling under high pressure with sodium carboxymethylcellulose (CMC) binder solution or a water vehicle. In another process (R106), the mixture, containing methylcellulose as a binder and carbon as a conducting medium, was compressed.

One relatively common technique, employed in the battery industry for over ten years, utilizes a thick paste of extrusion consistency prepared with AgO and an aqueous solution of a binder such as methylcellulose or CMC. This is applied by hand pasting to sections of silver grid, expanded mesh, or screen or by machine pasting to a continuous strip of grid material. Only moderate pressure is applied because of the fluidity of the paste. The finished plates are stamped out from the strips while still moist and then are dried. On drying, the binder sets up to hold the oxide together and to the grid. The plates are generally fragile and precautions are necessary during cell assembly to prevent edge fragments of AgO from creating shorting paths to the zinc anodes. Once assembled in a cell, sufficient pressure is maintained to prevent plate disintegration under normal handling conditions.

The fragility of this electrode and the resulting handling difficulties in battery production prompted the investigations discussed in this paper (R638, 639A, 649). These resulted in a mechanically sound structure capable of withstanding normal handling and severe environmental stresses in equipments with no appreciable damage. The plates retained the advantages of the earlier pasted plates in comparison to conventional electroformed plates, namely, higher utilization of active material, operation within narrower voltage tolerances, and, to some extent, lower cost.

## EXPERIMENTAL PROCEDURES

Theoretically, a bivalent AgO electrode should discharge in a two-step manner as shown by curve 1 of Figure 79. These steps correspond to the reduction of bivalent to univalent oxide

$$4AgO = 2Ag_2O + O_2$$

and then of univalent oxide to silver

$$2Ag_2O = 4Ag + O_2$$

The voltage levels at low rates of discharge in zinc-silver oxide cells can approach the open circuit values of 1.86 and 1.62 V. The theoretical two-step discharge is never seen in practice because of resistance effects within the silver oxide electrode. It can, however, be demonstrated in the laboratory with electroformed plates discharged at very low rates with almost monomolecular layers of AgO under conditions of utmost purity and refinement in technique.

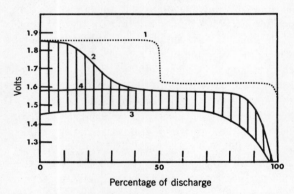

Figure 79     Representative voltage curves, zinc-silver oxide cells: (1) theoretical curve at infinitely low rate; (2) electroformed AgO plates, low-rate discharge; (3) electroformed or chemically prepared AgO plates, high-rate discharge; (4) chemically prepared AgO plates, low-rate discharge.

Electroformed plates will show part of the first discharge step or hump, as it is often called, with its extent and shape depending mainly on the discharge rate. A hump as large as that shown by curve 2 might be seen at a 50-100 hour rate. At higher rates the hump diminishes and finally disappears at very high rates due to *IR* drop as shown by curve 3, which might represent a 10-minute rate.

For most high rate discharge applications, the hump is not seen and therefore has no significance. However for an application which requires close voltage regulation with low current pulses, the hump can present a serious initial problem as shown by the generalized peaks under curve 2. This problem can be solved for the electroformed plate only by a reduction to the univalent form during the electrode fabrication process, by heat treatment or partial discharge. The electrode capacity is reduced by such treatment. For low rate applications, where close voltage regulation is required, the hump is a severe problem since weight restrictions generally require the full AgO capacity. In this case supplemental voltage regulation equipment is needed to maintain the required tolerances.

The chemically prepared bivalent silver oxide electrode may show a high open circuit voltage, but this drops rapidly, as shown by curve 4, to the univalent discharge level, even at very low rates of discharge. This is a notable advantage both for low rate purposes and for high rates where low rate periods are required.

The most probable explanation for the absence of the bivalent oxide potential in the chemically prepared oxide plate is the $IR$ factor which is considerably higher than in the electroformed plate. In the latter, metallic silver is pressed onto the silver grid. This gives a firm silver-to-silver conducting bond throughout the plate which is maintained due to incomplete electrolytic oxidation to AgO. This is illustrated in Figure 80, a photomicrograph of a section through an electrolytically formed plate with a magnification of 200. Apart from the expanded mesh silver grid, traces of unoxidized silver are seen throughout the plate.

In the chemically prepared oxide plate, particle-to-particle contact is made through the semiconductor AgO with a correspondingly higher $IR$ drop. The absence of silver, other than for the grid, is shown in the photomicrograph of Figure 81. Enlargements of the surface of the plates show similar evidence of the presence of highly conducting silver in one case and its absence in the other.

Bivalent silver oxide voltages may be seen, however, in chemically prepared oxide plates when using copper instead of silver grids. This has given rise to the so-called reservoir theory, which states that AgO in contact with silver grid in the presence of electrolyte will oxidize the silver to silver monoxide. This will be discharged directly to silver which, in contact with more of the higher oxide, will be oxidized and then reduced on discharge. The bivalent silver oxide thus acts as a reservoir for the formation of monoxide from silver, and the discharge is always at the univalent potential.

The term dry process was introduced (R638) to differentiate between the earlier paste process and the later method (R638) where the plate bonding forces were primarily dependent on high pressure compaction of the dry AgO powder. Small amounts of CMC binder were introduced in 1% aqueous solution to increase the subsequent bonding strength after drying, without affecting the compaction under typical pressures of 10,000 psi (700 kg cm$^{-2}$).

Preliminary pressure studies with dry AgO powder alone had shown that the powder would compact but would not adhere to either screen or expanded mesh silver grid material. A compressed layer would be formed on either side of the grid, which would separate from the grid in normal handling. A strong bonding structure was formed, however, when the grid was first coated with 1% CMC

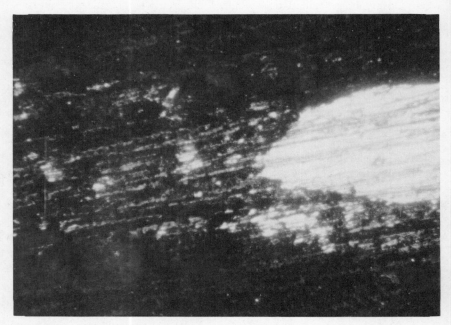

**Figure 80**   Photomicrograph of cross section of electroformed AgO Plate, ×200.

solution before being placed on a layer of dry powder, or between two layers, prior to pressing. It was believed that the binder solution was uniformly dispersed throughout the powder during pressing. On drying, a linkage was created by the binder to hold the plate together and to the grid. The weight ratio of CMC to AgO was small, in the order of 1:1000, derived from a normal ratio of 0.1 ml of 1% CMC solution per gram of AgO.

The electrodes were prepared in a jig with 4.0 g of AgO distributed in a cavity with dimensions, excluding terminals, of 4.13 x 5.08 cm, yielding a test electrode of 21.0 cm$^2$ area. Grids coated with CMC were added to the AgO, and pressure was applied ranging from 3,077 to 15,385 psi (224-1072 kg cm$^{-2}$). Various grid materials, mesh sizes, CMC concentrations, CMC-to-AgO ratios, pressing times, drying times, AgO sources, and particle sizes were studied as well as AgO weight densities ranging from 0.02 to 0.29 g cm$^{-2}$. Discharges were run, with sponge zinc anodes, primarily at 167 mA cm$^{-2}$ with initial pulses at 24 mA cm$^{-2}$ to check electrode impedance, which was normally found to be in the range of 0.03 to 0.04 ohm.

Coulombic efficiencies were calculated to various end voltages and were found to be, for plates in the 2-4 g range, about 74% to a 0.1 V drop below maximum voltage, 84% to a 0.2 V drop, and 88% to a 0.3 V drop.

Figure 81     Photomicrograph of cross section of chemically prepared AgO plate, x200.

Concurrent with the electrical studies, investigations were conducted on the mechanical strength of the plates, considering both the handling required in battery production and the environmental stresses which would be seen when the batteries were installed in typical equipments such as missiles. A test procedure was established based on a severe dry vibration applied to groups of nine plates assembled in test cells. The vibration shedule consisted of a 30-minute sinusoidal sweep from 5 to 2,000 cps at 40 g and a 30-minute sweep back to 5 cps in each of the three mutually perpendicular axes of the plates, all nine of which were covered with V-shaped Viscon separators and packed not too tightly in a plastic cell case mounted to the vibrator table. Each group of plates was examined after vibration. The separators adjacent to the plates were covered to some extent with powdery AgO, which occurs with any silver oxide plate during dry vibration. The plates had an excellent appearance with no signs of fracturing or breaking off at the edges. The difference in weight of each plate was recorded, before and after vibration, and the percentage loss was calculated. This was typically less than 1%, demonstrating that the chemically prepared silver oxide plates were capable of surviving normal handling in production and any environmental stresses in actual use. They were the equivalent of electroformed plates in this sense and vastly superior to the previous pasted plates.

The next phase of the investigations (R639A) resulted in eliminating the need for the CMC binder. It was postulated that the small amount of binder, in a 1:1000 weight ratio to the AgO, could not significantly affect the bonding of the oxide to the grid. Therefore the water vehicle in the binder solution was the major factor by permitting a proper orientation of the AgO particles under the intense pressures applied. This concept was verified experimentally in a series of plates made in the jig as before except that a small amount of water, about 0.1 cc per g of AgO, was applied to the grid or, by means of a vaporizer, to the layer of dry AgO powder. Plates were made and determined to have excellent electrical and physical properties. A series of plates was discharged under loads ranging from 1.75 to 28 A or current densities of 42 to 667 mA cm$^{-2}$. The results are given in Table 24.

These plates were made on expanded mesh silver grid ranging from 2/0 to 4/0 mesh size. The AgO weight was 4.0 g or 0.19 g cm$^{-2}$. In each case the maximum voltages are given and the coulombic efficiencies in descending order to a 0.3 V, 0.2 V, and 0.1 V drop below maximum voltage. For example, the 2/0 mesh plate discharged at 7 A or 167 mA cm$^{-2}$, gave a maximum voltage against zinc of 1.35 V, an efficiency of 82.5% to 1.25 V, 89.4% to 1.15 V, and 92.3% to 1.05 V, a 14-minute discharge. At 1.75 A, the approximate 2-hour rate, there was no evidence of the bivalent voltage and the efficiency even to a 0.1 V drop exceeded 90%. The low peak voltage at the highest current has no significance since thinner electrodes operating at much lower current densities would be used under such high rate conditions. However, the slight differences in efficiencies indicate quite flat discharges even under this high current density condition.

## DISCUSSION

The experimental work had established procedures for the fabrication of strong, high efficiency bivalent silver oxide electrodes, with good voltage regulation characteristics, by the direct application of the chemically prepared oxide to expanded metal silver grid. These electrodes were superior to the conventional electroformed plates in discharge efficiency and to pasted AgO plates in physical strength. The cost factor is also of considerable importance in the comparison with the electroformed plates. The higher cost of the chemically prepared AgO in comparison to silver powder or Ag$_2$O should be more than

TABLE 24.   PERFORMANCE OF
CHEMICALLY PREPARED PLATES
WITHOUT BINDER

| Amperes | Mesh Size | | |
|---|---|---|---|
| | 2/0 | 3/0 | 4/0 |
| 28 | 70.7 [a] | 76.6 | 78.4 |
| | 65.4 | 70.5 | 73.3 |
| | 58.1 | 55.7 | 67.1 |
| | 0.68 V [b] | 0.90 V | 0.79 V |
| 14 | 81.1 | 83.3 | 78.6 |
| | 75.8 | 78.9 | 70.1 |
| | 65.1 | 68.9 | 56.3 |
| | 1.12 V | 1.09 V | 1.19 V |
| 7 | 92.3 | 90.3 | 85.7 |
| | 89.4 | 85.3 | 78.6 |
| | 82.5 | 74.3 | 64.3 |
| | 1.35 V | 1.36 V | 1.36 V |
| 3.5 | 90.4 | 93.6 | 94.4 |
| | 86.9 | 91.9 | 93.0 |
| | 78.7 | 87.5 | 88.1 |
| | 1.44 V | 1.45 V | 1.44 V |
| 1.75 | 94.6 | 94.8 | 94.7 |
| | 93.4 | 93.7 | 93.8 |
| | 91.4 | 91.7 | 90.3 |
| | 1.54 V | 1.52 V | 1.50 V |

[a] Body of table, efficiency in %
[b] Maximum voltage on discharge.

offset by the simplicity of the production process. All of the conventional steps of electroformation, washing, and drying are eliminated as well as any special treatment for voltage control where needed. Apart from this, the substantially higher, 20 to 30%, coulombic efficiency of the chemically prepared oxide plate will decrease the quantity of silver required in any battery. The only exception to this latter would be an extremely low-rate battery where the bivalent oxide voltage can be tolerated. In this case efficiencies of over 90% can be obtained with electroformed plates if special charging procedures are used.

In spite of its advantages, the chemically prepared oxide plate cannot be produced by any interested company without an extensive and costly engineering program on either of the two possible methods, rolling or punch pressing. Recognizing the need for a program to establish the necessary techniques and facilities, the U.S. Army Electronics Command has instituted a contractual effort for the "Production of Dry Process Divalent Silver Oxide Electrodes by Continuous Rolling" (R649), a two-year program in effect since December 1967. It will establish techniques and facilities for the high speed, fully automated production of electrodes at a rate of at least one per second. These will be finished plates, stamped out with coined edges, and with welded terminal tabs, ready for cell assembly. The production equipment is being designed and fabricated and should be in operation within a few months. An important objective of the program is a detailed description of the procedures and equipment to enable any battery manufacturer to duplicate the process.

In the engineering studies thus far, important advances have been made over the previous laboratory effort. Much emphasis has been placed on the development of thin, low-weight electrodes for exceptionally high rate missile batteries. It has been found that strips for such plates can be rolled out with dry AgO powder and silver grid with no binders or moisture, thus simplifying the process considerably. It has also been found that uncoated areas are not necessary on the strips to provide for terminal tabs. The plates can be stamped out and corners can be subjected to intense flash heat to reduce the oxide to silver, as a base for welding on the tabs, without affecting the AgO elsewhere in the plate. This again simplifies the process.

In the process as developed to date, the bulk silver oxide is fed into a high speed micropulverizer which breaks up agglomerates and forces the oxide through a 100-mesh screen. The powder is then fed onto a paper-covered moving belt which feeds the powder to a pair of rapidly oscillating doctor blades (R383). The first blade levels the powder and distributes it the full width of the strip. The second blade smoothes out the layer and creates an exceptionally uniform surface. The powder is then fed to the roller where the expanded mesh silver grid material is introduced.

The property of satisfactory compaction without binders or moisture is believed to be attributable to the particle size resulting from the micro-pulverizing and the forced sieving through the 100-mesh screen. The resulting thin plates are structurally sound and will withstand the previously mentioned vibration regime. It has been determined that the slightly powdery surface can be effectively stabilized by spraying on a minute quantity of binding agent before compaction.

Thicker electrodes, which are also being investigated, will probably require a binder or the use of traces of water as a compacting agent. This has not been fully determined as yet. Efforts will be made to avoid the use of binders, except for surface treatment.

The successful completion of this program should see the realization of all of the original objectives with respect to chemically prepared silver oxide plates: high coulombic efficiency, optimum voltage regulation, structural soundness, and cost effectiveness through the simplicity of the process and full mechanization.

# PART V
## SEPARATORS

# 16.

## Polymeric Membranes as
## Effective Silver–Zinc Battery Separators

### Harry P. Gregor

Battery separators composed of synthetic polymers have been increasingly considered in the search for improved performance and life of batteries. For the most part, these separators have been commercially available films prepared for other purposes, such as cellophane. It is only in the past few years that serious attempts have been initiated to make films specifically for use as battery separators. Therefore, it is not surprising that, because of the exacting and manifold requirements of the zinc-silver oxide system, effective separators have not been prepared. On the other hand, the very real advances made in the last decade in the preparation of membranes for processes of separation and purification (for desalination, artificial kidneys, ultrafiltration) indicate that a truly practical separator should be forthcoming with appropriate effort.

This communication will summarize current knowledge of the requirements of a silver-zinc separator and then enumerate those known polymer systems which may find application.

In order to discuss the function of separators, it is useful to differentiate the important functions which a separator system must perform. At one time only a single separator material was used, usually in several wraps. With a view to leading to overall improvements, the special functions are examined here by subdividing the electrolyte space into four regions.

Starting at the electrode, the separator adjacent to the silver electrode will be designated as the Ag spacer. This is a porous, inert matrix of high chemical stability because it is in contact with the strongly oxidizing silver oxide electrode. Usually, nonwoven fabrics are employed because the Ag spacer's functions are to maintain an electrolyte layer adjacent to the electrode and to protect the next spacer system from oxidation. The argentistatic spacer has as its function the prevention or inhibition of the transport of soluble silver from the silver oxide to the zinc electrode. Next, the dendristatic spacer acts to inhibit or prevent the growth of zinc dendrites across the cell from the zinc electrode to the silver electrode. Adjacent to the zinc electrode is the Zn spacer, which acts

219

to retain electrolyte adjacent to this electrode, as a mechanical barrier to maintain the integrity of the electrode by maintaining the zinc active material in contact with the electrode and, possibly, by minimizing electroosmosis.

The failure of primary batteries when used in a wet-charged state is in large measure the result of loss in shelf life by self-discharge which in turn arises from a malfunction or inadequacy of the argentistatic separator. Those batteries which are stored in the dry state with delayed electrolyte addition fail primarily because of deficiencies in the Ag spacer caused by lack of wetting or wicking. Secondary batteries often fail at the dendristatic spacer because of its puncture and the subsequent short-circuiting of the battery by zinc dendrites.

To a large extent, the separator functions enumerated as special to the zinc-silver oxide system, relate to transport phenomena. Aside from the ionic current carried largely by the hydroxide ion and to lesser extents by the potassium, argentate usually designated as $Ag(OH)_2^-$ (R12, 107A, 453) and zincate ions, there is a substantial transfer of solvent and electrolyte during the charge and discharge processes. These transport phenomena are often the result of several phenomena. For example, on discharge, the volume of the silver oxide electrode is markedly reduced, and the hydrostatic pressure in the cell drives liquid from the zinc to the silver side of the battery. Simultaneously, $K^+$ (and $OH^-$) move toward the zinc electrode and away from the Ag electrode. Accordingly, there are hydrostatic forces, concentration and osmotic forces, and electroosmotic forces acting across the separator system. Each of these forces can produce a flow of liquid, this flow being proportional to the permeability of the separator. Very fine-pored or "tight" membranes minimize such flows.

## SILVER ELECTRODE SPACER

The function of the Ag spacer is to maintain a liquid film adjacent to the electrode. Those materials which have found most favor are the nonwoven fabrics, usually of polypropylene, nylon, or Dynel. Usually, one layer is employed with a fabric thickness ranging from 0.02 to 0.2 mm. These fabrics are of an open character with 80-90% void volumes and openings up to several microns in size.

In many military applications, activation is accomplished by inserting the electrolyte immediately prior to use. Under these conditions, it is particularly important that the spacer wet immediately.

The maintenance of wetting is particularly important also in space applications because of the high acceleration to which the battery is subjected and because of the special conditions of weightlessness which obtain. Even in ordinary use, spacers made of hydrophobic materials of poor wettability do not always maintain proper wicking. For example, polypropylene fiber separators are desirable because of their high thermal and chemical stability, but they wet poorly and show poor wicking. Fibers of finer denier aid in liquid retention and wicking, but a surface treatment to make the fabric hydrophilic is probably more effective. The latter can be accomplished by techniques employed in producing antistatic coatings. With polypropylene, a surface sulfonation can be effected by the use of chlorosulfonic acid, sulfur trioxide (the Turbak procedure is a mild yet effective procedure, (R595)), and a number of other agents. The treatment of polyolefin films to make fixed-charge membranes of the sulfonic-acid type is described in the patent literature (R44).

Probably Ag spacer membranes of a finer porosity and improved wicking characteristics will result from the use of woven or nonwoven fabrics made of plastids (R48, 431). Plastids are bundles of extremely fine fibers of one of many common polymers, prepared by dissolving the polymer in a liquid which is a solvent at elevated temperature and pressures but a nonsolvent at normal temperatures such that, when the solution is extruded through the spinneret, the solvent boils away leaving a multifilament fiber. While the overall diameter of the plastid fiber may be 25 $\mu$, the individual fibers often are in the form of fine ribbons having a thickness which varies from 0.1 $\mu$ to 1 $\mu$ and an overall width of 2-5 $\mu$. The surface developed by this procedure is quite high: An average plastid material has a surface area of 100 m$^2$ g$^{-1}$ compared with 5 m$^2$ g$^{-1}$ for cellulose. The cost for the production of plastids in appreciable amounts is relatively low, of the order of 50 cents per pound over and above the usual processing cost.


## ARGENTISTATIC SPACER

The argentistatic spacer must serve to inhibit or prevent the migration of $Ag(OH)_2^-$ at an approximately $10^{-4}$ $N$ solution in the alkaline electrolyte. Cellophane has been employed for this purpose since the discovery of its utility made the first practical silver-zinc cell possible. There is little doubt that this is a sacrificial membrane, with the reaction of silver with cellophane producing a reduced and insolubilized form of the metal and deterioration of the film. Studies on the permeability of silver across cellophane membranes as measured (R107B) by the use of [110]Ag show an initial low rate of permeation; after an appreciable period of time, the membrane capacity for Ag is apparently reached, and the rate of permeation increases (R341). This "sink" mechanism probably also obtains with separator materials which contain sulfur, probably in the mercaptan form because these show a similar argentistatic behavior. On the other hand, with apparently inert membranes one finds a reasonably good correlation between the ohmic resistance of a separator and its imperviousness to silver.

Among materials which have been employed as argentistatic spacers there are a number of copolymer or terpolymer systems wherein the active substituent is a carboxylic group, as with such materials as acrylic or methacrylic acid or maleic anhydride. Radiation or chemical grafting of monomers such as methacrylic acid to chemically stable matrices such as polyethylene has been employed, but there are often difficulties in nonuniformity of the product (R342).

An argentistatic separator can function in accordance with one or more of four mechanisms: (1) it can have such a fine-pored structure that the $K^+$ and $OH^-$ ion pass but the larger silver complex ion is retarded selectively; (2) it can form a weakly dissociated complex with silver or one of its complexes; (3) it can react irreversibly with the silver complex to form a nonmigrating species (the latter two are essentially sink mechanisms); (4) the membrane acts as a selective solvent, dissolving base but not the silver complex. The first and last mechanisms possess one important advantage. A simple calculation employing an estimated diffusion coefficient for the complex and the apparent path length reveals that, with a barrier which absorbs the complex irreversibly (or nearly so) over an appreciable period of time (about 6 months), a large portion of the silver electrode will dissolve making for poor battery shelf life.

The preparation of fine-pored membranes selectively permeable on the basis of ion size has been accomplished; for example, membranes (R650) permeable to monovalent cations or anions ($Na^+$, $K^+$, $Cl^-$) and quite impermeable to bivalent ones ($Ca^{2+}$, $Mg^{2+}$, $SO_4^{2-}$) by coating ordinary ion-exchange membranes (which do not show an appreciable selectivity) with a thin (5 $\mu$) film having a dense, very fine pore structure. Usually, condensates such as the phenolsulfonic acid-phenol-formaldehyde resins are employed. These coatings have a high specific resistance but are so thin that they add little to the overall membrane resistance. For example, a 10 $\mu$ coating adds only about 2 ohm cm$^2$ to the overall resistance of the membrane. A 1 $\mu$ coat is readily achieved and would add only 0.2 ohm cm$^2$ to the resistance, and a 0.1 $\mu$ coating is quite possible. The films which possess this property of size discrimination are not suitable for use in alkaline electrolytes because of their instability at high pH, but other condensates would appear to be usable.

Further encouragement for the concept of ion discrimination of the basis of their size can be obtained from the results of studies (R28) on the properties of valinomycin, a toroidal cyclic polypeptide of molecular weight about 1100. This molecule allows the hydrated $K^+$ ion (diameter of 4 Å) to pass while restricting passage of the $Na^+$ ion (diameter of 6 Å) by a factor of about 400:1.

As for the use of polymeric membranes, a prospective of this problem can be gained by considering the ionic dimensions that are involved. The solvated $K^+$ ion has a diameter of about 4 Å, and one can estimate that of the $Ag(OH)_2^-$ complex as being about 6 Å. The $OH^-$ ion migrates as part of the water structure, so that it is difficult to assign to it a specific diameter. Polymer chains have an average diameter ranging from 3-8 Å, so that their random arrangement in membrane gels allows for little selectivity on the basis of size.

However, many improvements are possible over the materials now employed. For example, the ion-exchange membranes composed of polystyrene-$p$-sulfonate and polystyrene-$p$-methylenetrimethylammonium exchange groups obviously cannot form very tight pore structures because of their shape. They usually have an effective pore diameter of about 10 Å, too large for ionic discrimination of the kind required for battery separators. However, cross-linked films of vinylsulfonic acid, $[CH_2CH(SO_3^-)]_n$, of aminated polyepichlorhydrin, $[CH_2CH(CH_2NMe_3^+)O]_n$, or polytrimethyl ammonium pentene-1, $[CH_2CH(CH_2)_3(NMe_3^+)]_n$, should lead to improved, tighter battery separator films. Polymeric films can be made to be very thin and thereby have a low resistance. A film having an overall thickness of 1000 Å would contain over 100 rods, so uniform film preparation should present no problems.

Other materials apparently capable of forming true, ionic sieve membranes are the inorganic exchangers such as the zeolites, faujasites, or molecular sieves and similar materials. The orthoclases, for example, show a $Rb^+/Na^+$ discrimination of 200:1 (R279). Even finer levels of selectivity (to organic molecules) have been demonstrated by the molecular sieves. The principal problem here appears to be the difficulty in making coherent, pinhole-free membranes of inorganic crystals, where the orientation of ionic passage ways is such as to make for a high conductivity. Inorganic materials can possess high chemical stability, and their thermal stability is unexcelled. There seems to be no fundamental reason that the kind of membrane properties exhibited by $\beta$-alumina at high temperatures (R146) cannot be exhibited by other inorganics at ordinary temperatures.

The most practical pathway to an ideal argentistatic separator, at least in terms of the technology available at hand, lies in the employment of polymers containing fixed groups capable of reacting irreversibly or nearly so with the silver complex. In general, the fixation of almost any complexing group to the surface of a film or fiber offers no serious technical problems. For example, a series of classical chemical reactions upon insolubilized polystyrene (R566) produced rather complex structures, some analogous to dipicrylamine. Almost any chemical reaction can be performed on polymers, but their by-products remain on the structure and the rate of these substitution reactions is usually much slower than those with low molecular weight materials.

In order to serve the purpose at hand, any fixed complexing group must have a particularly high formation constant with silver to achieve a very low concentration of $Ag^+$ in concentrated battery electrolytes. For example, many attempts have been made to construct a chelating resin which would abstract silver from photographic fixing solutions only to meet with indifferent success because silver is strongly bound in the high concentration of thiosulfate present. One could form a polymer-metal, low-molecular-weight ligand complex (such as the polyvinylimidazole-zinc-nitrilotriacetic acid complex (R249, 399), but, even so, the high rates of complex dissociation which obtain would probably allow the silver complex to traverse the membranes at appreciable rates.

Among the various polymeric materials which form Ag chelates which have been described in literature may be included in the complexes with polymeric acids and bases, thiol- and mercaptopolymers (R262, 470), as well as a host of others.

The formulation of polymeric chelates does not necessarily require the synthesis of a new monomers or polymeric reactions. For example, porous gels (R631B) have been prepared from usually impermeable condensates. The urea, thiourea, and phenylthiourea-formaldehyde resins are well known. In their usual form they are infusible and so dense as to prevent the well-known reaction with silver,

$$Ag^+ + C_6H_5NHCNH_2 \overset{S}{\overset{\|}{}} \rightleftharpoons C_6H_5=C-NH_2 \overset{SAg}{\overset{|}{}} + H^+$$

from occurring. Thiourea is known as a strong complexing agent for silver, even in acidic media. Preparation (R631B) of high capacity complexing resins by forming the thiourea condensate in the presence of an approximately equal amount of a hydrophilic polymer (polyvinylalcohol) in solution, yielded a material of good mechanical properties and of high porosity on a molecular scale. Figure 82 shows the titration of this resin in the absence and presence of $Ag^+$; results on the sorption of silver even in acidic media at pH 2 show a capacity of 0.2 g of Ag/g of resin. Under the highly alkaline conditions which obtain in the silver-zinc battery, one would expect to find a much higher capacity for silver and possibly an ability to fix practical amounts of silver even from its dilute solution as the complex ion.

The insolubilization of the silver complex is readily achieved by reduction, and here one can employ one of the many redox polymers which have been described (R90). For example, a stable pyrogallolformaldehyde condensate of high exchange capacity was prepared (R263) by carrying out the reaction under strongly alkaline conditions; the same desirable properties could also be obtained by performing this condensation in the presence of a hydrophilic polymer. The reduction potential of these materials can be varied within limits, and a high absorptive capacity could be realized from these systems.

High sterilizing temperatures for space batteries have led to an interest in the so-called high-temperature polymers. A few deserve mention; poly-1,3 oxadia-zole polymers are stable at temperatures of 250° and unaffected by 10% sodium hydroxide at reflux for two days. An analogous polymer contains sulfur in place of the oxygen and may have an additional advantage in being able to bind the silver complex. Indeed, some success in argentistatic properties has been achieved by the employment of sulfur-containing high-temperature polymers. The poly(phenylene)triazoles are stable at $300°$, and the nitrogen anion may also have silver-fixing properties.

One can also employ a solvent-type, pore-free membrane. The solvent-type membranes now used for desalination by reverse osmosis are made of cellulose acetate and nylon. These have an effective wall thickness of the order of $0.2 \mu$ and function as water permeable and salt impermeable because the water contained within the film is almost entirely in the "icelike" state, highly hydrogen bonded to the polymer chains. As a consequence, this form of water

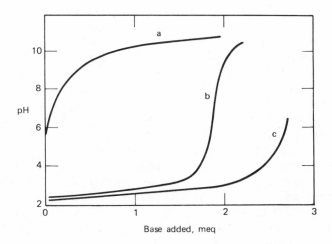

Figure 82  Titration of 0.3g of phenylthiourea-formaldehyde resin in $1N$ KNO$_3$: (a) no metal; (b) $0.005N$ Ag$^+$; (c) $0.02N$ Ag$^+$.

does not dissolve appreciable amounts of ionic substances; the more solvated ions are preferentially excluded. These types of membranes are capable of dissolving appreciable amounts of acids and bases, such as sulfuric acid and potassium hydroxide. Measurements (R532) of the solubility and conductivity of the electrolytes in cellulose acetate membranes have shown an appreciable conductivity for potassium hydroxide in the membrane phase. While one cannot employ cellulose acetate in the silver-zinc cell because it is not stable in alkaline media, some nylons are stable. There is every reason to believe that the latter or other solvent-type membranes should be capable of selectively dissolving potassium hydroxide while excluding the complexes of silver. These films would have to be very thin because of their high intrinsic ohmic resistance.

## DENDRISTATIC SEPARATORS

There appears to be general agreement that it is at the zinc electrode where dendrites form to short-circuit the battery and the majority of separator problems in the Ag-Zn battery arise. The development of a truly dendristatic separator would have manifold applicability because of the interest in the zinc electrodes for other battery systems.

The formation of zinc dendrites (R241) can be readily correlated with the surface current density (see the paper by Oswin and Blurton in this volume). As noted in the Oswin and Blurton paper, many charging regimes have been proposed to diminish the tendency for dendrite growth on recharging cells. However, the nonuniformities of current density distribution, restricted convection conditions, uneven plate alignments, and other design considerations have probably defeated the proposed countermeasures.

In batteries, these dendrites grow from the zinc electrode through the highly porous Zn spacer and to the dendristatic separator. Dendritic growth may continue parallel to the film for a short period of time, but the dendrites invariably seem to find a "weakness" in the membrane and grow through it to the Ag electrode. Dendritic growth proceeds during the recharge of secondary cells; dendritic trees grow at a more rapid rate when and where zincate ion has been largely depleted, especially near the end of the recharge cycle.

The best dendristatic separators developed thus far make use of methyl cellulose, and some employ this material together with copolymers of vinylmethylether and maleic anhydride, (R341-342). One does find penetration with membranes of these kinds but at longer times. Their resistances are also somewhat higher when compared to cellophane, and the mechanism of their function is not necessarily the same.

The time for the penetration of zinc dendrites across a membrane separator can be measured in a special cell (R107C) where the growth of the zinc dendrites across and through the membrane is detected by a short circuit. With cellophane separators, the dendrite grows out from the electrodes to the membrane and then may grow laterally across the surface of the membrane before penetrating. Most authors have assumed that this penetration is through micropores or pinhole imperfections in the membrane. Since the dendrite obviously grows by cathodic deposition at its point, one cannot speak of a dendrite's "punching through" a membrane, rather, it grows through it.

An important clue to the effectiveness of cellophane as a dendristatic separator is provided, particularly in the presence of zincate ion, by the fact that soluble material is leached from the separator and this soluble polymer modifies the growth of zinc dendrites. The modification of crystal growth by the presence of dissolved surface active agents is well known (R410). Surface active agents have been claimed to show dendristatic properties but ones of low molecular weight are not readily retained within the appropriate interspacer position because they diffuse or migrate readily. Polymeric surface active agents or ones of polyelectrolyte character would remain in place because many common membranes (including cellophane) are quite impermeable to polyelectrolytes of reasonable molecular weight.

The ability is well known of polymeric additives to mask rapidly growing crystals so as to inhibit their growth and produce finely divided and relatively soft deposits. Well known also is the utilization of polymeric additives to boiler feed waters to produce a calcium carbonate or calcium sulfate deposit which is flocculant and nonadherent to the metal surfaces. Some starches have traditionally been employed, but these are largely supplanted by synthetic polyacids.

For some time a considerable body of literature held that it was not possible to form a coherent crystal in a pore of molecular dimensions. A 10-Å pore of an ion-exchange membrane could accomodate a dendrite having a diameter formed by 5-10 atoms of zinc, but the surface energy of a structure having such a high surface-to-volume ratio has been assumed to be so great that the crystal structure could not be maintained.

A test of this particular theory was performed using the sulfonic acid and quaternary ammonium ion-exchange membranes of commerce. The properties of certain of these membranes have been studied extensively (R337) with calculation of their porosity from measurements of the rate of permeation of exchange ions and uncharged solutes, employing classical hydrodynamic equations to calculate pore diameters. Direct evidence on the rate of permeation of quaternary ammonium cations of different sizes into a membrane of this type shows that the rate falls off sharply as the size of the exchange ion approaches 10 Å, indicating that the pore diameter is of about this size. Solutes somewhat larger, such as vitamin $B_{12}$ of diameter 12 Å, do not permeate these membranes at all. Further, it is relatively easy to determine the absence of pinholes and other microscopic imperfections in membranes of this kind (R600).

Employing a modification of the standard zinc penetration test cell (R341), carefully selected samples of a number of membranes were tested for zinc penetration. Results are presented in Table 25. Figure 83 shows photomicrographs of dendrites grown through the membranes (taken of the membrane face away from the Zn cathode). In each case one observes a multiplicity of fine dendrites of typical metallic character. Since the membranes were apparently intact and since their pores are interconnecting over short distances, these dendritic crystals of zinc have an overall length of approximately 2 mm and, for each branch, a diameter of approximately 10 Å. The high surface free energy of these crystals is evidenced by the fact that, when current is no longer passed across the systems, gas begins to be evolved and the dendritic crystal within the pore apparently begins to dissolve spontaneously. At the end of each experiment

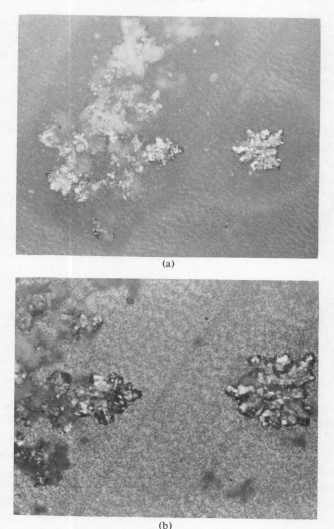

(a)

(b)

**Figure 83**   Dendritic growths through ion-exchange membranes:  (a) C 103,  ×50;  (b) C 103,  ×80;  (c) DA1,  ×50; (d) CMV-2, ×50.

all traces of zinc were removed by dissolution in dilute hydrochloric acid, and no pinhole or other visible evidence of dendritic growth remained in the stable membranes. These membranes are rather brittle, being highly crosslinked, so any "puncture" would be readily visible. Further, dyes and other tests for the presence of anything larger than molecular (10 Å) pores were negative.

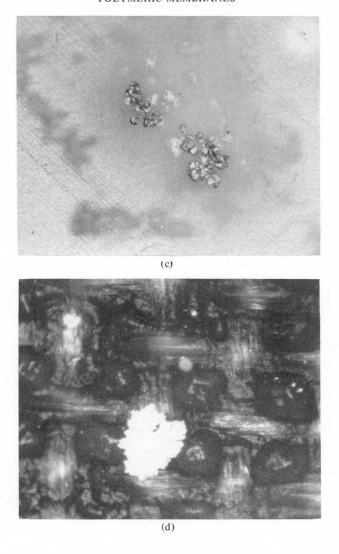

(c)

(d)

The data of Table 25 show many interesting effects. The AMF–C 103 and ACI–DK1 sulfonic acid cation-exchange membranes are quite stable in concentrated base; the same is true of their quaternary ammonium anion-exchange analogs A 103 and DA1. The sulfonic acid membranes favor the accumulation of positive ions such as $Zn^{2+}$, and their negative fixed charges would act to stabilize Zn metal dendrites. Their figures of merit are equal to or twice that for cellophane (the DK1 membrane is more homoporous than is C 103). On the other hand, the anion-exchange membranes would sorb $ZnO_2{}^{2-}$ preferentially while rejecting $Zn^{2+}$ ions; also, their fixed charges would have the same sign as that of the dendrite. Penetration times were 2 to 3 times as long for these anion-exchange films as for the cation-exchange membranes, and their figures of merit are 3.5 and 4.2 times that for cellophane.

TABLE 25.   ZINC PENETRATION RESISTANCE [a]

| Membrane[b] | Type[c] | Areal Resist. ohm-cm$^2$ | Wet Thickness microns | Specific Resist. ohm-cm | Zinc-Penetration Time hr | Figure of Merit hr/cm thickness |
|---|---|---|---|---|---|---|
| Cellophane 193 |   | 0.065 | 75  | 8.6  | 1.5  | 200 |
| AMF−C 103 | C | 4.10  | 300 | 136. | 7.2  | 240 |
| AMF−A 104 | A | 27.0  | 320 | 85.8 | 22.5 | 714 |
| ACI−DK1   | C | 2.40  | 230 | 105. | 9.2  | 400 |
| ACI−DA1   | A | 2.79  | 250 | 110. | 23.7 | 948 |
| CMV       | C | 1.96  | 280 | 70.4 | 1.4  | 50  |
| CMV−2     | C | 4.91  | 280 | 176. | 8.3  | 297 |
| AMT−10    | A | 0.522 | 170 | 15.4 | 0.93 | 54  |
| AST       | A | 2.46  | 230 | 109. | 3.2  | 142 |

[a] All values are the average of three determinations in 40% KOH. All membranes stable in 40% KOH except AMT−10 and AST.

[b] Sources of membranes: AMF, American Machine and Foundry; ACI, Asahi Chemical Industries; CMV, CMV−2, AMT−10, and AST from Asahi Glass Co.

[c] Cation-permeable C; anion-permeable A; all C membranes, sulfonate, all A membranes, quaternary ammonium, types.

Some of the membranes showed poor chemical resistance; the AMT−10 and AST films were partially dissolved in the battery electrolyte, and only the CMV and CMV−2 films remained intact, and their figures of merit were usually poorer than that for cellophane. Most of these films are preferentially permeable to monovalent cations and anions, but this effect was probably largely negated by a chemical instability and a lack of uniformity because all of these membranes are cast onto cloth backings.

Table 25 does show that a fine pore structure having fixed positive charges does lead to improved dendristatic properties; the substantial differences in ohmic resistance (contrasted with cellophane) can be reduced markedly by the use of thinner films.

## ZINC ELECTRODE SPACER

The zinc spacer has several functions in the cell. First, it imparts mechanical strength to an electrode composed largely of zinc oxide powder and therefore quite weak but, when properly wrapped, an electrode has considerably greater mechanical strength. Further, during the discharge process this electrode forms a nonadherent insoluble oxide, and it is essential that this insoluble zinc oxide be maintained in direct contact with the electrode during the charging process so that electron transfer can occur. This separator must also function to maintain electrolytic contact across the face of the electrode and also act as a wick. Because of electroosmotic dehydration of the zinc electrode region on discharge, the Zn spacer should show a greater water retention than the Ag spacer.

Another function of the Zn spacer is to inhibit dendritic growth by preventing the electrode from drying out so as to maintain a uniform current density over the plate. At the same time, the spacer makes for a relatively uniform, nonstirred boundary layer adjacent to the electrode so it acts to increase the throwing power of the electrolyte. One can ultimately relate throwing power of an electrolyte to a high slope of the Tafel curve, so that at the higher current densities which one obtains at protruberances of the electrode, a higher overvoltage obtains with less deposition on the "mountains" and more in the "valleys." Most separators do not maintain a uniform current density over the electrode; on recycle one observes more erosion at the edges and at the middle, with the electrode slowly changing in shape and becoming thicker at its bottom.

Many of the suggestions made earlier in this contribution are applicable to the Zn spacer. One could employ relatively fine pore membranes, and cation-selective membranes should inhibit $ZnO_2{}^{2-}$ migration across the battery. Because of the high dissolved ZnO content (8-9%) a complexing mechanism is probably inoperable on the basis of capacity considerations alone. The problem of electroosmotic solvent transfer can be minimized as suggested earlier.

Another mechanism deserving mention (R261) is to use membranes of good permselective character and low electroosmotic permeability (to ions of one sign) in a mosaic arrangement with a small area of membrane permeable to ions of the opposite sign and having high electroosmotic transport characteristics. Films which show zero or even negative electroosmotic flows could be prepared according to this general principle.

In summary, on the basis of our fundamental and applied knowledge, we are in a position to prepare membranes for the silver-zinc cell having markedly improved characteristics, particularly when we direct our attention to a composite film, each part of which performs a separate separator function in a specific and superior manner.

# I7.

## Mass Transfer Properties of Membranes and Their Effect on Alkaline Battery Performance

### Manny Shaw and Allen H. Remanick

Sealed silver oxide secondary batteries designed for long cycle life have had relatively poor performance at temperatures below ambient and at discharge depths greater than 25%. This is particularly true when the batteries are cycled under a combination of these conditions. Comparison of cells made with different membrane separator systems leads to the inference that major differences in performance can be associated with variations in separator properties and that in some manner the membrane is associated with changes in electrolyte concentration. Mass transport of ions and water, under conditions of discharge and charge, can result in concentration changes dependent on the transference number and diffusion coefficient of the dissolved species in the membrane as well as on the current and discharge or charge time. Concentration changes would, also, result from preferential absorption of water and KOH by the membrane. These changes could be a function of the initial concentration as well as the amount of membrane relative to the original solution.

Regarding the mass transfer effects through a membrane, the concentration of anolyte in a zinc-silver oxide cell would change during discharge as a result of the following events: (1) depletion of $OH^-$ by electrochemical reaction; (2) replenishment of $OH^-$ by migration from the catholyte; (3) replenishment of $OH^-$ by diffusion across the concentration gradient; and (4) transport of water including electroosmotic movement and depletion or replenishment by electrochemical reaction.

These events are shown schematically in Figure 84, together with the equations expressing the changes in the amounts of KOH and water occurring in the anolyte. In the figure, $Me$ represents any anode material; $D$ is the diffusion coefficient of the electrolyte in the membrane; $A$ and $\Delta x$ are the area and thickness respectively of the membrane; $t_-$ and $t_w$ are the transference number of the hydroxide ion and water respectively; $C_c$ and $C_a$ are the concentrations of the KOH in the catholyte and anolyte respectively; and $i$ is the current.

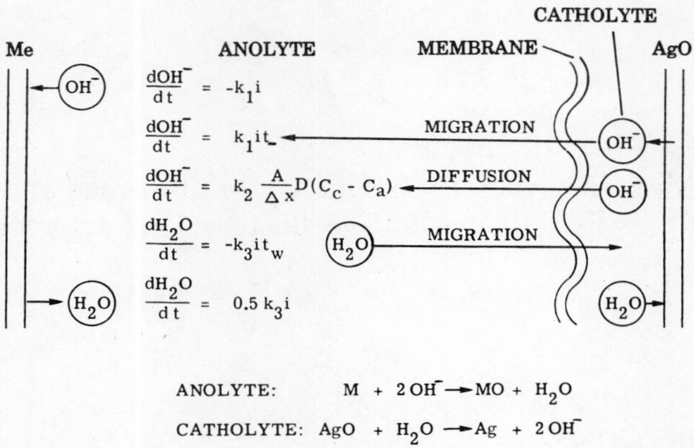

**Figure 84** Change of anolyte during discharge in zinc-silver oxide cell.

Combining these equations with those describing the effect of concentration on voltage should permit the development of a mathematical model which expresses cell voltage as a function of time, temperature, and current and which contains constants specific to the transport characteristics of specific membrane separators. This model, together with the effect of membrane absorption, can contribute to the establishment of optimization criteria for the design of zinc-silver oxide batteries having improved low temperature performance.

There will be described the task of measuring diffusion constants, transference numbers, and absorption properties of three commercially available membrane materials, as a function of concentration and temperature, and mathematical expressions will be evolved relating these values. The membranes used in this study were PUDO-600 cellophane, Visking V-7 sausage casing, and Permion 2.2XH (20%), an acrylic acid grafted polyethylene.

## DIFFUSION CONSTANTS FOR KOH IN MEMBRANES

Using a measuring cell (R107) in which the membrane separates two chambers, each filled with a known volume of solution of specified concentration, then, from Fick's First Law, the amount of material diffusing through a membrane of exposed area $A$ under steady-state conditions is given by

$$\frac{1}{A}\frac{d(Cv)}{dt} = D\frac{(C'- C)}{\Delta x} \tag{1}$$

where $C$ and $v$ are the concentration and volume respectively of the particular chamber under consideration, $D$ is the diffusion coefficient, and $\Delta x$ is the membrane thickness. Good mixing of the chamber contents is maintained such that the only possible concentration gradient $(C' - C)$ appears across the membrane.

If conditions are chosen such that $v$ remains essentially constant, Equation 1 may be rearranged to give

$$\frac{dC}{dt} = \frac{AD'}{v} (C' - C) \tag{2}$$

where $D' = D/\Delta x$ and is defined here as the *reduced diffusion coefficient.*

If the chamber containing solute at concentration $C'$ is made very large with respect to the other chamber, $C'$ will be essentially constant. Under these conditions integration of Equation 2 gives

$$\ln \frac{C_0' - C_t}{C_0' - C_0} = \frac{-AD'}{v} t \tag{3}$$

subscripts representing initial value and concentration at time $t$.

From the published data (R5), it can be shown that for small concentration differences

$$C = kd + a \tag{4}$$

where $C$ and $d$ are the concentration and density respectively of the KOH solution, and $k$ and $a$ are constants. Combining Equations 3 and 4 yields

$$\ln \frac{d_0' - d_t}{d_0' - d_0} = \frac{-AD'}{v} t \tag{5}$$

Using the buoyancy method of determining the weight of a sinker immersed in the liquid under measurement, the density may be expressed as

$$d = \frac{w_a - w_l}{v'} \tag{6}$$

where $w_a$ and $w_l$ are the weights of the sinker in air and liquid respectively, and $v'$ is the sinker volume. Combining Equations 5 and 6 yields

$$\ln \frac{w_0' - w_t}{w_0' - w_0} = \frac{-AD'}{v} t \tag{7}$$

Therefore, $D'$, can be calculated directly from $w_0'$ and $w_t$. In practice, the compartment corresponding to $w_t$ contained the higher concentration and smaller volume. Equation 7 in the form

$$\log (w_0' - w_t) - \log (w_0' - w_0) = -\frac{AD'}{2.303v}t \tag{8}$$

was used for the actual calculation of $D'$ (facilitated by a program written for an Olivetti-Underwood Programma 101 digital computer).

Figures 85, 86, and 87 show plots of the reduced diffusion coefficient against concentration for PUDO-600, Visking V-7, and 2.2XH membranes, respectively. For convenience, the results have been expressed in units of ml min$^{-1}$ in.$^{-2}$ The flux per unit area across one layer of membrane may be quite readily calculated by multiplying the value of $D'$ by the proper concentration difference in units of moles ml$^{-1}$. The average deviation of $D'$ was less than 5%, with a number of measurements showing less than 3% deviation.

As shown, concentrations are expressed as the initial concentration gradients

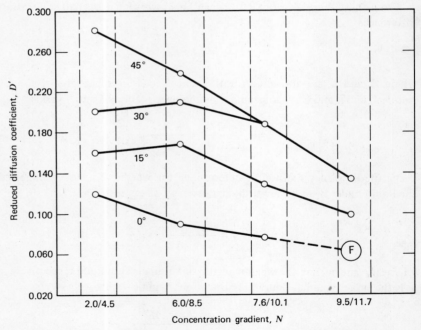

**Figure 85** Effect of KOH concentration and temperature on the reduced diffusion coefficient of KOH in PUDO-600 membrane.

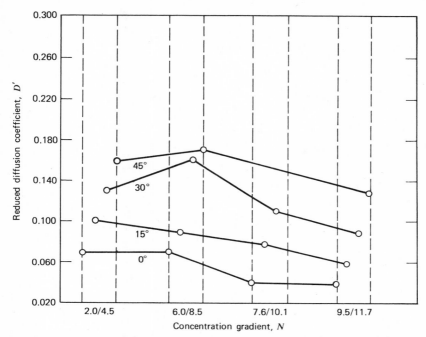

**Figure 86** Effect of KOH concentration and temperature on the reduced coefficient of Visking V−7 membrane.

across the membrane. Under the experimental conditions, the lower one remains almost invariant whereas the higher concentration decreases by about 10-15%. The values for $D'$ probably reflect an average concentration under a given set of conditions. According to Equation 3, the zero time may be arbitrarily selected thereby allowing the physical state (temperature, swelling, etc.) within the membrane to reach equilibrium. The membrane is exposed to reasonably steady-state conditions during the measurement since the change in KOH concentration in the compartment during the time of measurement, is less than 10%.

In general, $D'$ decreases with increasing concentration, the effect being most pronounced for PUDO-600. Except for the latter, values of $D'$ generally lie in the range of 0.06 to 0.18 ml in.$^{-2}$ min$^{-1}$, irrespective of temperature or concentration. It was not our purpose to relate the significance of these measurements to the membrane but rather to determine the $D'$ values as a function of temperature and concentration for later insertion into mathematical expressions. Values of $D'$ will depend on the effects of temperature and concentration on the diffusion coefficient of KOH and on membrane properties such as permeability, swelling and structure.

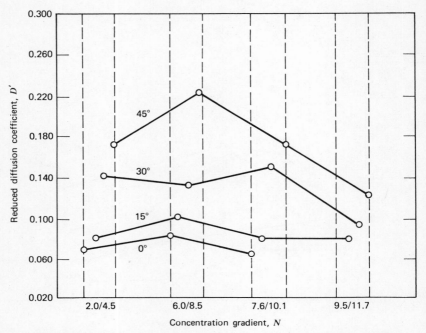

**Figure 87** Effect of KOH concentration and temperature on the reduced coefficient of 2.2XH membrane.

## EFFECT OF TEMPERATURE ON DIFFUSION

The diffusion constant can be calculated for any temperature from the kinetic relationship

$$D' = A^{\ddagger} \exp\left(-\frac{E^{\ddagger}}{RT}\right) \tag{9}$$

where $A^{\ddagger}$ is the preexponential term containing the entropy of activation, and $E^{\ddagger}$ is the energy of activation. Values for $A^{\ddagger}$ and $E^{\ddagger}$ are shown in Table 26.

For all membranes, both $A^{\ddagger}$ and $E^{\ddagger}$ increase with concentration reaching a maximum at the 7.6-10.0$N$ gradient. The change in $A^{\ddagger}$ is much more marked than the change in $E^{\ddagger}$. The change in values, as a function of concentration, may possibly indicate a change in mechanism of the diffusion process. One cannot be certain of this because a change in membrane thickness or self-diffusion or both would cause a change in $A^{\ddagger}$ or $E^{\ddagger}$ according to the effect on $D'$. It is probable, however, that the extent of change in the kinetic parameters is significantly greater than that expected from a change in the physical parameters of the membrane.

TABLE 26.   MEMBRANE DIFFUSION RATE PARAMETERS

| Membrane | 2-4.5$N$[a] | | 6-8.5$N$ | | 7.6-10.1$N$ | | 9.5-11.7$N$ | |
|---|---|---|---|---|---|---|---|---|
| | $A^{\pm}$ | $E^{\pm}$ | $A^{\pm}$ | $E^{\pm}$ | $A^{\pm}$ | $E^{\pm}$ | $A^{\pm}$ | $E^{\pm}$ |
| Visking V-7 | 33.6 | 3.36 | 66.3 | 3.72 | 731 | 5.24 | 218 | 4.68 |
| PUDO-600 | 56.8 | 3.35 | 89.4 | 3.67 | 736 | 4.93 | 60.4 | 3.71 |
| 2.2XH | 36.0 | 3.36 | 74.5 | 3.73 | 347 | 4.73 | 4.20 | 2.24 |

[a] Values of $A^{\pm}$ are in ml min$^{-1}$ in.$^{-2}$ Values of $E^{\pm}$ are in kilocalories per mole.

Further evidence for a possible mechanism change is derived from a comparison of $A^{\pm}$ and $E^{\pm}$ among each of the membranes at the various concentrations. In the lower concentration ranges both $A^{\pm}$ and $E^{\pm}$ vary only slightly among the membranes at a given concentration. This indicates that some solution property controls the diffusion rate since the only common factor among the three membranes is the concentration of the diffusing solution. At higher concentrations significant differences in $A^{\pm}$ and $E^{\pm}$ occur among the membranes, becoming most pronounced at the highest measured range. The membrane properties, therefore, appear to control the diffusion process at the higher concentration ranges.

## TRANSPORT NUMBERS FOR KOH AND WATER IN MEMBRANES

Another area of effort important to the definition of electrolyte-membrane relationship, is the measurement of cation and anion transference numbers through membranes, together with the apparent transport number of solvent. One cannot strictly define the transport of solvent per faraday as a transference number since the solvent is an uncharged material not directly entering into the current-carrying process. It has become common practice, however, to designate the change in the number of moles of solvent associated with the anolyte or catholyte per faraday as the solvent transport number. Since the direction of solvent movement is not defined by the stoichiometry of a reaction, a positive value for the solvent transport number is arbitrarily chosen to correspond to net solvent transport in the direction of cation movement.

The transference numbers were determined by a method (R356) differing from the usual Hittorf method in that not only is the change of concentration after electrolysis measured but also the total weight of material remaining in the electrode compartment. In this manner, both the cation transference number $t_+$ and the water transport number $t_w$ can be determined. The measurement of transport through membranes required other modifications of the Hittorf method (R356) in order to facilitate the experimental procedure. The middle compartment was eliminated, and the anolyte compartment was made much larger than that for the catholyte. The latter modification was necessary since $t_+$ was found to be dependent on the concentration in the receiving compartment. The effects of diffusion and concentration polarization were eliminated by determining a range of time and current density in which $t_+$ remained constant. In our own work, sufficient layers of membrane were used to eliminate the effects of back diffusion.

Measurements of cationic transport numbers were made at $0°$ and $30°$ for PUDO-600 and Visking V-7 over a wide range of concentration. No significant difference in values, as a function of membrane type or temperature, was evident. Figure 88 shows a plot of all values as a function of concentration, the straight line representing the data average according to the method of least squares. Values of $t_+$ decrease about 20% with increasing concentration from 2.6 to 15.0$N$ KOH. Cation transport numbers in KOH solution have been previously determined (R356). Values decrease from 0.26 at 1.0$N$ to 0.22 at 3.0$N$, remaining constant at 0.22 until 12.0$N$, and then increase to 0.26 at 17.0$N$. It thus appears that the values of $t_+$ in the membranes are slightly higher than in free solution. That is, the apparent mobility of the potassium ion, relative to the hydroxide ion, is slightly greater in the membranes. It is not known whether this increased mobility results from change of ionic hydration, ion aggregation, or environmental effects (direct membrane-ion interaction).

Values for the water transport number, $t_w$, are shown in Figure 89 for PUDO-600 and Visking V-7. Nearly identical values were obtained for each membrane. The water transport number shows a marked dependency on

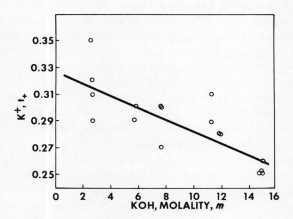

**Figure** 88 Effect of KOH concentration on the transference number of $K^+$ ion in PUDO–600 and Visking V–7 membranes.

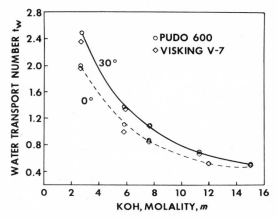

**Figure 89** Effect of KOH concentration on the water transport number in PUDO-600 and Visking V-7 membranes.

concentration, decreasing from 2.5 at 2.6$N$ KOH to 0.5 at 15.0$N$ KOH at 30°. Values are 10-20% lower at 0° than at 30°. The effect of concentration appears to be similar at both temperatures.

## KOH ABSORPTION BY MEMBRANES

In order to describe the change in electrolyte distribution in the anolyte and catholyte chambers of a cell on charge or discharge, it is necessary to know the initial distribution of electrolyte. Since it is known that membranes commonly employed in batteries absorb electrolyte, determination of both quantity and concentration of electrolyte absorbed was necessary. The method used to measure absorption (R107D) was modified in that, after the gross weight of absorbed electrolyte had been measured, the electrolyte in the membrane was extracted and titrated in order to determine the distribution of KOH and water in the absorbed electrolyte.

Absorption measurements were made by placing a 5 x 5 cm (2 in. square) section of membrane of known weight, $w_m$, into a polypropylene test tube. About 50 ml of KOH solution of specified molality, $m_s$, and weight, $w_s$, was added. After being sealed with a stopper, the test tube was placed in a constant temperature bath and the system allowed to equilibrate for a minimum of four hours. Longer soaking times did not affect the measurements. The membrane was then removed from the solution, lightly wiped on a glass plate, and weighed. The wet weight made up the sum of the weight of water contained in the membrane, $W_i$, plus the weight of KOH in the membrane, $w_i$, plus the weight of

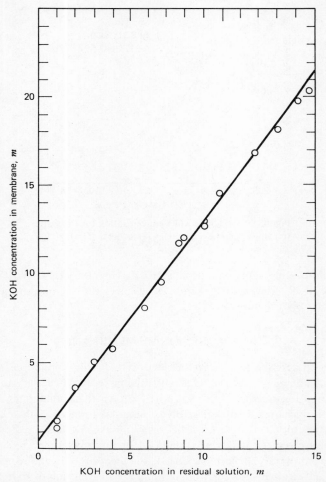

**Figure 90** Relationship of membrane KOH concentration with residual solution concentration, Visking V-7, $30^{\circ}$.

the membrane, $w_m$. The wet membrane was then placed in a Soxhlet extractor and extracted for a minimum of four hours with water that was free from $CO_2$. Longer extraction times had no effect. After being cooled, the contents of the boiling flask were titrated with standard acid, giving the weight of the KOH in the membrane. Knowing $w_i$ and $w_m$ gives the weight of the water in the membrane, $W_i$. These measurements were repeated over a range of concentrations.

In the present work, the ratio of membrane weight to solution weight was selected at a quite low value so that there would be a relatively small change in the electrolyte concentration as a result of the absorption. The absorption measurements, however, permitted the calculation of electrolyte distribution between the membrane and the solution at equilibrium as a function of membrane weight by the use of derived relationships involving $w_e$, the weight of KOH in the solution after the absorption; $W_e$, the weight of water in the solution after absorption; $w_i$, $W_i$, and $w_m$. Also, for the calculations it was convenient and necessary to have the values of the original weight of KOH, $w_0$, in the solution taken for the experiment, and of the original weight of water, $W_0$, in this solution; these were determined by the following relationships,

$$w_0 = \frac{m_s w_s}{17.8 + m_s} \quad \text{and} \quad W_0 = \frac{17.8 w_s}{17.8 + m_s}$$

derived by linear combination of $w_0 + W_0 = w_s$, and $m_s = 1000 w_0 / 56.1 W_0$ (mol. wt. of KOH is 56.1).

From the absorption measurements, two relationships could be obtained. These were: the internal-external molality and the fractional absorption of KOH and water, $w_i/w_m$ and $W_i/w_m$, respectively.

INTERNAL-EXTERNAL MOLALITY

Figure 90 shows a typical relationship between the concentration of KOH within the membrane (internal molality) and that in the residual solution (external molality), which can be expressed in the form:

$$\text{internal molality} = N (\text{external molality}) + P \qquad (10)$$

where $N$ and $P$ are functions of the membrane and the temperature, as shown in Table 27. Since the plotted data do not pass through the origin, values for $P$ are not applicable at very low molality.

TABLE 27. INTERNAL-EXTERNAL MOLALITY PARAMETERS

| Membrane | At 0° | | At 30° | |
|---|---|---|---|---|
| | $N$ | $P$ | $N$ | $P$ |
| Visking V-7 | 1.20 | 1.04 | 1.38 | 0.535 |
| PUDO-600 | 1.13 | 1.49 | 1.12 | 2.22 |

## FRACTIONAL ABSORPTION OF KOH OR WATER

The fractional absorption of KOH by the membrane has been correlated with the external molality. This quantity is defined by the equation,

$$\text{fractional absorption KOH} = \frac{w_i}{w_m} \qquad (11)$$

Although the absorption data were reproducible for two of the membrane types at a given temperature and given concentration, when taken over the entire range of concentration, the relationship between fractional absorption and external molality was complex.

For mathematical simplicity in subsequent derivations, the total concentration range was therefore divided into a number of distinct linear regions. Within each region, the data could then be expressed according to the following relationship:

$$\text{fractional absorption KOH} = J \,(\text{external molality}) + K \qquad (12)$$

with equilibrium conditions implied. The terms $J$ and $K$ are constants for a given region of concentration.

A similar relationship exists for the fraction of water absorbed as a function of external molality, that is,

$$\text{fractional absorption water} = L \,(\text{external molality}) + M \qquad (13)$$

Equations 12 and 13 are empirical representations of the data on membrane absorption. Values of $J$, $K$, $L$, and $M$, as a function of molality, are given in Tables 28 and 29.

The results of the absorption measurements for 2.2XH were very erratic and did not afford sufficient reproducibility for any type of correlation. In general, the 2.2XH material absorbed much less electrolyte than the other types of membranes.

The following relationships hold at equilibrium:

$$w_e + w_i = w_0 \qquad (14)$$

$$W_e + W_i = W_0 \qquad (15)$$

$$\frac{w_i}{w_m} = J\,\frac{1000 w_e}{56.1 W_e} + K \qquad (16)$$

$$\frac{W_i}{w_m} = L \frac{1000w_e}{56.1W_e} + M \qquad (17)$$

Equations 14 and 15 do not involve Equation 10, the relationship between internal and external molality, but are expressions of mass balance only.

TABLE 28.  ABSORPTION PARAMETERS, PUDO-600

| | Fractional Absorption KOH | | | Fractional Absorption Water | | |
|---|---|---|---|---|---|---|
| | $m$ | $J$ | $K$ | $m$ | $L$ | $M$ |
| At 30° | 0.0- 3.5 | 0.105 | 0.0283 | 0.0- 1.0 | 0.267 | 0.873 |
| | 3.5- 6.6 | 0.168 | −0.170 | 1.0- 2.1 | −0.185 | 1.32 |
| | 6.6- 8.7 | −0.0546 | 1.29 | 2.1- 6.1 | 0.257 | 0.412 |
| | 8.7-15.0 | 0.0510 | 0.365 | 6.1- 7.6 | −0.450 | 4.70 |
| | | | | 7.6-10.0 | −0.0888 | 1.96 |
| | | | | 10.0-13.0 | −0.0122 | 1.18 |
| | | | | 13.0-15.0 | 0.0737 | 0.0653 |
| At 0° | 1.0- 6.2 | 0.2186 | −0.141 | 1.0- 2.1 | −0.165 | 1.49 |
| | 6.2- 8.4 | −0.184 | 2.37 | 2.1- 6.1 | 0.459 | 0.177 |
| | 8.4-15.0 | 0.0874 | 0.0822 | 6.1- 8.0 | −0.849 | 8.08 |
| | | | | 8.0-10.2 | −0.0645 | 1.85 |
| | | | | 10.2-11.9 | 0.0030 | 1.15 |
| | | | | 11.9-15.0 | 0.100 | −0.0057 |

TABLE 29.  ABSORPTION PARAMETERS, VISKING V-7

| | Fractional Absorption KOH | | | Fractional Absorption Water | | |
|---|---|---|---|---|---|---|
| | $m$ | $J$ | $K$ | $m$ | $L$ | $M$ |
| At 30° | 0.0- 6.7 | 0.1330 | −0.0114 | 0.0- 1.0 | 0.908 | 0.823 |
| | 6.7- 8.3 | −0.0552 | 1.28 | 1.0- 2.1 | −0.488 | 2.208 |
| | 8.3-15.0 | 0.0623 | 0.262 | 2.1- 6.6 | 0.110 | 0.957 |
| | | | | 6.6- 7.6 | −0.456 | 4.700 |
| | | | | 7.6-10.0 | −0.0668 | 1.746 |
| | | | | 10.0-15.0 | −0.00092 | 1.064 |
| At 0° | 0.0- 6.4 | 0.183 | −0.0744 | 0.0- 3.0 | 0.188 | 1.01 |
| | 6.4- 8.1 | −0.133 | 1.89 | 3.0- 5.7 | 0.3506 | 0.547 |
| | 8.1-14.8 | 0.0787 | 0.179 | 5.7- 7.8 | −0.628 | 6.23 |
| | | | | 7.8-10.7 | −0.876 | 2.05 |
| | | | | 10.7-14.8 | 0.0654 | 0.406 |

Equations 16 and 17 are restatements of Equations 12 and 13, respectively, and where

$$\frac{1000 w_e}{56.1 W_e} = 17.8 \frac{w_e}{W_e} = m_s$$

the molality of the original solution. These four equations permit us to solve for the four unknowns, $w_e$, $w_i$, $W_e$, and $W_i$.

Although the values of $J$, $K$, $L$, and $M$ are dependent on the particular value of the equilibrium external molality, only the final numerical solution will be dependent on the proper choice of $J$, $K$. $L$, and $M$. A mathematical solution for $w_e$, $W_e$, $w_i$, and $W_i$ is dependent only on the form of Equations 12 and 13, as will be seen.

From equations 14 through 17, the weight of potassium hydroxide and water in both the membrane and residual solution, under equilibrium conditions, can be calculated. The weight of KOH in solution at equilibrium is given as

$$w_e = \frac{w_0 W_e - w_m W_e \quad K}{W_e + 17.8 \, w_m \, J} \tag{18}$$

and the weight of water in solution at equilibrium is given as the quadratic

$$W_e = \tfrac{1}{2} \{ -17.8 w_m \, J - w_m M + W_0 \pm \quad [(17.8 w_m J + w_m M - W_0)^2$$
$$- 71.2 w_m (w_0 \, L - w_m KL + w_m JM - W_0 \, J)]^{\frac{1}{2}} \} \tag{19}$$

Equation 19 expresses $W_e$ in terms of known quantities. From the value of $W_e$, $w_e$ can be found by Equation 18. Similarly, having $w_e$ and $W_e$, $w_i$ and $W_i$ may then be determined from Equations 16 and 17. However, since substitution of $W_e$ in Equation 19 leads to a rather involved expression for $w_e$, numerical solution for $W_e$ is made before substitution in Equation 18. Since the expression for $W_e$ is itself rather cumbersome, a computer was used for the mathematical solution. The computer was programmed to solve for $W_e$ using different choices of $J$, $K$, $L$, and $M$ values taken from Tables 28 and 29. Certain restrictions on the final value of $W_e$ aided in the choice of the proper values of $J$, $K$, $L$, and $M$. A value of $W_e$ which is imaginary, negative, or greater than $W_0$ is a physical impossibility. Therefore, an alternate choice of $J$, $K$, $L$, and $M$ was necessary to establish a value of $W_e$ within the limits $W_0 < W_e > 0$.

Since it was anticipated that alternate choices of $J$, $K$, $L$, and $M$ might be necessary, the computer program was written so as to allow facile substitution of such values. Having decided on the proper value of $W_e$, it was then possible to solve for $w_e$, $W_i$, $w_i$, the equilibrium internal molality and the equilibrium external molality. As a final check, the internal molality was determined from the external molality and Equation 10. Since $N$ and $P$ in Equation 10 are the most exact parameters, the value of $W_e$ was adjusted until the value of the internal molality derived from $w_i$ and $W_i$ agreed within $0.1m$ with that found using the computer value of equilibrium molality. This adjustment procedure was also written into the computer program. The successful solution to Equations 14, 15, 16, and 17 means that for a given membrane it is possible to describe the distribution and concentration of electrolyte within any portion of the cell under equilibrium conditions and for any given weight of membrane.

Tables 30, 31, 32, and 33 present some typical calculations of electrolyte

TABLE 30.   ELECTROLYTE DISTRIBUTION, PUDO-600[a]

| | $w_m$ (g) | $w_e$ (g) | $W_e$ (g) | External Molality ($m$) | $w_i$ (g) | $W_i$ (g) | Internal Molality ($m$) |
|---|---|---|---|---|---|---|---|
| $m_s = 9.0m$ | 10 | 25.3 | 53.7 | 8.4 | 8.3 | 12.8 | 11.6 |
| | 20 | 16.1 | 37.7 | 7.6 | 17.4 | 28.8 | 10.8 |
| | 30 | 5.9 | 15.6 | 6.7 | 27.7 | 50.8 | 9.7 |
| | 35 | 2.1 | 5.9 | 6.4 | 31.5 | 60.5 | 9.3 |
| | 37.5 | 0.83 | 2.4 | 6.2 | 32.8 | 64.1 | 9.1 |
| $m_s = 14.0m$ | 10 | 33.6 | 45.2 | 13.3 | 10.4 | 10.8 | 17.1 |
| | 20 | 23.9 | 33.9 | 12.6 | 20.1 | 22.1 | 16.2 |
| | 30 | 15.0 | 22.5 | 11.8 | 29.1 | 33.5 | 15.5 |
| | 45 | 2.9 | 4.9 | 10.7 | 41.1 | 51.1 | 14.3 |
| | 47.5 | 1.1 | 1.8 | 10.6 | 42.9 | 54.2 | 14.1 |

[a] Weight of original solution, $w_s$ = 100 g; temperature = 30°.

TABLE 31.   ELECTROLYTE DISTRIBUTION, PUDO-600[a]

| | $w_m$ (g) | $w_e$ (g) | $W_e$ (g) | External Molality ($m$) | $w_i$ (g) | $W_i$ (g) | Internal Molality ($m$) |
|---|---|---|---|---|---|---|---|
| $m_s = 9.0m$ | 10 | 25.3 | 53.2 | 8.5 | 8.2 | 13.2 | 11.1 |
| | 20 | 18.0 | 40.2 | 8.0 | 15.6 | 26.2 | 10.6 |
| | 30 | 10.9 | 25.2 | 7.7 | 22.7 | 41.2 | 9.8 |
| | 40 | 5.6 | 14.0 | 7.1 | 28.0 | 52.4 | 9.5 |
| | 45 | 2.7 | 6.9 | 6.9 | 30.9 | 59.5 | 9.2 |
| | 50 | 0.45 | 1.2 | 6.6 | 33.1 | 65.2 | 9.0 |
| $m_s = 14.0m$ | 10 | 31.6 | 42.5 | 13.2 | 12.4 | 13.5 | 16.4 |
| | 20 | 20.5 | 29.2 | 12.5 | 23.5 | 26.8 | 15.6 |
| | 30 | 10.6 | 15.9 | 11.8 | 33.5 | 40.1 | 14.9 |
| | 40 | 1.5 | 2.4 | 11.2 | 42.5 | 53.6 | 14.1 |

[a] Weight of original solution, $w_s$ = 100 g; temperature = 0°

TABLE 32.   ELECTROLYTE DISTRIBUTION – VISKING V-7 [a]

|  | $w_m$ (g) | $w_e$ (g) | $W_e$ (g) | External Molality $(m)$ | $w_i$ (g) | $W_i$ (g) | Internal Molality $(m)$ |
|---|---|---|---|---|---|---|---|
| $m_s = 9.0m$ | 10 | 25.7 | 54.6 | 8.3 | 7.8 | 11.9 | 11.8 |
|  | 20 | 16.4 | 38.5 | 7.6 | 17.2 | 28.0 | 10.9 |
|  | 30 | 6.31 | 16.8 | 6.7 | 27.7 | 49.6 | 9.8 |
|  | 35 | 3.59 | 9.78 | 6.5 | 30.0 | 56.6 | 9.4 |
|  | 40 | 0.60 | 1.87 | 6.3 | 32.9 | 64.6 | 9.1 |
| $m_s = 14.0m$ | 10 | 33.3 | 45.7 | 13.0 | 10.7 | 10.4 | 18.5 |
|  | 20 | 23.7 | 34.9 | 12.1 | 20.3 | 21.1 | 17.2 |
|  | 30 | 15.2 | 24.0 | 11.3 | 28.9 | 32.0 | 16.0 |
|  | 40 | 7.33 | 12.4 | 10.5 | 36.7 | 43.6 | 15.0 |
|  | 47 | 2.49 | 4.43 | 10.0 | 41.5 | 51.5 | 14.3 |
|  | 50 | 0.48 | 0.88 | 9.80 | 43.5 | 55.1 | 14.1 |

[a] Weight of original solution, $w_s = 100$ g; temperature = 30°.

TABLE 33.   ELECTROLYTE DISTRIBUTION – VISKING V-7 [a]

|  | $w_m$ (g) | $w_e$ (g) | $W_e$ (g) | External Molality $(m)$ | $w_i$ (g) | $W_i$ (g) | Internal Molality $(m)$ |
|---|---|---|---|---|---|---|---|
| $m_s = 9.0m$ | 10 | 25.1 | 53.0 | 8.5 | 8.4 | 13.5 | 11.2 |
|  | 20 | 17.5 | 39.4 | 7.9 | 16.1 | 27.1 | 10.6 |
|  | 30 | 5.0 | 12.5 | 7.1 | 28.6 | 53.9 | 9.4 |
|  | 33 | 0.44 | 1.2 | 6.7 | 33.1 | 65.2 | 9.0 |
| $m_s = 14.0m$ | 10 | 31.9 | 43.1 | 13.2 | 12.2 | 12.9 | 16.8 |
|  | 20 | 20.9 | 30.0 | 12.4 | 23.1 | 26.0 | 15.8 |
|  | 30 | 11.2 | 17.1 | 11.6 | 32.9 | 38.9 | 15.0 |
|  | 35 | 6.7 | 10.5 | 11.3 | 37.4 | 45.5 | 14.6 |
|  | 40 | 2.4 | 4.0 | 10.9 | 41.6 | 52.0 | 14.2 |
|  | 42 | 0.7 | 1.2 | 10.8 | 43.3 | 54.8 | 14.1 |

[a] Weight of original solution, $w_s = 100$ g; temperature = 0°.

distribution between membrane and solution for various weights of membrane in a given weight of original solution. Original KOH concentrations of 9.0 and 14.0$m$ have been chosen as representative examples. From the eight sets of calculations (2 membranes x 2 temperatures x 2 concentrations) the following observations were made:

1. The amount of residual solution $(W_e + w_e)$ is greatly reduced with increasing membrane to initial solution weight ratio. This was observed for both

membranes at both temperatures. A typical example is shown in Figure 91 for PUDO-600 in 14.0*m* KOH.

2. Membrane absorption is greater at the lower concentration. This was observed for both Visking V-7 and PUDO-600 at 30°. The one set of calculations for PUDO-600 in 9.0*m* KOH at 0° showed a lower membrane absorption than PUDO-600 in 14.0*m* KOH.

3. Membrane absorption is greater at the lower temperature. Again, the one set of calculations (PUDO-600 in 9.0*m* KOH at 0°) did not fall into line. The lower amount of residual solution at a lower temperature could mean cell failure in certain cases where the temperature is dropped from 30° to 0°. As an example, using Visking V-7 membrane in 9*m* KOH at a membrane-to-solution weight ratio of 0.35, a change in temperature from 30° to 0° eliminates the residual solution (13.4 g at 30°). Decreasing the temperature to 0° results in complete loss of residual solution, as shown in Table 33. Therefore, what may constitute a sufficient amount of electrolyte at 30° may be insufficient at 0°, depending on the membrane/solution weight ratio. (Neglected in this discussion is the amount of electrolyte absorbed in the plates, which may still be adequate to support the reaction. Measurements of plate absorption are needed for the total picture of electrolyte distribution).

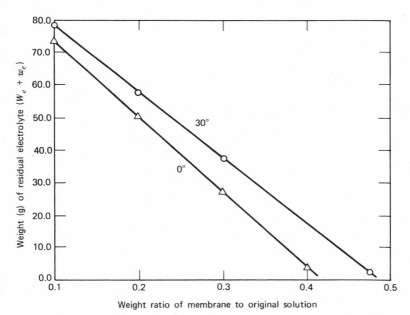

**Figure 91** Effect of membrane absorption as a function of membrane/solution weight ratio and temperature, illustrated by a decrease in the amount of residual electrolyte, PUDO-600 membrane in 14.0*m* KOH.

4. The electrolyte concentration is always greater within the membrane than in the residual solution.

5. The concentration of both the membrane-contained electrolyte and that of the residual electrolyte decrease with increase in the membrane to original solution weight ratio.

6. The electrolyte concentration within the membrane approaches the value for the original solution with increasing membrane-to-solution weight ratio. This is a natural consequence of items 4 and 5. These last three observations are graphically presented in Figure 92. As shown, if the original weight ratio of membrane to solution is 0.1, the KOH concentration at equilibrium is significantly higher than in the original solution, whereas the concentration in the residual solution is only slightly lower. On the other hand, for a ratio of about 0.3-0.4 the concentration of the residual solution is markedly lower than the original solution, whereas the membrane concentration is approximately the same as the original concentration.

The possibility of solution depletion occurring in cells operating at a relatively high membrane-to-solution weight ratio was referred to above. Operating a cell at low membrane-to-solution weight ratios may in some cases result in freezing of the electrolyte contained in the membrane. Figure 92 shows that, in the case of Visking V-7, in an initial concentration of $14.0m$ KOH, the concentration within the membrane is increased to $17.0m$.

A $17m$ KOH solution freezes at $-5°$, so that operating a cell at $0°$ could result in freezing of the electrolyte in the membrane if the membrane-to-solution ratio is less than 0.1. This actually occurred (see Figure 85) during measurement

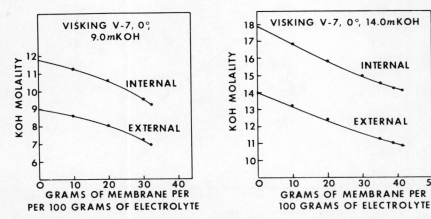

**Figure 92** Electrolyte concentration inside the membrane and in the residual solution (external) as a function of membrane/solution weight ratio for Visking V-7 membrane.

of the diffusion coefficient of KOH in PUDO-600 at $0°$, in which one side of the membrane was in contact with $11.7N$ KOH (equivalent to $14.5m$).

The method used to determine the absorption parameters involved a small weight of membrane in a much greater weight of KOH solution. Since these experiments formed the basis for calculating the data presented in Tables 30-33, it was important to confirm the absorption correlation. For Visking V-7, a membrane-to-solution ratio of 0.3 and an initial concentration of $8.97m$, the equilibrium concentration of KOH in the solution was found to be $7.12m$, compared to the calculated concentration of 6.82. It therefore appears that the mathematical correlation of absorption is reasonably accurate.

## FORMULATION OF CONCENTRATION-TIME EQUATIONS

The overall anolyte reaction in a silver-zinc battery may be formally described as

$$Zn + 2OH^- = ZnO + H_2O + 2e^-$$

The corresponding overall reaction in the catholyte compartment is

$$AgO + H_2O + 2e^- = Ag + 2OH^-$$

Use of an alternate stoichiometry (involving zincate ion) would not change the method of derivation but only the final terms of the equations. This type of derivation should therefore be applicable to any type of alkaline battery or fuel cell, if proper adjustment is made for reaction stoichiometry.

As described in Figure 84, the anolyte concentration changes during discharge and charge as a result of $OH^-$ depletion by faradaic reaction, combined with $OH^-$ replenishment by migration and diffusion effects, as well as by transport of water. These changes are now expressed mathematically. Faradaic depletion occurs according to

$$\frac{d(OH^-)_a^*}{dt} = k_1 i \tag{20}$$

The quantity of $OH^-$ replaced by migration is

$$\frac{d(OH^-)_a}{dt} = k_1 it \tag{21}$$

---

* In these and subsequent equations, use of parentheses for chemical species denotes total *quantity* in grams of material rather than concentration.

Replenishment by diffusion may be expressed as

$$\frac{d(OH^-)_a}{dt} = k_2 \frac{A}{\Delta x} D(C_c - C_a) \tag{22}$$

Combination of Equations 20, 21, and 22 yields

$$\frac{d(OH^-)_a}{dt} = -k_1 i(1 - t_-) + k_2 \frac{A}{\Delta x} D(C_c - C_a) \tag{23}$$

Using the definition of $D'$ as previously indicated and expressing the concentration units in molality as defined by

$$m = \frac{1000(KOH)}{56.1(H_2O)} = 17.8 \frac{(KOH)}{(H_2O)} \tag{24}$$

together with the necessity of maintaining ionic balance in the anolyte, we may transform Equation 23 to

$$\frac{d(KOH)_a}{dt} = -k_1 i t_+ + 17.8 k_2 \frac{A}{n} D' \left[ \frac{(KOH)_c}{(H_2O)_c} - \frac{(KOH)_a}{(H_2O)_a} \right] \tag{25}$$

Similarly, the rate of change of water with time is

$$\frac{d(H_2O)_a}{dt} = -k_3 i(t_w - 0.5) \tag{26}$$

The symbols represent the following:

| | |
|---|---|
| $(KOH)_a$ | = Total grams KOH in anolyte |
| $(H_2O)_a$ | = Total grams $H_2O$ in anolyte |
| $(KOH)_c$ | = Total grams KOH in catholyte |
| $(H_2O)_c$ | = Total grams $H_2O$ in catholyte |
| $t_+$ | = Transference number of $K^+$ in the membrane |
| $i$ | = Current |
| $A$ | = Area of membrane |
| $n$ | = Number of layers of membrane |
| $D'$ | = Reduced diffusion coefficient per layer of membrane |
| $t_w$ | = Transference number of water in membrane |
| $k_1, k_2, k_3$ | = Conversion constants, to adjust each of the terms to g sec$^{-1}$ in Equations 25 and 26 |

Numerical values of $D'$ as given in Figures 85, 86, and 87 cannot be used directly in Equation 25 because of incorrect dimensions. Conversion of $D'$ from units of ml min$^{-1}$in.$^{-2}$ to the correct units of g $H_2O$ min$^{-1}$in.$^{-2}$ may be made by use of known density percentage relationships (R5).

The amount of KOH and water initially absorbed by the membrane may be calculated from the previously determined absorption measurements. Since the total amount of solution in the cell is known and remains constant, $(KOH)_c$ and $(H_2O)_c$ may be expressed as

and

$$(KOH)_c = (KOH)_{tot} - (KOH)_m - (KOH)_a \qquad (27)$$

$$(H_2O)_c = (H_2O)_{tot} - (H_2O)_m - (H_2O)_a \qquad (28)$$

where the subscripts tot and $m$ refer to total quantity in cell and membrane respectively. Furthermore, it has been experimentally found that

$$D' = a_2 m^3 + b_2 m^2 + c_2 m + d_2 \qquad (29)$$

$$t_+ = a_1 m + b_1 \qquad (30)$$

$$t_w = a_5 m + b_5 \qquad (31)$$

within ranges of molality $m$ of interest to the program. The values of the constants in Equations 29, 30, and 31 may be readily established from the data plotted in Figures 85, 86 and 87 (Equation 29), Figure 88 (Equation 30), and Figure 89 (Equation 31). Substitution of Equations 24 and 27-31 in Equations 25 and 26 yields the following two simultaneous differential equations, in which the variables are time, $(KOH)_a$, and $(H_2O)_a$.

$$\frac{d(KOH)_a}{dt} = W + XY \qquad (32)$$

$$\frac{d(H_2O)_a}{dt} = Z \qquad (33)$$

where

$$W = -k_1 i \left( \frac{17.8 a_1 (KOH)_a}{(H_2O)_a} + b_1 \right)$$

$$X = 17.8 k_2 \frac{A}{n} \left[ a_2 \left( 17.8 \frac{(KOH)_a}{(H_2O)_a} \right)^3 + b_2 \left( 17.8 \frac{(KOH)_a}{(H_2O)_a} \right)^2 + 17.8 c_2 \frac{(KOH)_a}{(H_2O)_a} + d_2 \right]$$

$$Y = 17.8 \left[ \left( \frac{(KOH)_{tot} - (KOH)_m - (KOH)_a}{(H_2O)_{tot} - (H_2O)_m - (H_2O)_a} \right) - \frac{(KOH)_a}{(H_2O)_a} \right]$$

$$Z = k_3 i \left( 17.8 a_5 \frac{(KOH)_a}{(H_2O)_a} + b_5 \right)$$

Since direct integration of Equations 32 and 33 was deemed a rather difficult task, an iterative method of solution was chosen. Since the iterative method involved a great number of numerical and graphical operations, the computer was programmed to carry out this task. Solution of Equations 32 and 33 gives the concentration-time relationship for the anolyte. Sign reversal in these same equations results in similar relationships for the catholyte.

## APPLICATION OF CONCENTRATION-TIME EQUATIONS TO A TYPICAL CELL

The concentration-time equations derived above, permit the calculation of the anolyte (or catholyte) concentration at any time $t$ for a given cell containing a given weight of a given membrane at a given temperature, if also the amount and concentration of the initial electrolyte is known. Using the absorption, diffusion, and transference data obtained from measurements on Visking V-7 membrane and applying the concentration-time equations, anolyte concentration-time curves have been plotted for a typical silver-zinc cell.

These curves are not to be construed as representing the actual change in concentration with charge or discharge time which can be expected in a typical silver-zinc cell. Membrane measurements were made in zincate-free KOH solutions, and the equations were derived on the basis of Zn/ZnO stoichiometry involving 2 moles of $OH^-$ per mole of zinc. Further, no data were available concerning plate absorption of electrolyte, which could affect the equilibrium of membrane absorption. Also, changes in the properties of the membrane caused by the electrode species, as a result of continued cycling, have been neglected. These curves are, nevertheless, presented to show how the anolyte concentration may be expected to change as a function of time, current density, temperature, and number of membrane layers.

Concentration-time curves have been calculated for a 25 Ah silver-zinc cell having the following design and operating parameters:

| | |
|---|---|
| Total negative plate area: | 7.73 dm$^2$ (120 in.$^2$) |
| Current density: | 21 mA cm$^{-2}$ (133 mA in.$^{-2}$) |
| Type of membrane: | Visking V-7 |
| Layers of membrane: | 4 |
| Temperature: | 0° |
| Amount of electrolyte: | 140 g |
| Concentration of added electrolyte: | 9.0$m$ |

Calculations were carried out to 30 minutes of discharge (or charge), which represents 30% depth of discharge at the 21 mA cm$^{-2}$ rate. The curves are shown as initiating at less than 6.7$m$ KOH at zero time. This represents the effect of membrane absorption, whereby the residual electrolyte concentration has been decreased from the original 9.0$m$ solution. For the sake of clarity, the temperature effect of membrane absorption was not considered in describing the effect of temperature on concentration changes. Although the temperature-membrane absorption relationship may be one of the most important features of cell design, the lack of plate absorption data prevents a more definitive correlation.

The effect of current density on anolyte concentration is shown in Figure 93. The increased slope of the 31.5 mA cm$^{-2}$ curve is caused by the increased water transport according to Equation 33. The corresponding catholyte curves, not

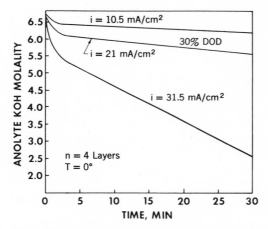

**Figure 93** Anolyte concentration-time curves as a function of current density, as calculated using Equations 32 and 33.

shown here, would approximate mirror-images of the anolyte curves were it not for the effect of $t_w$, which tends to inhibit the change in catholyte concentration. This involves the assumption that the water is transported wholly by cations so that the increase in catholyte concentration is offset by the dilution effect of water transported into the catholyte region.

Figure 94 shows the effect of increasing the number of membrane layers. The more layers of membrane used, the less is the replenishment of OH$^-$ to the anolyte by diffusion, so the greater is the change in anolyte concentration. Figure 95 shows the effect of decreasing the temperature. Diffusion and transport values at $-30°$ were obtained by extrapolation of measured data. Similar calculations made for a cell containing initially 14.0$m$ KOH showed a much greater change in OH$^-$ concentration in going from $0°$ to $-30°$. In the latter case there was a decrease of nearly 2 molal units.

Examining the overall effect of membrane-induced electrolyte changes and

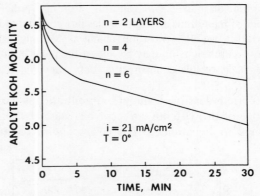

**Figure 94** Anolyte concentration-time curves as a function of number of membrane layers, as calculated using Equations 32 and 33.

current-induced electrolyte changes, an overall change in concentration of 4-5 molal units could occur, attributed to 2.5 to 3.5 molal units change as a result of membrane absorption, plus 1-2 units change as a result of current flow. For instance, a cell filled with 30% KOH might possess a final anolyte concentration of 13-17% after discharge. The possibility of localized freezing within the anolyte becomes immediately apparent even though the concentration of the initially added electrolyte would not be expected to allow freezing. The change in anolyte concentration becomes even more marked at lower temperatures, since the diffusion coefficient, a primary factor in the electrolyte restoring mechanism, markedly decreases with temperature, whereas the other factors are relatively unaffected.

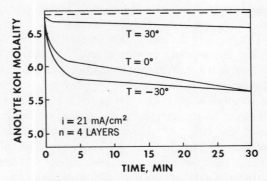

**Figure 95** Anolyte concentration-time curves as a function of temperature, as calculated using Equations 32 and 33.

In addition to cell failure or considerably reduced capacity due to electrolyte freezing, is the effect of KOH concentration on plate voltage and cell capacity. Subsequent work has been involved with this effect. Having equations relating concentration with time and the effect of concentration on cell voltage and cell capacity permits the development of a mathematical model that expresses cell voltage as a function of time, temperature, current, and concentration and that contain constants specific to the mass transfer and absorption characteristics of membrane separators and cell geometry.

## EXPERIMENTAL PROCEDURES

### DETERMINATION OF DIFFUSION RATES

As described earlier, diffusion rates of KOH in the membranes were determined by a buoyancy method of density measurement in which the weight of a sinker is directly translatable into a value for the diffusion constant.

Measurements were made in a dual chamber cell constructed of Plexiglas, the two parts being joined through a membrane holder, the purpose of which was to vary the orifice size between the two chambers. The small compartment of the cell was filled with 100.0 ml KOH of the higher concentration to be used for the particular run. The larger compartment was filled to the same liquid level with KOH of the lower concentration. The volume of solution was usually 1300 ml in the larger compartment. Magnetic stirring bars were placed in each compartment. The cell was contained in a constant temperature bath.

A Plexiglas sinker, weighed with lead to adjust apparent density, was suspended by means of a platinum wire from a hook extended below the balance. (A Mettler B5 analytical balance had been modified by means of a Mettler GD adaptor for determining weights of material suspended below the balance.) The cell was positioned and raised so that the top of the sinker was about 2.5 cm below the liquid level in the large compartment. Stirring of both solutions was then started.

After about two hours equilibration, readings of the sinker weight were taken. (The stirrer was stopped during the readings, and the solution was allowed to come to rest for about 30 seconds.) Readings were continued at 10-15 minute intervals until a constancy of about 0.3 mg was reached. The sinker was then transferred to the smaller compartment. After an equilibration period of 30 minutes, measurements of the sinker weight at fixed time intervals were made. Usually 13 points were then taken over a three-hour period.

For laboratory convenience, Equation 8 was converted to

$$\log\left[10(w_1 - w_s)_t\right] = \log\left[10(w_1 - w_s)_0\right] - bt \qquad (34)$$

where $b = AD'/2.303\ V$. Since $(w_1 - w_s)_0$ is a constant, Equation 34 represents a straight line, the slope of which permits the calculation of $D'$, the reduced diffusion coefficient. Table 34 shows a typical run, with the calculated data being plotted in Figure 96. A computer program was established to determine the straightline constants $a$ and $b$, as well as the average error for the diffusion coefficient determination.

The fact that the KOH concentration in the large compartments remains invariant can readily be seen by the following relationships:

maximum total flux
$$= D' \times \text{area} \times \text{total time} \times \text{maximum gradient}$$
$$= 0.2 \text{ ml min}^{-1} \text{ in.}^{-2} \times 0.5 \text{ in.}^2 \times 180 \text{ min}$$
$$\times 2.5 \times 10^{-3} \text{ moles ml}^{-1}$$
$$= 0.045 \text{ mole}$$

The total amount of KOH in the large compartment is 1.3 liters × 6.0 moles per liter or 7.2 moles. Therefore, the percent of change in the large compartment is

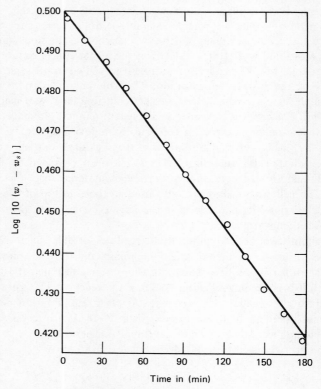

Figure 96 Diffusion rate measurements for KOH in PUDO-600.

TABLE 34.   EXPERIMENTAL DATA FOR MEMBRANE (PUDO-600)
DIFFUSION OF KOH[a]

| Large Compartment[b] | | Small Compartment[c] | |
|---|---|---|---|
| Elapsed Time, min | Sinker Weight $w_1$, grams | Elapsed Time, min | Sinker Weight $w_s$, grams |
| 0 | 2.7540 | 0 | 2.4371 |
| 15 | 2.7530 | 18 | 2.4417 |
| 30 | 2.7528 | 30 | 2.4457 |
| 45 | 2.7528 | 45 | 2.4501 |
| 60 | 2.7526 | 60 | 2.4550 |
| | | 75 | 2.4599 |
| | | 90 | 2.4647 |
| where | | 105 | 2.4688 |
| $y = a + bt$ | | 120 | 2.4730 |
| $y = \log [10(w_e - w_s)_t]$ | | 135 | 2.4777 |
| $b = AD'/2.303v$ | | 150 | 2.4825 |
| $b = 4.63 \times 10^{-4}$ | | 165 | 2.4864 |
| $D' = 0.213$ ml in.$^{-2}$ min$^{-1}$ | | 180 | 2.4904 |

[a] Orifice area, $A = 0.5$ in.$^2$; temperature = 29.9°
[b] 6.0$N$ KOH
[c] 8.5$N$ KOH

(0.045/7.2) x 100 or 0.6%. A similar analysis indicates that the change in concentration in the smaller compartment is of the order of 5%.

## DETERMINATION OF TRANSPORT NUMBERS

Transport numbers were measured according to the method of Kressman (R356). The cell was similar to that used for the diffusion measurements, except for reduction in size of the smaller compartment. Smooth platinum electrodes and stirrers were employed in both compartments. Measurement of electrolyte change was made in the anolyte chamber. Temperature control was maintained by an external bath and, at currents of 300 mA or less, remained within ±0.3°. Usually a current of 200 mA was used. Reproducibility of measurements ranged from 1-10%. All determinations were made in duplicate. Details of a typical determination are given below:

| | |
|---|---|
| Membrane: | PUDO-600 |
| Temperature: | 30° |
| Initial concentration of anolyte: | 5.72$m$ |
| Initial weight of anolyte: | 12.012 g |
| | 2.921 g KOH |
| | 9.091 g H$_2$O |
| Final concentration of anolyte: | 4.83$m$ |

Final weight of anolyte:                          10.808 g
                                                  2.305 g KOH
                                                  8.503 g $H_2O$
Loss of KOH from anolyte:                         0.616 g
Loss of $H_2O$ from anolyte:                      0.588 g
Electrolysis time:                                18,070 sec
Current:                                          0.200 A
$t_+$:                                            0.293
$t_w$:                                            1.38

MEMBRANE ABSORPTION MEASUREMENTS

Experimental procedure for determining membrane absorption of electrolyte was described earlier. A typical experimental run is shown below:

Membrane:                                         Visking V-7
Temperature:                                      $30°$
KOH concentration in solution:                    $7.92m$
Dry weight of membrane:                           0.1231 g
Wet weight of membrane:                           0.3601 g
Weight of KOH solution absorbed:                  0.2370 g
Volume of $0.1001N$ $H_2SO_4$ to titrate extracted KOH:17.75 ml
Weight of KOH in membrane:                        0.0997 g
Fraction KOH in membrane:                         0.810
Weight of water in membrane:                      0.1373 g
Fraction water in membrane:                       1.12
KOH concentration in membrane:                    $12.94m$

## SUMMARY AND CONCLUSIONS

Measurements of KOH diffusion and transference through some typical cell membranes, and of KOH absorption by the latter, have shown that changes in electrolyte concentration can occur, which could have adverse effect on cell performance. These experiments were initiated for the purpose of determining the quantitative changes in KOH concentration as related to the mass transfer characteristics of membranes as a function of current, time, and temperature. Such information, combined with the effect of concentration on cell voltage and capacity, could ultimately be applied to the design of batteries having improved low temperature performance.

Diffusion coefficients of PUDO-600 cellophane, Visking V-7, and Permion 2.2 XH (20%) were determined as a function of temperature and KOH concentration under a constant initial concentration gradient ($2.5N$) across the membrane. The concentration range was from 2.0 to $11.7N$, and the temperature range from 0-45°. The concentration changes from which calculation of diffusion coefficients was made, were measured by a density-buoyancy method using an analytical balance suspended above a 2-chamber cell. Analysis of the data indicated that in the lower concentration ranges, the diffusion coefficient was controlled by solution properties whereas at higher concentrations the membrane properties appeared to be the controlling factor. Transference numbers for both ionic species and solvent were determined at 0° and 30°. Values of the cation transport number indicated a slightly higher mobility of $K^+$ (relative to $OH^-$) in the membranes as opposed to free solution. The water transport number showed an expected concentration dependence.

It was suspected that the quantity of membrane used in the cell could alter the initial electrolyte distribution by preferential absorption. Therefore, the relationship between internal (within the membrane) and external concentration of KOH was determined. Equations were derived which allowed the calculation of the equilibrium values of the weights of water and KOH in the membrane and in the residual solution, as well as the internal and external molalities. These were calculated from the known weights of membrane, original solution, and original molality. Membrane absorption appeared to be greater at 0° than at 30° and for 9.0m than for 14.0m KOH. The KOH concentration was always greater within the membrane than in the residual solution. Under certain conditions, the increase of KOH concentration on the membrane would be sufficient to cause freezing within the membrane.

The ratio of the membrane weight to the solution weight was found to have profound effect on electrolyte distribution. The amount of residual solution is markedly reduced with increasing weight ratio of membrane to initial solution. Both the internal and external KOH concentration (relative to the membrane) decrease with increasing membrane-to-solution weight ratio. Also, the electrolyte concentration within the membrane approaches the value for the original solution with increasing membrane-to-solution weight ratio.

Differential equations were derived describing the changes in KOH and water within the anode and cathode compartments. Graphical solution to these equations was obtained by an iterative method using typical battery design parameters. Curves of anolyte concentration against time were obtained as a function of current density, number of layers of membrane, and temperature.

The experiments and data reported here are to be considered only as an initial attempt in correlating electrolyte concentration changes with the mass transfer properties of membranes. Further work should include measurements in zincate solution and the effect of plate absorption of KOH.

# 18.

## The Nature of Cellulose as Related to Its Application as a Zinc–Silver Oxide Battery Separator

### Robert E. Post

The importance of cellulose as a battery separator material has best been indicated by the statement (R107), "The recognition of the utility of regenerated cellulose film as a separator was a forward step which advanced the silver-zinc system to the status of a practical secondary battery." However, in spite of the desirable features of commercial cellulose films, these films do exhibit severe limitations when used in batteries. In particular, cellulose films suffer significant degradation as a result of reaction with battery constituents. This degradation limits battery life and performance.

It is the purpose of this paper to identify and describe the properties of cellulose that relate to separator requirements, then, based on this information, to suggest modifications to cellulose films that should lead to improved separator performance.

### REQUIREMENTS OF SEPARATORS

The requirements for separators have been set forth in the Screening Methods handbook (R107) that includes data on properties that are representative of films presently in use. In this treatment only barrier properties as opposed to absorption capability will be considered. As a barrier, the separator should prevent the migration of soluble zinc and silver species between the electrodes. In particular, the growth of metallic zinc dendrites from the surface of the zinc electrode should be prevented.

However, the transport of water and ionic components of the electrolyte (KOH) must not be restricted to the point where the electrical resistance is so high as to overcome the energy density advantage of the silver-zinc system.

The separator performs in an alkaline environment in which it must maintain dimensional stability and structural integrity for extended periods of time. It is also desirable that it be able to withstand the temperature required for sterilization while immersed in the electrolyte.

Both bivalent silver and oxygen will be present in the battery, which gives rise to the requirement of stability towards oxidation. However, it is preferred that the separator suffer some attack by bivalent silver rather than permitting it to reach the zinc electrode.

## STRUCTURE OF CELLULOSE

This section is for the most part a battery-oriented summary of material from standard reference works (R411-412) except where other specific references are noted.

Cellulose is a natural high polymer. In useful forms of regenerated films and fibers, the mean degree of polymerization (DP) is at least 200 and in natural fibers, such as cotton, the DP runs into the thousands. The crystallizing tendency can be controlled so as to obtain manufactured products having a wide rage of distributions of order ("pore size").

The crystal lattice is shown in Figure 97 where some idea of the nature of the molecular geometry and the arrangement of the bonds can be obtained. The tendency to crystallize arises from the chemical nature and the unique geometry of the polymer chain. The monomer unit of cellulose is glucose, which in the polymerized state contains three reactive hydroxyl groups. Hydroxyl groups of neighboring chains fall into just the right position to form hydrogen bonds in

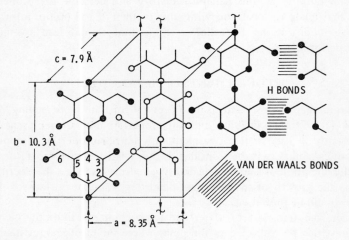

Figure 97 Schematic crystal lattice of native cellulose. ⌒, Line junctions denote carbon. Ring at lower left shows conventional numbering system. ○, Circles denote oxygen. Oxygen atoms in and between rings are part of the acetal configuration. Pendant oxygen atoms belong to hydroxyl groups. Hydrogen, not shown, completes all omitted valencies.

one plane, while their fit in the other plane is conducive to a maximum of van der Waals bonding strength. However, hydrogen bonds are only one-tenth as strong as the chemical bonds in the polymer, and van der Waals bonds are only one-tenth as strong as the hydrogen bonds, so that the forces that hold the structure together arise from the reinforcing effect of large numbers of these lateral bonds. Conversely, the structure is weakened by agents that specifically weaken these bonds, and in the case of cellulose the list of weakening agents for hydrogen bonds includes water, alkali, and various specific complexing entities including soluble zinc species.

Crystallinity is a term that can be loosely applied to describe the extent to which the structure comprises bonds that mutually reinforce each other against attack. However, with polymers, unlike inorganic crystals, no macroscopic single crystal exists, all structures being to some extent disordered. Various concepts of disorder in cellulose are depicted in Figure 98, where the extent of disorder ranges from minor deviations in the crystal regularity (lateral disorder) to disconnected perambulations (network).

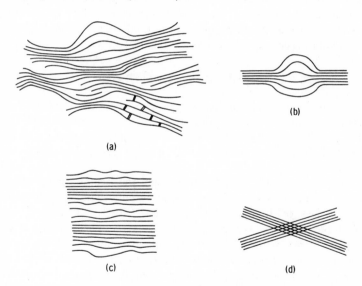

(a)

(b)

(c)

(d)

Figure 98 Concepts related to distribution of crystallinity. (a) Network of crystalline regions connected by polymer chains in varying degrees of order, showing restrictive effect of cross-links ( // ) in lower portion as compared to unrestricted swelling in upper portion. (b) Longitudinal disorder, a limitation on the DP of crystalline regions. (c) Lateral disorder, a limitation on the number of chains that can form a stable aggregate. (d) Crossover disorder, well-developed but unaligned crystalline regions—typical of cotton—also a limitation on orientation in regenerated cellulose.

The degree of order determines the amount of swelling in the presence of alkali. Because of the specific interaction of hydroxyl ion and the hydroxyl groups of cellulose, alkali aids in the penetration of water through a complex process involving the interplay of forces of chemisorption and osmosis (R412F). As a consequence, the amount of swelling in alkali depends on the degree of crystallinity. In well-developed crystals, the structure parts only far enough to permit penetration of the alkali species and water. The crystal lattice is maintained with enlarged spacing between the planes formed by van der Waals bonds. The enlargement comes about as the hydrogen bonds are spread apart by the intervention of water and alkali. For example, sodium hydroxide causes an increase of 4.0 Å in the $a$ dimension of Figure 97 (R412B). The spacing of the van der Waals bonds changes very little, if at all.

In amorphous regions, structure is lost completely, and the polymer behaves as though dissolved. The concept of pore size gives way to one of a contribution to the viscosity of a solution (R596).

A high degree of crystallinity also contributes to stability against chemical attack because of decreased accessibility (R418A). Thus crystallinity is a desirable factor in that it contributes to structural integrity, size selectivity with respect to migrating species, and chemical stability.

## CHEMISTRY OF CELLULOSE

The subject of chemical stability is a complex one. For example, degradation processes are sensitized by certain functional groups on the cellulose. Oxidation can either produce or remove some of these sensitizing groups. In the following paragraphs the degradation reactions will be considered and preventive measures will be suggested.

The cellulose polymer is linked through the acetal configuration. This structure is subject to alkaline hydrolysis at elevated temperatures. This attack involves cleavage of terminal monomer units that still possess unreacted aldehyde groups. Strong oxidizing agents (oxygen, bivalent silver, bleaching agents applied during manufacture) are able to oxidize hydroxyl groups to aldehyde or ketone groups. Whenever this occurs, the chain cleaves readily in alkali even at room temperature. Also the aldehyde groups seem to initiate the process of oxygen oxidation (R418B).

Alkaline hydrolysis can be greatly inhibited by partial oxidation (mild oxidizing agents) of terminal aldehyde groups to carboxyl groups, by the reduction of these aldehyde groups to alcohol groups or by the rearrangement involving the aldehyde.

It is of interest to note that metallic silver results from oxidation of aldehyde by monovalent silver. Not only does this silver serve as an inhibitor of the reaction with oxygen (R418B) but it will compete with hydroxyl groups as a reducing agent for bivalent silver with the production of more monovalent silver. Therefore a judicous pretreatment of the separator with monovalent silver has logical merit (R421).

A means of preserving structural integrity as well as reducing the effects of chemical degradation is chemical crosslinking. This process can be carried out with a variety of acid catalysts using formaldehyde as the cross link. This reaction ties the chain together with the same (acetal) type of bond that forms the natural polymer chain. This bond is highly resistant to the action of alkali and complexing agents (R411C). Crosslinking therefore tends to preserve the order that existed beforehand. On the one hand, swelling of noncrystalline regions can be suppressed (R412D), and, on the other, spontaneous recrystallization of amorphous material can be prevented (R621).

In any disruptive or degradation process, a high DP will prolong the time to the onset of failure. In a sense, crosslinking increases the DP by joining molecular chains of low DP.

## PROPERTIES OF CELLOPHANE AND SAUSAGE CASING

A minimum of crystallinity is desired in commercial cellophane because crystallinity detracts from toughness and flexibility, which are the desirable qualities (R411D). A tight barrier type of structure is not necessary, since specific barrier properties such as moisture resistance are commonly applied as a coating. Also, with sausage casing a barrier to the penetration of smoke and water vapor would not be wanted, so that a minimum of crystallinity is again desirable. Hence, cellophane and sausage casing are subject to much swelling. This fact and the presence of submicroscopic cavities and spongy material (R315) within the structure preclude outstanding barrier qualities. However, the electrical resistance is also relatively low.

The cellophane industry is a highly competitive tonnage industry where cost factors weigh temendously. The manufacturing begins with a relatively degraded pulp because of a number of economies that are gained. The product does not have a high DP, and the number of available aldehyde chain ends is large. Sausage casing manufacturers generally use a higher grade of pulp and obtain a film of somewhat better quality, in particular regarding wet strength. Both processes employ a bleaching stage, which can generate additional aldehyde groups.

Because of a structure of minimum crystallinity that even for sausage casing is somewhat degraded in chain length to begin with, commercial cellulosic films do not hold up well under further degradation. A striking example of the extremes is seen in the markedly superior durability under adverse conditions of the natural fiber in fibrous sausage casing (R107). The presence of aldehyde groups (from bleaching) will make the film susceptible to immediate alkaline chain cleavage when exposed to the battery environment and to hot alkaline hydrolysis if sterilization is attempted.


## PROSPECTS FOR BETTER MATERIALS

### THE CELLOPHANE CUTICLE

The description given thus far of cellophane applies to most of the film. A possibly highly significant item that makes up only about 0.1 $\mu$ thickness of the film is the cuticle layer found on viscose cellophane (R316). This layer is compact and apparently highly crystalline, because its permeability is very much less than that of the rest (R633). It may, in fact, contribute a substantial part of the barrier effect. Imperfections exist in this layer (R315) which would be expected to constitute preferred paths of penetration. It may be possible, using this cuticle as a substrate, to regenerate cellulose upon it from solution so as to build up a thicker initial layer.

Encouraging prospects exist for the manufacture of a cellulosic film especially adapted for battery separator applications. Manufacturing conditions, which for present films are aimed at minimizing crystallinity, could be modified to emphasize it. In particular, those conditions that lead to cuticle formation deserve particular attention. The existing cuticle might be augmented by employing it as a substrate, as was mentioned earlier. Higher electrical resistances should be compensable by decreased total thickness. Particularly for films of decreased thickness, the elimination of imperfections through quality control is especially important. A high DP would be in order, and efforts should be made to obtain an optimum balance of reducing property so as to stop silver migration with a minimal catalytic activity with respect to oxidation and hot alkaline degradation. Crosslinking might further increase the structural integrity and barrier properties of such a film or of ordinary cellophane for that matter.

### HIGH WET MODULUS REGENERATED CELLULOSE

The rayon industry has in recent years seen the upturn of a market for rayons that simulate cotton in their properties. No film of this type has been reported. However, inspection of the fiber characteristics indicates compatibility with separator requirements. These high wet modulus (HWM) rayons (R413) have a high degree of crystallinity, which tends to suppress the reduction of elastic modulus when wet. The manufacturing process emphasizes high DP, and the best obtainable pulps are used. Resistance to alkali is remarkable, especially of crosslinked varieties. Solubility in 6.5 NaOH is as low as 1% compared with 20% for regular rayon, and the amount of water absorbed is reduced to 60% from 100%. A HWM film should possess superior barrier properties compared with ordinary cellophane. Because of the use of highest quality pulp and minimum degradation in processing, chemical resistance should be improved. The higher crystallinity and higher DP should permit an HWM film to take far more punishment than can cellophane before reaching the point of failure associated with short chain lengths. The only possible drawback that appears at this time might occur in battery fabrication. The HWM process produces a material that is not only highly crystalline but also highly oriented. A film of the material would be more brittle than cellophane is when dry and highly directional in its mechanical properties. However, though such a film would be more brittle than cellophane, it is unlikely that it would remain brittle in the alkaline electrolyte.

It will be observed that HWM technology embodies those factors suggested earlier for adapting existing films to separator requirements, except that stretching is fundamental in the production of HWM and no cuticle layer has been reported.

## CONCLUSIONS

The essential weaknesses of cellophane as a battery separator result from the need to meet the requirements of the packaging market which are rather far removed from those of a battery separator. However, as a barrier film, regenerated cellulose can be improved by introduction of highest quality pulp, emphasis on high DP and crystallinity in manufacturing, crosslinking, and chemical pretreatment to reduce aldehyde content.

In particular, the technology of high wet modulus rayon bears investigation. Another primary area is in the exploitation of the tendency for a compact cuticle to be formed on viscose cellophane. This might be accomplished by emphasizing those manufacturing conditions which cause it to occur. Another possibility is to use the cuticle as a substrate and build it up to increase thickness by regenerating cellulose upon it from solution.

# 19.

## Grafted Membranes

### Vincent D'Agostino, Joseph Lee and Guyla Orban

The word *membrane* is used herein interchangeably with the term *separator*, which is the more commonly used term in the battery industry. The separators or membranes discussed in this paper were designed primarily for use in secondary zinc-silver cells and are semipermeable. The dry pore size of these membranes is about 25-40 Å.

In preparing these membranes, a resin such as polyethylene is selected, extruded as a one mil film.

The term *grafting* refers to a chemical or radiation process whereby the base polymer, in this case polyethylene, is modified by adding onto the long polyethylene molecules side chains which have a chemical characteristic different from that of the polyethylene. In this way a polyethylene which has certain desired characteristics such as oxidation resistance, but which is not electrolytically conductive, may be modified, by grafting, to yield a membrane which has a low resistance in 40% KOH. The grafting step is initiated by either atomic radiation (i.e., gamma ray) or by chemical initiators. This is shown schematically for radiation grafting in Figure 99.

Figure 99 shows that a highly energetic gamma ray will cleave a carbon-hydrogen bond on polyethylene to give a free radical (i.e., a chemical species with an odd electron). This free radical state is unstable, and three principal reactions can take place. If no other chemicals are present during irradiation, the radicals on the polyethylene undergo simultaneously crosslinking and degradation. *Crosslinking* changes the polyethylene by modifying the chains to give a three-dimensional tight network structure. It is a desired reaction. *Degradation* is not desirable because it leads to a decrease in the polyethylene chain length or molecular weight of the polyethylene. Both reactions occur simultaneously during radiation. *The ratio of the extent of crosslinking* to scission is what is measured macroscopically. Fortunately, with polyethylene the rate of crosslinking is much greater than scission thereby yielding a crosslinked three-dimensional structure.

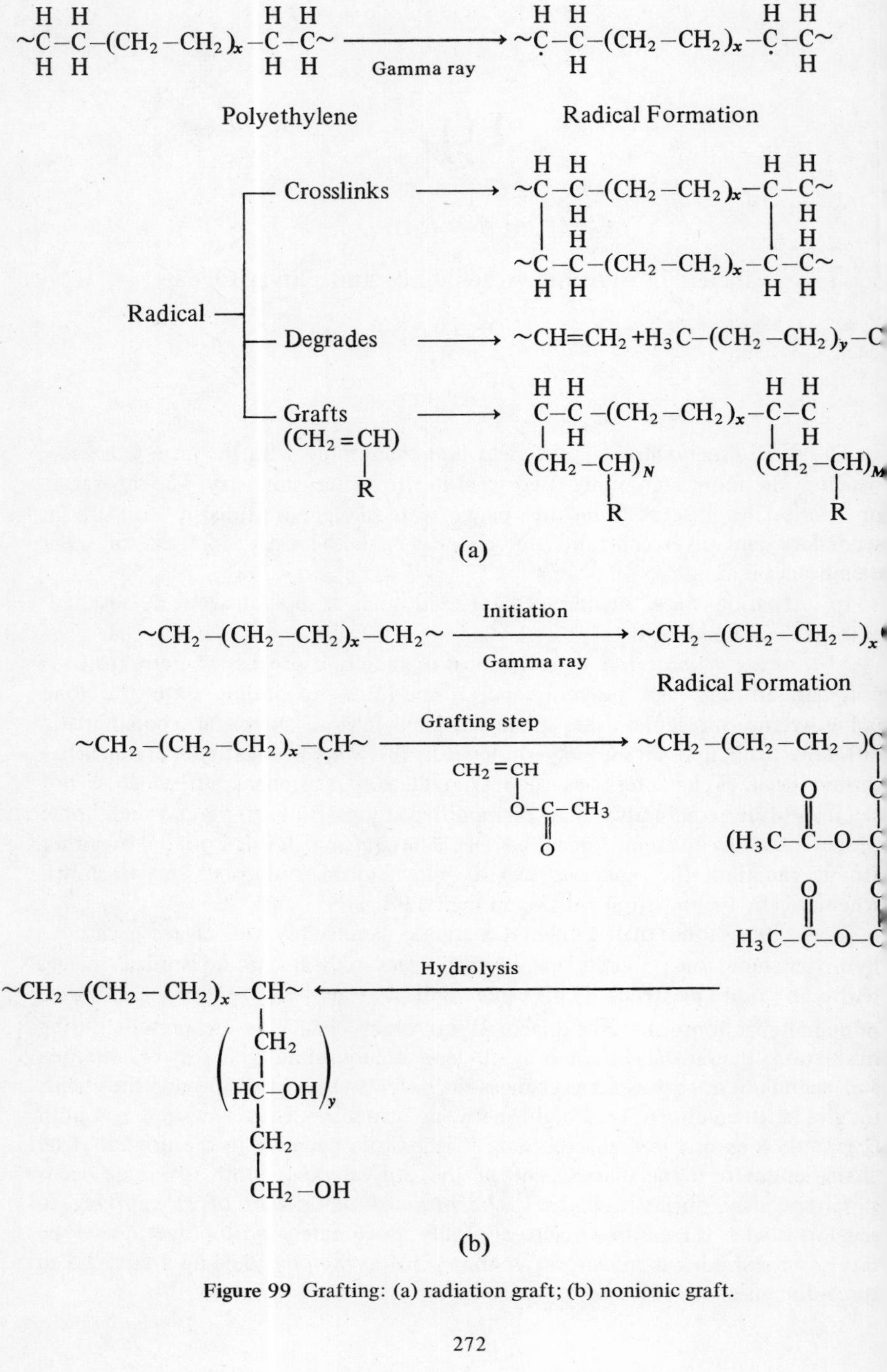

$$\underset{\text{H H}}{\overset{\text{H H}}{\sim C-C}}-(CH_2-CH_2)_x-\underset{\text{H H}}{\overset{\text{H H}}{C-C}}\sim \xrightarrow[\text{Gamma ray}]{} \underset{\text{H}}{\overset{\text{H H}}{\sim \overset{\cdot}{C}-C}}-(CH_2-CH_2)_x-\underset{\text{H}}{\overset{\text{H H}}{\overset{\cdot}{C}-C}}\sim$$

Polyethylene        Radical Formation

Radical

— Crosslinks ⟶ 
$$\sim C-C-(CH_2-CH_2)_x-C-C\sim$$
$$\sim C-C-(CH_2-CH_2)_x-C-C\sim$$

— Degrades ⟶ $\sim CH=CH_2 + H_3C-(CH_2-CH_2)_y-C$

— Grafts ⟶
$(CH_2=CH)$
$|$
$R$

$$C-C-(CH_2-CH_2)_x-C-C$$
$$(CH_2-CH)_N \qquad (CH_2-CH)_M$$
$$R \qquad\qquad R$$

(a)

$$\sim CH_2-(CH_2-CH_2)_x-CH_2\sim \xrightarrow[\text{Gamma ray}]{\text{Initiation}} \sim CH_2-(CH_2-CH_2-)_x$$

Radical Formation

$$\sim CH_2-(CH_2-CH_2)_x-\overset{\cdot}{C}H\sim \xrightarrow[\underset{\underset{O}{O-C-CH_3}}{CH_2=CH}]{\text{Grafting step}} \sim CH_2-(CH_2-CH_2-)C$$

$$(H_3C-\overset{O}{\overset{\|}{C}}-O-C$$

$$H_3C-\overset{O}{\overset{\|}{C}}-O-C$$

$$\sim CH_2-(CH_2-CH_2)_x-CH\sim \xleftarrow{\text{Hydrolysis}}$$
$$\left(\begin{array}{c} CH_2 \\ | \\ HC-OH \\ | \\ CH_2 \\ | \\ CH_2-OH \end{array}\right)_y$$

(b)

**Figure 99** Grafting: (a) radiation graft; (b) nonionic graft.

272

When a reactive monomer is present during irradiation as shown in Figure 99a, it may react with the free radicals generated giving a *grafted* side chain. This reactive monomer can be either a weak or a strong acid, a weak or a strong base or a nonionic monomer. Examples of these are acrylic acid as a weak acid, sulfonic acid as a strong acid and amines and quaternized amines as weak and strong basic exchangers. Nonionic grafts, currently under development, are made by grafting vinylacetate and subsequently hydrolyzing or by grafting a vinylpyrrolidone monomer. Schematically this is shown for vinylacetate in Figure 99b.

In Figure 99b the polyethylene is irradiated to give a radical formation that reacts with the vinylacetate monomer to give a grafted polyethylene membrane. This nonionic ester is then hydrolyzed chemically to produce a grafted polyalcohol chain. Normally polyvinylalcohol is water soluble, but, when it is grafted to an insoluble polyethylene backbone, it becomes insoluble. The resistance of some membranes made with (vinylalcohol) grafts are extremely low, being of the order of 10-15 milliohm in.$^2$ (65-100 mohm cm$^{-2}$) in 40% KOH.

To develop membranes which will have extended cycle life in secondary Zn-Ag systems, it is of primary importance to have a membrane which is low in resistance, is resistant to oxidation, preferably is thermally stable, and prevents zinc diffusion. Based on previous operating experience, three basic and critical parameters important in formulating secondary battery membranes are the following: (1) the base resin used in making the polyethylene film; (2) the extent to which the base resin, in film form, is crosslinked; and (3) the type of monomer grafted to the crosslinked film.

The order of performing the two operations is important. Although it is easier to graft first and then to crosslink, test results indicate that a preferred membrane is made by crosslinking first, then grafting. Initially, the effect of the base film on cycle life shall be considered. The molecular properties of the base resin which are of importance in making membranes include the crystallinity, the molecular weight distribution, and the absence of low molecular weight fractions. The distribution is actually the ratio of the weight average molecular weight of the resin to the number average molecular weight of the resin. In this study three different base films, all low density polyethylenes, were selected after screening numerous resins from various manufacturers. These films were made from resins which differed in melt index. The molecular weights of each resin was determined by gel permeation chromatography (gpc). The results of these determinations are given in Table 35.

The important numbers in Table 35 are given in the last column. These figures indicate that the molecular weight distribution of the polymer is very narrow for the B-6 resin and very wide for P-1 resin. The gpc curves also indicate a large low molecular weight tail for the P-1 resin.

TABLE 35.   PROPERTIES OF POLYETHYLENES USED TO
PREPARE BASE FILMS

| Base Film | Density | Melt Index | $M_n$ No. Average[a] | $M_w$ Wt. Average[a] | $M_w/M_n$ |
|-----------|---------|------------|----------------------|----------------------|-----------|
| P-1 | 0.917 | 1.20 | 29,261 | 223,000 | 7.6 |
| U-2 | 0.922 | 1.40 | 22,362 | 58,300 | 2.6 |
| B-6 | 0.922 | 0.75 | 17,130 | 34,400 | 2.0 |

[a]Determined by gel permeation chromatography.

These three base films were next crosslinked to various extents using electron bombardment of 30, 50, 70, and 90 megarads of energy. As the amount of crosslinking energy increases, the extent of crosslinking increases. The crosslinked films were then grafted under identical conditions. These membranes were cycle life tested using 3-plate, 1.1 Ah cells at a 40% depth of discharge. One layer of membrane was used to separate the element consisting of 2 silver and 1 zinc plates. The 2-hour cycle regime comprised a charge of 350 mA for 85 minutes and a discharge of 800 mA for 35 minutes. The charging cutoff voltage was set at 2.00 V and the discharge was terminated at 1.3 V. The results of the cycle life testing are given in Figure 100.

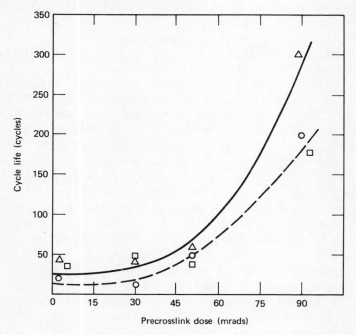

Figure 100  Effect of base film and precrosslink dose on the cycle life.

Figure 100 illustrates two very important points. Plotting cycle life vs. crosslinking dose for each of the three different low density polyethylenes shows that (1) as the crosslinking dose is increased the cycle life of the grafted film increases for all of the polyethylenes and that (2) the cycle life at a high crosslinking dose appears to be related to the base resin properties.

The reason for the improvement in cycle life at high crosslinking dose has now been shown to be dependent on the efficiency with which each of the films is crosslinked. This leads to the second important parameter in preparing grafted membranes, the extent of crosslinking of the base film.

The molecular weight distribution of the three films differed. The B-6 film had a distribution of 2.0, the U-2 film was 2.6, and the P-1 film was 7.6. Previous researchers have shown that for polyethylene (above a certain molecular weight) the narrower the molecular weight distribution, the more efficient is the crosslinking of the film per unit energy expended. This behavior has been substantiated by the gel-dose curve (Figure 101) developed in this study (R377).

When a polyethylene film is crosslinked and extracted with hot xylene, that fraction which is three-dimensional in character is insoluble. This insoluble

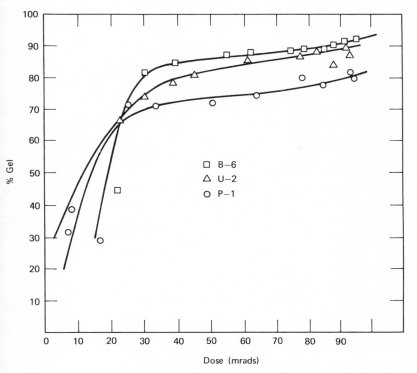

**Figure 101** Gel fractions of irradiated polythylene films versus radiation dose.

fraction is expressed in Figure 101 as the % gel. The more efficiently a film is crosslinked the greater will be the % gel per unit radiation dose. These curves indicate that the efficiency of crosslinking the B-6 film is greater than that of the U-2 which in turn is greater than that of the P-1 film.

Also it should be noted that above 30 Mrads the change in % gel with dose is an insensitive measure of the extent of crosslinking. This is true because a three-dimensional structure having either one or 100 crosslinks will be insoluble, that is, the gel fraction does not distinguish between a lightly crosslinked film and a highly crosslinked film. A more sensitive parameter for characterizing the tightness of the crosslinks in an irradiated film is given by the $M_c$ value, the molecular weight between crosslinks. The significance of this is illustrated in Figure 102. This validates the importance of the numbers in the last column in Table 35.

| % Gel | Polyethylene | Number of Crosslinks | Solubility | $M_c$ |
|---|---|---|---|---|
| 0 | | 0 | Soluble | — |
| 100 | | 2 | Insoluble | $X$ |
| 100 | | 10 | Insoluble | $0.1X$ |

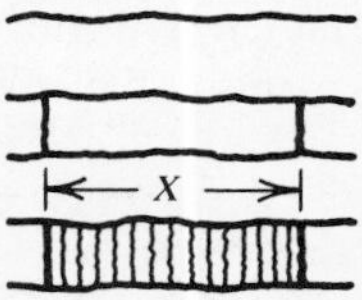

Figure 102 Comparisons of gel fraction and $M_c$ values. The gel fraction of a structure with two crosslinks is 100%, and this gives an $M_c$ value of $X$. Increasing the radiation dosage would increase the number of crosslinks. The gel fraction would not reflect this increase; the $M_c$ value does.

It should be noted that in Figure 102 the gel fraction of a structure with 2 crosslinks is 100% and this gives an $M_c$ value of $X$. An increase in radiation dose would increase the number of crosslinks. The gel fraction would not reflect this increase; the Mc value does.

The importance of the $M_c$ value is that it indicates on a semiquantitative basis the tightness of the structure and it is also a measure of the decrease in pore size with crosslinking. Two differences noted between crosslinked and uncrosslinked films are: (1) highly crosslinked grafted films swell much less than do identically grafted films which are not crosslinked; and, (2) highly crosslinked films are more difficult to graft.

Certain monomers graft crosslinked films or uncrosslinked films, or both, more readily than others. Table 36 gives the $M_c$ values determined for the three films evaluated in this program.

Table 36 indicates that, although the percentage of increase in the gel content is only 2% greater for the B-6 than for the U-2 film, the crosslink density of this film is just about twice that of the U-2 film, that is, the $M_c$ value or distance

TABLE 36.   $M_c$ VALUES FOR IRRADIATED BASE POLYMERS

| Base Film | Crosslink in Mrad | % Gel | $M_c \times 10^3$ | $M_c/1.77$ | $M_w/M_n$ |
|-----------|-------------------|-------|-------------------|------------|-----------|
| P-1 | 90 | 75 | 6.22 | 3.5 | 7.6 |
| U-2 | 90 | 89 | 3.31 | 1.9 | 2.6 |
| B-6 | 90 | 91 | 1.77 | 1.0 | 2.0 |

between crosslinks is ½. This tightness of structure is probably related to a decrease in zinc diffusion through the membrane and the noted increase cycle life for membranes made with this polymer.

The third important characteristic to be discussed relates to the type of monomer grafted to the crosslinked resins. Two monomers, both weak acids, have been thoroughly evaluated to date. The results indicate that the methacrylic acid graft gives better cycle life than the acrylic acid graft, and this poses the question of the effect of acid strength of the grafted chain on cycle life. The mode of failure caused by going from a cellulosics to a crosslinked ion-exchange membrane seems to lead to a greater tendency for capacity loss. This may be caused by the higher resistance of the ion exchange membranes over cellulosics or by the ionic character of the membrane. This point is not established firmly and needs further research work. Non-ionic grafted membranes with resistances of 10-15 mohm in.$^2$ (65-100 mohm cm$^2$) are being prepared currently on a laboratory scale; membranes with resistances of 30-35 mohm in.$^2$ (195-225 mohm cm$^2$) have been prepared recently by grafting methacrylic acid on

TABLE 37.   CYCLE LIFE vs. MONOMER TYPE

| Sample | Monomer Type | Cycle Life at 40% Depth | |
|--------|--------------|--------------|--------------|
| | | at 25° | at 50° |
| 101A | AA | 284 | |
| 101B | AA | 236 | |
| 104A | MA | 372 | |
| 104B | MA | 386 | |
| Control-Visking[a] | | 90 | |
| Control-Visking | | 60 | |
| 50A | MA | | 310 |
| 50B | MA | | 474 |
| 49A | AA | | 232 |
| 49B | AA | | 126 |
| Control-Visking | | | 50 |
| Control-Visking | | | 104 |

[a]Fiber-strengthened sausage casing.

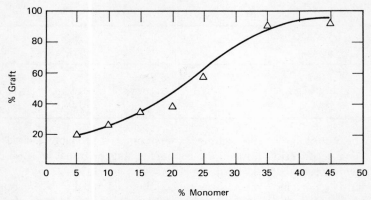

**Figure 103** Effect of monomer concentration on percent graft. Dose rate, 9214 r/hr; dose, 1.55 Mrads; base film, B-6.

90 Mrad crosslinked film. Evaluation of these membranes should give some insight into the effect of monomer type on capacity loss and of resistance on capacity loss.

Table 37 compares the effect of monomer type on cycle life. The base film used in preparing the membranes shown in this table was made from the B-6 resin. The crosslinking energy level was 90 Mrad. Cells were cycled at 25° and 50°. The results indicate that cycle life to a shorting failure mode is not any lower at the higher temperature. Additional testing is needed to confirm this. The data also indicate that methacrylic acid grafted membranes gave better cycle

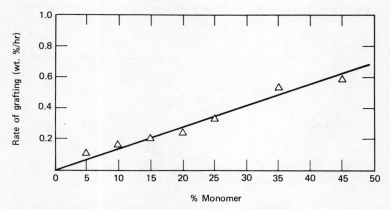

**Figure 104** Relationship between rate of grafting and monomer concentration. Crosslink dose, 90 Mrads; base film, B-6; monomer, methacrylic acid.

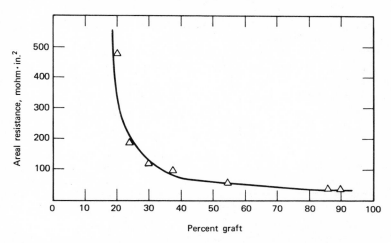

**Figure 105** Effect of percent graft on resistance in 40% KOH. Base film, B-6; crosslink dose, 90 Mrads; monomer, methacrylic acid.

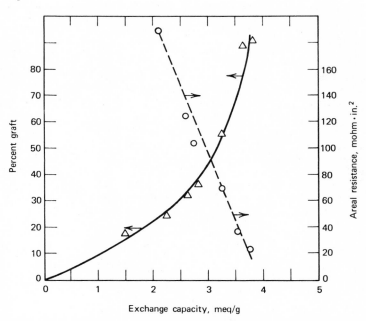

**Figure 106** Exchange capacity as a function of percent graft and areal resistance.

life than acrylic acid grafted membranes. The control cells used were made with one layer of fiber-strengthened sausage casing.

Figures 103 and 104 give the result of studies on grafting procedures. Figure 103 shows the effect of monomer concentration on % graft at a constant radiation dose and dose rate when B-6 film, crosslinked at 90 Mrad, is grafted with methacrylic acid. Under these conditions the % graft increases up to 90-95% with increase in monomer concentration and then levels off. This is

| Polymer Film as Received | Lay-Up Film for Crosslinking | Crosslinking | Prepare Film for Grafting |
|---|---|---|---|
| Control | Control | Control | Control |
| 1. Density 2. Gpc ($M_w$/$M_n$) 3. Thickness 4. IR | 1. Nitrogen 2. No. of films in pack | 1. Dosimetry on run 2. Gel, $M_c$ | 1. Weight of roll 2. Diameter and tension of roll |
| (1) | (2) | (3) | (4) |
| Precondition | Radiation | Wash to Remove Homopolymer | Wash to Convert Acid to Salt |
| Control | Control | Control | Control |
| 1. Swelling time 2. Solution | 1. Dosimetry of $^{60}$Co ensures dose, dose rate 2. Temperature | 1. Temperature 2. Time | 1. Temperature 2. Time |
| (5) | (6) | (7) | (8) |
| Wash to Remove Excess KOH | Dry Film | Testing | |
| Control | Control | Control | Control |
| | 1. Moisture content | 1. Resistance 2. IR spectra 3. Swelling 4. Tensile 5. Hull test 6. Exchange value | |
| (9) | (10) | (11) | (12) |

**Figure 107** Membrane processing and controls.

shown more vividly in Figure 104, which is a plot of the rate of grafting as a function of monomer concentration.

With the grafting system used, the rate of grafting increases with monomer concentration to 35% monomer concentration. Figure 105 gives the resistance of the grafted films as a function of percentage graft.

At low % graft the resistance drops rapidly and then at above 30% graft the rate of decrease in resistance with a change in % graft is very much decreased. It is believed that the reason for this change in rate is the inability to convert the acid completely to the salt form at the higher graft levels. This is also shown by plotting % graft and resistance as a function of exchange capacity, as in Figure 106.

The plot of % graft as a function of exchange capacity yields a curve approaching an asymptotic value (about 4 meq/g) with increased % graft. The resistance falls linearly for any one set of grafted samples. These data support the idea that the resistance decreases as a function of the graft in the salt form and not simply as a function of the % graft.

Therefore, lower resistance values might be expected with complete ionization of the graft or with nonionic grafts. This is currently being evaluated. This may also be a clue to why ionic membranes cycled at higher temperature perform better.

Based on the results obtained in this investigation 5000 feet (1500 m) of a new membrane was made for evaluation in 25 Ah batteries. This membrane was delivered to Delco-Remy and is currently under evaluation. Figure 107 indicates schematically the process and control steps which were used in preparing this film.

# 20.

## Heat Sterilizable Separators

### Ralph Lutwack

This paper presents the results of research carried out at the Jet Propulsion Laboratory, California Institute of Technology, under contract NAS7-100 sponsored by the National Aeronautics and Space Administration. Its purpose was to develop heat sterilizable batteries capable of meeting the requirement that capsules landing on other planets not contaminate their surfaces. This requirement placed a severe environmental condition on the battery and consequently on the separator system.

The major efforts of the NASA-JPL program, begun in 1962, have been directed to study, develop, fabricate, and test Zn-Ag oxide batteries and to synthesize and test new separator and case materials. During the course of this program, the heat sterilization program, which all capsule components must be capable of surviving with subsequent satisfactory performance, has been changed twice. The original specification required three 36-hour cycles at 125°; the temperature was changed to 145°; the testing condition is now 135° for 72 hours.

From observations made in the first investigations of Ag-Zn batteries (R55) in this program it was concluded that the imposition of the heat sterilization condition on the Ag-Zn battery caused gross material failures and changes in the electrical performance. Considering only the results related to the separator, these conclusions regarding the effects of sterilization were reached: fibrous sausage casing (FSC); Dynel 470, a nonwoven fabric made from acrylonitrile monomer; Permion 600, a modified cellophane; and Teflon grafted with sulfonated styrene became too fragile or too brittle to handle; Teflon grafted with acrylic (AA) or methacrylic (MAA) acid resulted in capacity loss or short-circuiting; and high density polyethylene (PE), crosslinked and then grafted with AA or MAA, could be used as separator wraps. However when cells were built using the latter materials and then sterilized at 145° the cell capacity decreased by about 50%. There was some improvement when the sterilization of similar cells was done at 135°. It was shown in a cell component evaluation that portions of the separator were soluble in the KOH solution and may have affected plate performance. Thus, from these material evaluations it was concluded that the successful development of a heat sterilizable Ag-Zn cell depended in a crucial sense upon the development of a heat sterilizable separator material, that is, a material which is hydrolytically, oxidatively, and thermally stable in 40% aqueous KOH containing soluble silver and zinc ions during the sterilization cycle. As a consequence, a top priority was given to the separator development, and several different approaches have been used.

## DEVELOPMENT OF GRAFTED AND CROSSLINKED POLYETHYLENE AS SEPARATOR MATERIALS (R2-4, 540)

The first phase related to polyethylene-based materials because the early results indicated this type to be the best of the materials tested; it consisted of utilizing various methods of crosslinking to stabilize the film that was rendered hydrophilic by grafting with AA. Crosslinking was done with electron beam irradiation or with divinylbenzene (DVB) and $^{60}$Co irradiation. Grafting of AA was accomplished with $^{60}$Co irradiation. A total of 51 materials were fabricated and tested for dimensional stability, tensile strength, exchange capacity, resistivity, and performance in cells. The specimens were prepared in 51 distinct ways, combining variations in the density of the PE, in the levels of grafting and crosslinking, and in molecular weight distributions.

Density is an important factor because lower density corresponds to an increase in the content of amorphous material where crosslinking and grafting reactions occur rather than in crystalline regions. However, low density PE melts at lower temperatures so three density grades were investigated. The following preparation scheme was used:

(1.) Electron beam radiation crosslinking before grafting. Four levels of radiation were used on each of the three base polymers. Since radiation-crosslinked films have higher dimensional and thermal stability than have uncrosslinked films and increased degradation also takes place with increased exposure to radiation, the dependence of film properties on the level of crosslinking was studied with these samples.

2. Electron beam radiation crosslinking of the grafted copolymer. One level of irradiation was used on samples of each density grade equilibrated at three different levels of relative humidity. The effect of the extent of water swelling of the film was studied.

3. Crosslinking with DVB before grafting with AA. Three levels of DVB were used on each density grade. With these samples precrosslinking was done with DVB using lower radiation doses than for the electron beam irradiation, thus lessening the radiation-induced degradation.

4. Crosslinking with DVB after grafting with AA. Nine levels of DVB were

used on one density grade. With these samples postcrosslinking results were compared with those for precrosslinking.

5. Simultaneous crosslinking and grafting. Nine ratios of AA : DVB were used on one density grade. Here the efficacy of a single step for both crosslinking and grafting was studied.

6. Fractionated PE as the base material. One sample of each density grade was prepared using the best crosslinking procedure. The films were pressed from resins which were fractionated to remove low molecular weight fractions. The objective was to remove the lower molecular weight fraction and thereby reduce or prevent the leaching of soluble fragments by KOH electrolyte.

The most promising procedure for grafting and crosslinking, as established in the program, was used. Each material was tested for stability and properties before and after sterilization.

The properties of 7 materials exceeded the criteria used to characterize the films, namely, an electrical resistance in 40% KOH less than 150 ohm-cm (60 ohm-in.), a wet tensile strength greater than 50 kg $cm^{-2}$ (700 psi), and a maintenance of at least 90% of initial capacity after 5 cycles at 100% depth-of-discharge on testing in 3-plate cells. All seven materials were prepared from the lowest density PE, 0.917, with two being precrosslinked with DVB, two being postcrosslinked with DVB, and one being simultaneously grafted and crosslinked. The two materials selected for more extensive evaluation differed in precrosslinking treatment, one being given a radiation dose of 70 Mrads and the other treated with DVB, and both grafted with AA. Their properties were as follows:

| Material Crosslinking | Radiation | DVB |
|---|---|---|
| Properties | | |
| Thickness    mm | 0.053 | 0.041 |
| Tensile strength    kg $cm^{-2}$ | 70 | 60 |
| Tensile strength˙   psi | 1000 | 850 |
| Exchange capacity    meq $g^{-1}$ | 4.25 | 6.67 |
| Resistivity    ohm-cm | 50 | 38 |
| Maintenance of capacity   % | 96-99 | 97-100 |

These two materials and a material containing DVB have been used in all the experimental cell developments of the sterilizable battery program.

The following general conclusions were derived in this phase of the program: no detectable difference in degradative effects of sterilizations at 137 and 145°; no effect in sterilizing in the presence of the unformed Ag electrode; practically no change in the physical and electrical properties after the first sterilization cycle (of 3 cycles of 36 hours each); the seven samples that met requirements were based on the lowest density PE and high levels of crosslinking and grafting;

and the crosslinking technique had no apparent effect in the precrosslinked materials provided that a minimum level was exceeded.

The second phase had two general objectives, the development of a processing technique for producing uniform DVB crosslinked material and of techniques for preparing better materials. To achieve the first of these objectives the processing variables of irradiation dose rate and total dose, continuous and discontinuous irradiation, temperature during irradiation, and the presence or absence of oxygen were investigated using a fractional factorial experiment. The results of this experiment were the attaining of lower resistivities by using higher temperatures in the absence of oxygen, finding of minor effects of irradiation dose rate and total dose, and improved films by crosslinking after grafting. Subsequent studies showed that the amount of crosslinking with DVB after grafting increased with DVB concentration up to 16% by weight, the highest concentration used, and that the dimensional changes which took place during sterilization were dependent partly on the temperature of the KOH wash solution. For example, changing from 80° to the boiling point of the 5% wash solution resulted in increasing the length in contrast to decreasing it for the 80° wash. The basic process has been modified accordingly, and very uniform material has been produced.

During this second phase, preliminary experiments indicated that different methods for grafting and crosslinking lead to materials with good physical and electrical properties. For example, some inorganic salts prevent large amounts of homopolymerization; strong and uniform films can be prepared using a grafting solution of acrylic acid and N-vinyl pyrrolidone; uniform films with good electrical properties were obtained using acetone in place of carbon tetrachloride in the grafting solution.

The separator material being used in the cell development was characterized at JPL by physical and chemical means, the purpose being to attempt correlations between the properties of the separator and the electrochemical performance of a cell. The following information was obtained:

1. A crystallinity content of 50% in the base polymer was calculated from the density.

2. Using *IR* data it was concluded that methyl side chains are present in a concentration of 2.1 methyl groups per 100 backbone carbon atoms, a value of 1.7 being expected for 50% crystallinity. Longer side chains are not present.

3. It was concluded from *IR* data also that pendant vinylidine groups are the most common form of unsaturation in the polymer.

4. All of the DVB was reacted with the polyethylene since none could be removed by extraction with hot benzene. Some unreacted vinyl groups are present.

5. A very small change in crystallinity occurs as a result of crosslinking with

DVB, and it was concluded that all crosslinking took place in the amorphous phase.

6. It was concluded from the differences in solubility properties between $^{60}$Co-irradiated polyethylene and the DVB- $^{60}$Co-crosslinked material that crosslinking had taken place mainly by DVB crosslinks rather than by bond formation between backbone carbon atoms.

7. Since the absence of free acid groups was shown with *IR* data, the total content of grafted AA chains is 43%, which was determined from the K content determined by flame photometry and ash titration. Thus the carboxyl groups had been completely neutralized in the wash process.

8. A decrease in crystallinity from 50 to 43% occurs during the grafting reaction or the subsequent washing at 80°.

9. The material is hygroscopic, absorbing 13 to 14% by weight $H_2O$ at a relative humidity of 40 to 50%.

This information is being used to formulate quality control criteria and to serve as an aid in determining which properties can be changed to yield material with improved properties.

Examples of data which describe the properties and uniformity of the modified polyethylene material are given in Tables 38 and 39. The data shown in Table 38 are physical measurements made on four rolls of the material. The data in Table 39 show the resistivity values for a sample run. The areal resistances of the unsterilized and sterilized RAI-116 material (R193) as a function of the number of layers, up to 6, may be expressed by the linear equations, $R$ (ohm-cm$^2$) = 0.4x and $R$ = 0.133x, respectively, where x is the number of layers.

## DEVELOPMENT OF ETHYLENE/ACRYLIC ACID COPOLYMERS AS SEPARATOR MATERIALS (R533)

Films comprised essentially of ethylene/AA copolymers have been synthesized to simulate the radiation-grafted material. Since ethylene and AA do not copolymerize readily, a precursor copolymer of ethylene/methyl acrylate was prepared and subsequently hydrolyzed in aqueous KOH.

Several steps were involved in this fabrication: (1) ethylene and methyl acrylate were copolymerized using benzoyl peroxide as the initiator and chain transfer agent to control the molecular weight, the copolymer being characterized by density, infrared spectroscopy, and melt index measurements; (2) films were prepared by casting from toluene solutions onto glass surfaces and by pressing between heated plates in a hydraulic press; (3) the films were vulcanized by additional heat treatment in the hydraulic press, the vulcanizing agent being dicumyl peroxide; and (4) a two-step hydrolysis procedure was necessary in which a treatment in 2.5$N$ alcoholic KOH for 24 hours at 60-70° is followed by

TABLE 38.   ACCEPTANCE TEST RESULTS, SWRI-GX
SEPARATOR MATERIAL (R193)

| Lot No. | Spec- imen No. | Thickness | | Length Change +% (Roll) | Width Change +% | Thick- ness Change % | Weight | | Unit Electrolyte Absorption g/g |
|---|---|---|---|---|---|---|---|---|---|
| | | Dry | Wet | | | | Dry | Wet | |
| | | mils | | | | | mg/in.$^2$ | | |
| | 1 | 1.3 | 2.5 | 5.0 | 7.0 | 100 | 21 | 48 | 1.5 |
| GX | 2 | 1.3 | 2.5 | 5.0 | 6.0 | 100 | 21 | 47 | 1.5 |
| 85 | 3 | 1.4 | 2.5 | 5.0 | 7.0 | 79 | 21 | 47 | 1.5 |
| | 4 | 1.8 | 2.5 | 5.0 | 7.0 | 43 | 21 | 48 | 1.6 |
| | $\overline{X}$ | 1.4 | 2.5 | 5.0 | 6.8 | 81 | 21 | 48 | 1.5 |
| | 1 | 1.4 | 1.5 | 6.0 | 7.0 | 7 | 22 | 46 | 1.4 |
| GX | 2 | 1.4 | 2.0 | 4.0 | 6.0 | 43 | 22 | 45 | 1.2 |
| 86 | 3 | 1.8 | 2.3 | 5.0 | 7.0 | 29 | 23 | 47 | 1.3 |
| | 4 | 1.3 | 2.0 | 4.0 | 6.0 | 60 | 22 | 47 | 1.4 |
| | $\overline{X}$ | 1.5 | 1.9 | 4.8 | 6.5 | 35 | 22 | 46 | 1.3 |
| | 1 | 1.5 | 1.8 | 8.0 | 6.0 | 20 | 26 | 52 | 1.3 |
| GX | 2 | 1.5 | 2.0 | 7.0 | 6.0 | 33 | 24 | 49 | 1.3 |
| 87 | 3 | 1.5 | 2.0 | 6.0 | 6.0 | 33 | 23 | 47 | 1.3 |
| | 4 | 1.5 | 2.3 | 5.0 | 5.0 | 53 | 24 | 48 | 1.2 |
| | $\overline{X}$ | 1.5 | 2.0 | 6.5 | 5.8 | 35 | 24 | 49 | 1.3 |
| | 1 | 2.0 | 2.3 | 7.0 | 8.0 | 13 | 26 | 53 | 1.4 |
| GX | 2 | 2.0 | 2.3 | 6.0 | 7.0 | 13 | 25 | 53 | 1.4 |
| 88 | 3 | 2.0 | 2.3 | 6.0 | 6.0 | 13 | 24 | 52 | 1.4 |
| | 4 | 2.0 | 2.5 | 4.0 | 6.0 | 25 | 23 | 52 | 1.5 |
| | $\overline{X}$ | 2.0 | 2.3 | 4.8 | 7.0 | 16 | 24 | 53 | 1.4 |
| All | $\overline{\mu}$ | 1.6 | 2.2 | 5.5 | 6.4 | 42 | 23 | 49 | 1.4 |
| | $\rho$ | .12 | .30 | 1.2 | 0.7 | 30 | 1.7 | 2.7 | 0.12 |
| ± | Max | 2.0 | 3.1 | 9.1 | 8.5 | 132 | 28 | 57 | 1.8 |
| $3\rho$ | Min | 1.2 | 1.3 | 1.9 | 4.3 | 0 | 18 | 41 | 1.0 |

one in 2.5$N$ aqueous KOH under the same conditions. The characteristics of the product depend upon the degree of hydrolysis and the state of the film. The completely hydrolyzed form, which is the copolymer of ethylene and the potassium salt of acrylic acid, is hygroscopic. Conversion to the completely acidified form resulted in a tough film which is less sensitive to $H_2O$ and somewhat stiff. Surface-acidified films are ductile, pliable, not hygroscopic, and easiest to handle.

Studies were made in efforts to improve the properties of resistivity, tensile strength, and ease of handling. It was shown that the electrical resistances of the films vary inversely with the acrylic acid content. However, the limit of acrylic

TABLE 39. UNIFORMITY OF GX SEPARATOR MATERIAL[a]

| Grafting Solution Composition: | | Results—Sample 118 GX | | |
|---|---|---|---|---|
| 25 wt % Acrylic acid | | Resistance | | Resistance |
| 70 wt % Benzene | Footage | milliohm-in.$^2$ | Footage | milliohm-in.$^2$ |
| 5 wt % Carbon tetrachloride | 10 | 14 | 260 | 10 |
| | 20 | 18 | 270 | 8 |
| Crosslinking Solution Composition: | 28 | 13 | 280 | 12 |
| | 30 | 15 | 290 | 7 |
| 1.0 vol % Divinylbenzene | 40 | 13 | 300 | 4 |
| 1.0 vol % Benzene | 50 | 12 | 310 | 8 |
| 98.0 vol % Methanol | 60 | 12 | 320 | 5 |
| | 70 | 15 | 330 | 5 |
| Conditions for Grafting: | 80 | 9 | 340 | 7 |
| | 90 | 10 | 350 | 9 |
| Dose rate: 0.0012 Mrad/hr | 100 | 13 | 360 | 5 |
| Total dose: 0.815 Mrad | 110 | 6 | 370 | 6 |
| Temperature: 16° | 120 | 7 | 380 | 12 |
| Atmosphere: Nitrogen | 130 | 12 | 390 | 8 |
| Roll Length: 500 feet | 140 | 10 | 400 | 7 |
| | 160 | 12 | 410 | 11 |
| Conditions for Crosslinking: | 170 | 6 | 420 | 15 |
| | 180 | 9 | 430 | 7 |
| Dose rate: 0.022 Mrad/hr | 190 | 13 | 440 | 20 |
| Total dose: 0.550 Mrad | 200 | 6 | 450 | 14 |
| Temperature: 26° | 210 | 7 | 460 | 11 |
| Atmosphere: Nitrogen | 220 | 10 | 470 | 9 |
| | 230 | 9 | 480 | 17 |
| | 240 | 5 | 490 | 12 |
| | 250 | 11 | 500 | 20 |
| | | | 510 | 20 |

[a]Average of 51 samples taken at 10 ft. intervals = 0.010 ohm in.$^2$; standard deviation 0.004 ohm in.$^2$

acid content for usable films seems to be 50%, at which level the films become tacky and too difficult to handle. Consequently, it was necessary to compromise between the handling properties and resistance values, and the material selected as the most suitable was that containing 48% methyl acrylate (43.7% AA) which gives a product having a resistivity of 25 to 38 ohm-cm.

Two methods for modifying the basic copolymer were investigated. In one, the copolymers were blended with carbon black before the vulcanizing step. The resultant films are less tacky, are somewhat thicker, and have lower resistivity values than those of a similar composition without carbon. The incorporation of carbon black increases the ionic conductance of the films only. However, these films have slightly decreased tensile strength and elongation values in contrast to the polydiolefin polymers. Since heat sterilization does not appreciably affect these properties, these films will withstand normal handling during cell fabrication.

Polyblending, the mixing of copolymers with widely variable monomer ratios so as to achieve a higher acrylic acid content without making the corresponding copolymer, was the other modification used to improve the properties of the films. Attempts to hot press films by incorporating a homopolymer of ethyl acrylate by mixing, forming into a film, and vulcanizing led to films with unsatisfactory resistance values. However, films with lowered resistances were made by solution casting.

Polyblending of low acrylate and high acrylate content copolymers resulted in a product which is of lower resistance and is less tacky than a homogeneous film of similar average acrylate composition. When ethylene/methyl acrylate and ethylene/vinyl acetate copolymers were blended, low resistance, easily handled films resulted. Thus, it seems that films with satisfactory properties can be made using the polyblending techniques.

The evaluation of these films has begun at JPL. The film resistivities are lowered by the addition of 10% carbon black to 17 to 25 ohm-cm.

## DEVELOPMENT OF HETEROCYCLIC POLYMERS AS SEPARATOR MATERIALS (R594)

An investigation was conducted based upon the premise that the introduction of aliphatic moieties into heterocyclic polymers, which have excellent oxidative and thermal stability characteristics, would provide flexible films with the properties necessary for separator materials. Aliphatic polybenzimidazole (PBI), polybenzothiazole (PBT), polybenzoxazole (PBO), and polyquinoxaline (PQ) polymers were synthesized and tested for compatibility with 40% aqueous KOH under sterilization conditions; films were fabricated from the polymer with the best properties; various methods were used in attempts to improve the film characteristics; and the final product was evaluated.

The polymers which were synthesized, were oxidatively, hydrolytically, and thermally stable under heat sterilization conditions but films made from them had unacceptably high resistances. It was concluded that films with satisfactory resistance values could be prepared only by modifying the polymers or films or both. Only the PBIs were suitable for modification.

The investigations for lowering the resistivity included the incorporation of leachable additives in films and chemical modifications of the polymers to introduce hydrophilic groups. It was hypothesized that the films would be rendered more permeable to the KOH solution and consequently have a lower resistivity by the technique of leaching and that the inclusion of hydrophilic groups into the polymers would make possible the fabrication of films which are more easily wet by aqueous KOH solution.

The leaching of soluble additives dispersed throughout the film was not effective in reducing the resistivity. Apparently, without concomitant lessening of the hydrophobic characteristics of the polymer, the increase in porosity which resulted from the leaching did not increase the wettability of the films sufficiently.

Several chemical modifications, which introduced hydrophilic groups into the polymer, were then made. In these attempts, the sebacic acid based **PBI** was hydroxylated in the aliphatic chain, sulfonated in the aromatic ring, and *N*-hydroxyethylated and *N*-carboxyethylated in the imidazole ring. Only the *N*-carboxyethylated polymers could be formed into films with substantial improvements in the electrical resistivity.

To increase the effectiveness of this procedure, the density of the hydrophilic groups, the number of carboxyl moieties per equivalent weight of the repeating unit of the polymer was increased by using aliphatic acids with shorter chains. In this manner an aliphatic **PBI** based on suberic acid was synthesized, cyano-ethylated, cast into films, and converted to the carboxyl form by sterilization. The resultant films were not uniform in resistivity and tensile strength properties, and efforts to overcome this difficulty were intensified. Additional work revealed that several of the processing steps were characterized by critical operating ranges. The films that were made by the best available technique were not uniform and were not suitable for use as separator material. The average resistivity of PBI-6 film, 0.08 mm in thickness, was 136 ohm-cm before sterilization and 75 ohm-cm after sterilization.

## DEVELOPMENT OF  LIGAND-CONTAINING POLYMERS
## AS SEPARATOR MATERIALS (R465A, B)

Films prepared from ligand-containing vinyl polymers were investigated. The program consisted of four phases: (1) preparation of model ligand-containing polymers; (2) screening of these polymers for hydrolytic, oxidative, and thermal stability under sterilization conditions; (3) preparation of membranes using polymer units with high stability; and (4) evaluation of the best membranes. Of the 32 types of polymers tested in a screening program, copolymers of styrene/maleic anhydride, 2-vinylpyridine (2VP)/methyl methacrylate (MMA), and 2VP/methyl acrylate (MA) had the best stabilities in 40% aqueous KOH. In further evaluations of films of these polymers, in which there were used as criteria the results of measurements (R107) of resistance to oxidation dimensional stability, electrolyte absorption, resistivity, appearance, and tensile strength, it was concluded that the 2 VP/MMA films were more durable than those of 2 **VP**/MA; films of 2 VP/MA with the acrylate component above 65% were too soluble under sterilization conditions; and, the styrene/maleic anhydride copolymers are easily hydrolyzed at room temperature and have low electrical resistivity but the dry films are too brittle to be usable. The improvements in flexibility which were made by plasticizing the films with $H_2O$ and by using the styrene/maleic anhydride half ester or salt copolymers were only marginal. In view of these results, the 2 VP/MMA and styrene/maleic anhydride systems were selected for further development.

A conclusion of the first phase of the development of the 2 VP/MMA system was that an increase in the copolymer molecular weight was necessary to obtain more durable films. Subsequently, it was shown that the degree of polymerization increased at high reactant concentrations, decreased in the presence of chain terminators such as oxygen, and was relatively insensitive to catalyst concentration and temperature. Employing the derived experimental conditions, copolymers with intrinsic viscosities as high as 4.2 were prepared. Films prepared from the new high molecular weight copolymers had improved mechanical properties and the resultant films were much lower in electrical resistivity, those containing 67-69% methacrylate being under 25 ohm-cm after sterilization. A study of the film-forming characteristics of the high molecular weight copolymers was then made conmitantly with evaluations of the electrical and mechanical properties of these films.

Since the copolymer molecular weight affects two very important characteristics of the film preparation, namely, the polymer solubility and the time necessary to saponify the esters under sterilization conditions, the optimization of these two factors was studied. The conclusions were that polymers of moderate molecular weight, having a relative viscosity of 2 to 2.7 in 0.5% dimethylformamide solutions, had better solubility characteristics and were saponified at higher rates than polymers with very high molecular weights. Films prepared from the moderate molecular weight polymers had tensile strengths of 500 kg $cm^{-2}$ in the dry state and electrical resistivities of 25 ohm-cm after sterilization for 120 hours. However, saponification was not adequate under the same conditions for 64 hours, an interval representing one sterilization cycle, to yield low resistance films. To increase the rate of saponification it was necessary to pretreat the films; pretreatment by soaking in 1-octanol to attain a weight gain of at least 1% lowered the resistivity of a 2VP/MMA separator, 0.094 mm thick, to 7.1 ohm-cm.

In the styrene/maleic anhydride system several polymer modifications were prepared and evaluated. This study revealed the strong dependence of the properties of the copolymer on molecular weight. Low molecular weight specimens were hard and brittle in a dry atmosphere. Copolymers of intermediate molecular weights (intrinsic viscosity of 1.8) had better film-forming properties than those of the highest molecular weight (intrinsic viscosity of 4.3) because of higher solubility in the casting solution. Terpolymers of styrene/maleic anhydride/MMA gave films with good mechanical properties but unsatisfactory resistance values. Tetrapolymers that incorporated MA in the terpolymers were modified further by crosslinking to reduce the solubility in aqueous KOH. The films from the crosslinked product which had dry tensile strengths of 400 kg cm$^{-2}$ withstood sterilization and had resistivities of about 50 ohm-cm after one 64-hour cycle. However, the preparation of uniform film from the tetrapolymers was so difficult that the material produced for JPL evaluation was made from the 33:67 2VP/MMA copolymer.

## DEVELOPMENT OF COMPOSITE MEMBRANES AS SEPARATOR MATERIALS (R385)

Composite membranes, which consist of a porous matrix, a binder, and an inorganic substance having ion exchange properties, are being investigated. This program includes surveys of suitable materials, establishment of fabrication techniques, testing of the parameters that determine the conditions for the preparation of satisfactory films, and devising process controls for the achievement of film with uniform properties.

In the first phase several filler and matrix materials were identified as promising components, using as the criterion compatibility with 40% aqueous KOH at 135°; experimental composites were prepared by impregnation. These membranes were characterized by conductivity, diffusion rates of KOH and soluble Ag and Zn species, resistance to Zn dendrite growth, mechanical properties (including microscopic and electrographic inspection for pinholes and cracks), and thickness determinations. After extensive evaluations, a polypropylene felt was selected as the matrix material, polysulfone as the binder, and zirconium oxide as the inorganic filler.

The first films were nonuniform and had high ohmic resistances. Since the filler layer was uneven, ball milling was used to obtain more uniform mixtures. To increase porosity and wettability, the method of leaching soluble additives was also used but without measurable effect except to adversely affect filler distribution. After considerable investigation it was shown that the variables which had appreciable effects on the uniformity and electrical resistivity were the solvent-to-filler ratio, the extraction technique, the pressure applied to the film, the coating procedure, and the drying process. Each of these variables was examined as to its effect on the product, and optimum conditions were obtained for a procedure using a dip-coating tank and a drying tower. Films, 0.18 mm thick, were prepared having a satisfactory resistivity of 25 ohm-cm after sterilization and satisfactory hydroxyl ion diffusion. None of the material has been evaluated at JPL yet.

## STERILIZABLE Zn-Ag CELL TEST RESULTS (R193)

The separator materials processed from modified polyethylene have been used in all of the sterilizable cell developments. In the earlier work both the electron-beam crosslinked and the DVB-$^{60}$Co crosslinked films were tested, but most of the data were obtained for cells with the DVB-crosslinked film. The films were first characterized by measurements, such as tensile strength and resistance, to ensure applicability and quality control, and then the various aspects of electrical performances of the cells such as plateau voltage, capacity, stand life, and cycle life were determined. Although the testing has not been complete enough to demonstrate that all of the development goals for the sterilizable batteries can be met, the test results indicate that the modified polyethylene survives sterilization and then functions as a separator.

# 21.

## Inorganic Separators

### G. Moe and F. C. Arrance

Research and development programs aiming at the achievement of a truly satisfactory separator and the improvement of zinc electrode performance have been carried out in both company-sponsored and government-funded programs since 1962. One of the basic objectives of this research has been the development of inorganic separators (R107, 225).

The function of the interelectrode separator is to prevent contact and shorting between the electrodes, to provide an electrolyte-retaining structure, and to prevent migration of undesired ions and materials between the electrode compartments. In order to meet these requirements, the separator must combine adequate physical strength with a high level of overall porosity and a pore structure sufficiently small to effect the required ionic screening. In addition, the separator must be highly resistant to concentrated KOH and to the oxidizing forces of the silver oxides.

The experimental work related to separators was based on the recognition of the limitations of organic separators and the premise that significant advances would result from the development of separators based on inorganic formulations. These advanced research and development efforts were devoted to the identification of suitable inorganic materials, establishment of the technology necessary to fabricate them into separator configurations, and evaluation of these separators in cells and batteries. Although the major thrust of activity was related to separators, it was recognized that work on the zinc electrode and on overall cell design was necessary to establish the full advantage of the new separator technology. The results are discussed in the following sections.

## DEVELOPMENT OF INORGANIC SEPARATOR MATERIALS

### RIGID INORGANIC SEPARATORS

The early research work started out with several assumptions and objectives based on the cited properties of separators to which were added the requirement of an average pore diameter in the range of 500 to 200 Å, and a capability for manufacture at a reasonable cost with a high degree of reproducibility and uniformity After extensive screening tests in which a large number of inorganic substances was evaluated as candidate materials for development, it was determined that certain synthetic and naturally occurring minerals were worthy of investigation.

One type of material chosen for an in-depth evaluation was a family of aluminosilicate compositions (R43B). These compositions were fabricated into separators for testing using compaction and sintering techniques. Typically, the formulations were ground in a ball mill to an ultimate particle size finer than 10 microns. After the grinding, the slurry was removed from the mill, passed through a 325-mesh sieve, and deironed by means of a powerful magnet.

The slurry was placed in a shallow stainless steel pan and dried to less than 0.5% moisture as determined by the weight loss at 95-105° for about 24 hours. After being dried, the material was pulverized by passing it through a special grinder and screened to produce a granulation satisfactory for compaction into separator shapes.

The separators were then formed by compaction at about 10,000 psi (700 kg/cm²). The pressed separators were then stacked on flat, refractory support plates for sintering. The firing schedule, which was established by experimentation, was cam controlled to provide optimum time/temperature relationships for the desired degree of sintering.

Development of rigid inorganic separators made it possible to build batteries having substantially improved performance characteristics. For example, cells were fabricated which were capable of withstanding thermal sterilization at 145°, operation at temperatures as high as 150°, and long life in both orbital-cycle and deep discharge tests. In typical orbital-cycle tests, the cells were discharged to moderate depths of discharge, such as 15-30%, in ½ hour and then fully recharged in 1 hour. Deep discharge regimes consisted of fully discharging the cells to 1.0 V in 1 to 5 hours and then recharging the cells in about 15 hours.

## FLEXIBLE INORGANIC SEPARATORS

Although rigid separators are highly satisfactory for many cell designs, it was recognized that inorganic separators would have even greater value and wider application if they could be produced as thin, flexible films. Flexible separators, for example, could be used in existing cell designs as a direct replacement for cellophane and other organic separators without major changes in the design of other components or manufacturing processes. Flexible separators might also be made in thin sheets or strips by a continuous process, thereby lowering the unit cost.

The flexible inorganic separator formulations were based on the technology developed for the fabrication of rigid separators. The inorganic ingredients in the ready-for-compaction stage were combined with a small amount of a suitable polymer material. This material was selected for its chemical resistance to concentrated KOH at both ambient and elevated temperatures, compatibility with the inorganic constituents, and adaptability to practical fabrication technology. After the polymer had been dissolved in a suitable solvent and incorporation of the inorganic filler, the slurry was brought to the desired viscosity for film formulation by solvent adjustments.

Flexible inorganic separator films were fabricated by casting and dipping techniques. The prepared slurry is cast on a smooth plate using a doctor blade to control film thickness. Films made by this method can be varied in thickness from about 0.1 to 0.5 mm with a thickness tolerance of about 10%. The arrangement produced films approximately 25 x 40 cm in size, which is adequate for fabrication of test cells ranging up to about 100 Ah capacity. This method can be used to produce larger film sizes by scaling up the equipment. For producing large quantities of film, a continuous film-making device can be developed.

## CHARACTERISTICS OF INORGANIC SEPARATOR MATERIALS

### PHYSICAL AND CHEMICAL EVALUATION

*Strength and Porosity.* The transverse strength at various porosity levels for typical rigid inorganic separators used in silver-zinc cells is shown in Table 40. The transverse strength values reported are expressed as the modulus of rupture in $kg/cm^2$. These data were obtained by cutting test specimens 1.2 cm wide from sintered separators using a diamond saw. The samples tested were approximately 0.7 mm thick. The breaking load was determined using a testing device which provides for three-point loading at uniform rate. The modulus of rupture was calculated as $kg/cm^2$ according to the equation $M = (3pl)/(2bd^2)$, in which $p$ is the breaking load in kg, $l$ is the span in cm, $b$ is the width of the sample in cm, and $d$ is the thickness in cm.

As shown, the transverse strength of rigid inorganic separators used in silver-zinc cells is generally quite high, ranging from 490 to 1120 $kg/cm^2$ for water absorptions from 10 to 18%. Although experimental separator materials have been developed which have substantially higher strength, experience has shown that separators having transverse strength above 350 $kg/cm^2$ are sufficiently strong for assembly handling and application in silver-zinc cells. Several of the materials shown in Table 40 have been assembled into 5 Ah silver-zinc cells and subjected to typical NASA shock, vibration, acoustic noise, and acceleration environmental tests. These tests resulted in no discernable separator or electrode damage.

Since the performance characteristics of inorganic battery separators are directly related to their ultrastructure, separator porosity is also an important

TABLE 40.   TRANSVERSE STRENGTH AND ABSORPTION OF
INORGANIC SEPARATORS FOR SILVER-ZINC BATTERIES

| Designation | Modulus of Rupture kg/cm$^2$ | Water Absorption % |
|---|---|---|
| Rigid inorganic: | | |
| 036-11 | 490 | 12.5 |
| 3420-09 | 700 | 10.0 |
| 3420-09S | 1000 | 10.0 |
| 3420-25 | 900 | 9.5 |
| 3355-25 | 850 | 11.6 |
| 4669-31 | 770 | 25.0 |
| 4669-35 | 875 | 9.5 |
| 7498-13 | 850 | 10.0 |
| 7498-29 | 900 | 10.0 |
| 7520-37 | 1120 | 9.5 |
| Flexible inorganic films: | | |
| 3420-09 | N/A | 13.0 |
| 3420-09S | | 15.0 |
| 3420-25 | | 18.0 |
| 4669-31 | | 20.0 |
| 4669-35 | | 14.0 |
| 7498-13 | | 17.0 |
| 7488-29 | | 16.0 |
| 7520-37 | | 12.0 |
| 8352-20 | | 243.0 |
| 8352-21 | | 150.0 |

parameter. Percent water absorption can be used as an estimation of porosity because the functional porosity of the separator is related to pores which are surface connected. Water absorption data also provide a sensitive quality control measurement because variations in raw materials or processing technology are reflected in absorption values after separator sintering. The percentage of absorption was determined by measurement of separator weight before and after saturation with distilled water. Although the absorption of rigid inorganic separators can be varied by formulation or process changes to meet specific cell requirements, the values shown in Table 40 which range from 10% to 18% are typical. Separators in this absorption range have been successfully used in Zn/Ag cells ranging in capacity from 1 to 400 Ah designed for low and moderate rate applications.

As shown in Table 40 the porosity of flexible inorganic separators can be varied to tailor separator characteristics to cell design. This control is effected by modification of the separator formulation or processing technology or combinations thereof. For example, separators designed for application in long-life secondary cells for operation in orbital cycle regimes are usually fabricated with porosities ranging from about 9 to 15% water absorption. However, the same basic formulation may also be fabricated into separators for high rate cells, in which case porosities ranging from 15 to 30% water absorption are desirable to provide favorable voltage performance. In the case of primary applications where long wet-stand life and high energy density are the significant parameters and only a few discharge-charge cycles are required, higher porosities may be used. Separators of this type generally range in absorption from 30 to 200% or higher.

*Structure.* In order to provide a better understanding of separator structure and operating mechanisms, electron microscopic studies of inorganic separators were made. Figures 108 and 109 show the typical ultrastructure of Type 3420-09 separators. It can be noted that the separator is comprised of particles which contain very fine pores or passages which interconnect, thus providing a large number of microchannels through the separator. The size of these channels is such that, while the resistance to transfer of hydroxyl ion is small, migration of electrode species is prevented or minimized. Although the effective ion-screening action of inorganic separators is evident from diffusion tests and the long cycle life obtained, the exact mechanism is not fully understood. It is believed that in addition to microporosity and tortuosity, ion exchange and double layer effects may play important roles. The electroviscous effect, in particular, may be significant in view of the nature of the separator materials, electrolyte, and the potential differential between the two electrode compartments (R603). This phenomenon may account for the apparent discrepancy between measured pore sizes of inorganic separators reported by other investigators and their effectiveness in preventing electrode species migration in silver-zinc cells.

*Compatibility with KOH.* There are space application requirements for silver-zinc secondary batteries which can be thermally sterilized so that the capability of separators for withstanding heat sterilization is of particular interest to NASA. It is believed that temperatures will range from 125 to 145° with total time at temperature ranging from 108 to 160 hours.

A large number of Astropower inorganic separators have been tested in accordance with JPL sterilization procedures (R318). Typical test results are shown in Table 41, which compares separator characteristics before and after heat sterilization at 135° for 180 hours in KOH (R290). As shown, the inorganic separators are unaffected by heat sterilization whereas organic materials, such as cellophane, specially treated cellophane, sausage casing, nylon, and Kynar, are either dissolved or substantially destroyed in the same test.

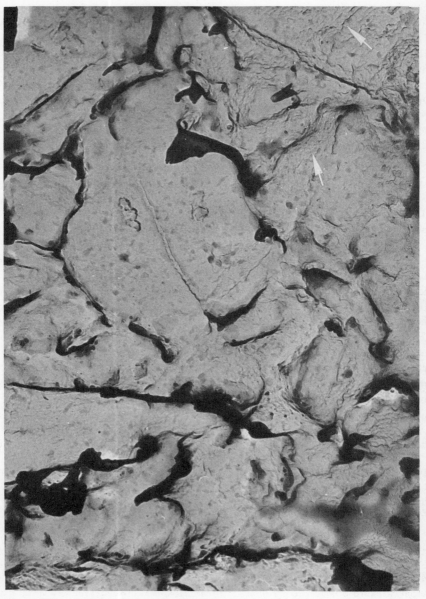

**Figure 108** Electron micrograph of a polished surface of the interior in separator 3420-09. Magnification 11,000x.

**Figure 109** Photomicrograph of surface structure in 3420-09 separators with estimated porosity of 10 and 20%, respectively. Note large grain size and intermediate material of low reflectivity that is indicative of high porosity in sample 20%. Note the significantly smaller grain size and a high light reflectivity over the entire surface in sample with 10% reported porosity. Magnification 450x.

*Resistance to Oxidation.* The resistance of inorganic separators to oxidation (silver pick-up) was measured by the Lander technique (R107E). Table 42 shows the results obtained in these tests at room temperature for both rigid and flexible inorganic separators. As shown, the amount of silver pick-up is negligible after 30 days.

*Resistivity in KOH.* The resistivity of inorganic separators was determined in accordance with the a-c method (R107F). The test procedure was modified in that the resistance of electrolyte alone is determined by measurement of voltages

TABLE 41.   EFFECT OF HEAT STERILIZATION IN 30% KOH
ON SILVER-ZINC INORGANIC SEPARATORS

| Designation | Before Sterilization Resistivity, ohm-cm | After Sterilization for 180 hr at 135° Resistivity, ohm-cm |
|---|---|---|
| Rigid inorganic: | | |
| 3420-25 | 10.0 | 16.8 |
| 4669-35 | 15.8 | 8.0 |
| 7498-13 | 19.8 | 9.0 |
| 7520-35 | 10.5 | 4.5 |
| Flexible inorganic: | | |
| 7498-13 | 10.0 | 11.0 |
| 3420-09 | 3.5 | 16.0 |
| 3420-25 | 4.5 | 5.0 |
| 4669-35 | 9.0 | 11.0 |
| Organic: | | |
| Cellophane | 7.2 | Dissolved |
| RAI 116 [a] | 68-220 [a] | 13-35 [a] |
| 2-Vinylpyridine | | |
| methyl acrylate | 330.0 (ohms) | 54.0 (16 hr) |
| Polybenzimidazole | $16 \times 10^5$ | $20 \times 10^4$ |

[a] This represents, in the view of the reviewer, a high batch-to-batch variability in early samples. An SWRI material, basically the same as RA 116 had a presterilization resistivity of 30 ohm-cm dropping to about 20 on sterilization.

at various currents and plotting $E$ vs $I$. The resistance of the electrolyte, $R_e$, is the slope of the curve. The measurements are then repeated with the separator sample inserted between the electrodes. Plotting these data determines $R_s$, the overall resistance. Separator resistance, $R$, is then the difference, $R = R_s - R_e$. Separator resistivity, $\rho$, in ohm-cm is then calculated by the usual equation $\rho = (Ra)/t$, where $a$ is the cross-sectional area of the separator in cm$^2$ and $t$ is the thickness in cm.

Typical resistivity data for inorganic separators are given in Table 41. These data show that the resistivity of Type 3420-09 rigid separators ranges from about 10 to 20 ohm-cm in 31% KOH depending upon porosity. Long exposure to KOH (up to 15 weeks) at both room temperature and 75° has little effect on separator characteristics. Table 41 also lists the resistivity of several flexible inorganic separators. As shown, resistivity ranges from 3.5 to 10 ohm-cm. In the same test, battery grade cellophane exhibits a resistivity of about 7.2 ohm-cm.

*Ionic Screening and Resistance to Penetration.* The separator interposes a gap between the electrodes while at the same time providing an electrolyte path. This gap between the plates prevents a direct electronic short circuit, whereas the electrolyte held within the pores of the separator provides a low-resistance ionic path for transfer of hydroxyl ions. In the silver-zinc cell, the separator must also prevent migration of silver colloids or ions to the negative plate and penetration by zinc trails or dendritic growth toward the silver plate.

*Silver Diffusion.* The impermeability of rigid and flexible inorganic separators to silver ions was measured using the Harris apparatus (R107G). This test is carried out by placing the test separators in the apparatus after vacuum impregnation with 31% KOH. After sealing the separator in place, 31% KOH saturated with silver oxide is placed in the primary compartment. The secondary compartment is then filled with silver-free, 31% KOH solution.

TABLE 42.   RESISTANCE TO OXIDATION
OF SILVER-ZINC SEPARATORS

| Separator | Silver Pick-Up at 25° after 30 Days mg/cm$^2$ |
|---|---|
| Rigid inorganic: | |
| 3420-09 | 0.0 |
| 3355-25 | 0.0 |
| 3420-09S | 0.0 |
| Flexible inorganic: | |
| 3420-09 | 1.30 |
| 3420-25 | 1.95 |
| 3420-09S | 1.22 |
| Organic: | |
| Cellophane | 31.2 (after 6 hr) |

At two-hour intervals, a 1-ml sample is removed from the secondary compartment and checked for silver content by colorimetric analysis. As shown in Table 43, flexible inorganic separators effectively control silver diffusion at room temperature.

*Zincate Diffusion.* Zincate diffusion was determined with the same apparatus except that the secondary compartment had a volume of 100 cc. Briefly, the procedure consists of assembling a separator sample into the apparatus and detecting zincate transfer between compartments by potential measurements.

The rate of zincate diffusion is the amount of diffusion per minute through one cm$^2$ of separator.

Values obtained for zincate diffusion for both rigid and flexible inorganic separators are shown in Table 44. It can be noted that the rate of diffusion is small in all cases.

## TABLE 43. SILVER DIFFUSION OF SILVER-ZINC SEPARATORS

| | Diffusion at 25° | |
|---|---|---|
| Designation | Absolute ($\mu$g/cm²hr) | Specific ($\mu$g mm/cm²hr) |
| Rigid inorganic: | | |
| 3420-09 | 0.37 | 0.23 |
| 3420-25 | 0.43 | 0.27 |
| Flexible inorganic: | | |
| 3420-09 | 0.28 | 0.03 |
| 3420-25 | 0.15 | 0.02 |
| 3420-09S | 0.25 | 0.03 |
| Organic: | | |
| Cellophane | 0.39 | 0.08 |

## TABLE 44. ZINCATE DIFFUSION OF SILVER-ZINC SEPARATORS

| | | Diffusion at 25° | |
|---|---|---|---|
| Designation | | (millimoles/cm² min) | Specific (millimoles mm/cm² min) |
| Rigid inorganic: | | | |
| 3420-09 | 0.6 mm | 0.17 | 0.10 |
| 3420-25 | 0.6 | 0.65 | 0.39 |
| Flexible inorganic: | | | |
| 3420-09 | 0.125 mm | 10.0 | 1.25 |
| Organic: | | | |
| Cellophane | 0.2 mm | 6.3 | 1.26 |

## TABLE 45. ZINC PENETRATION OF SILVER-ZINC SEPARATORS

| Designation | Thickness mm | Absolute (hr) | Specific hr/mm |
|---|---|---|---|
| Rigid inorganic: | | | |
| 3420-09 | 0.6 | 34-43 | 66 |
| Flexible inorganic: | | | |
| 3420-25 | 0.075 | 3.4 | 45.5 |
| 3420-09 | 0.075 | 4.1 | 55.0 |
| Organic: | | | |
| Cellophane | 0.2 | 8.8 | 45 |

*Zinc Penetration.* The resistance of both rigid and flexible inorganic separators to zinc penetration was determined using a cell assembly (R107C) that holds the separator under study between two zinc electrodes. Current is passed between these electrodes at controlled current density until a zinc trail traverses the separator. The experimental result is a time reading in hours or minutes.

Typical results for inorganic separators are given in Table 45. For reference purposes, cellophane samples having a wet thickness of 0.2 mm were penetrated in an average time of 8.8 hours.

## APPLICATION OF INORGANIC SEPARATORS

Although measurement of specific characteristics of separators provides much useful information, the practical value of a separator can best be determined when it is applied in batteries. *In situ*, the separator is simultaneously subjected to all of the various destructive forces which exist during charging and discharging.

In addition, the battery may be subjected to a variety of environmental stresses during service such as shock, vibration, acceleration, acoustic noise, and high ambient temperature. Accordingly, a substantial part of both the sponsored and company-funded effort has been devoted to evaluation of inorganic separators in silver-zinc cells over a broad spectrum of operating and service conditions.

### SINGLE ELECTRODE COUPLE TESTS

Single electrode couple tests provide a valuable vehicle for evaluating inorganic separators and demonstrating their merit. The unit test cell used for the evaluation consists of a pair of electrodes, the inorganic separator being tested, and a suitable case to permit testing up to temperatures of $150°$.

Inorganic separators have been evaluated in test cells of this type on both orbital regimes at adequate depths of discharge and in deep discharge tests in which the cells are totally discharged to 1.0 V during each cycle. A typical orbital regime is ½ hour discharge at 20-30 mA/cm$^2$ to 20-30% depth of discharge followed by a one-hour recharge.

In these tests, silver-zinc cells employing rigid inorganic separators were capable of cycling for 2500 times at 20 mA/cm$^2$ at $25°$ and 2200 times at $100°$ (R23).

### MULTIPLE-ELECTRODE CELL TESTS

Although satisfactory for preliminary separator evaluations, a more adequate test unit is a multiplate cell design. Typical 5-Ah silver-zinc cells employing rigid inorganic separators were cycled on a 90-minute regime at 20 and 30% depths of

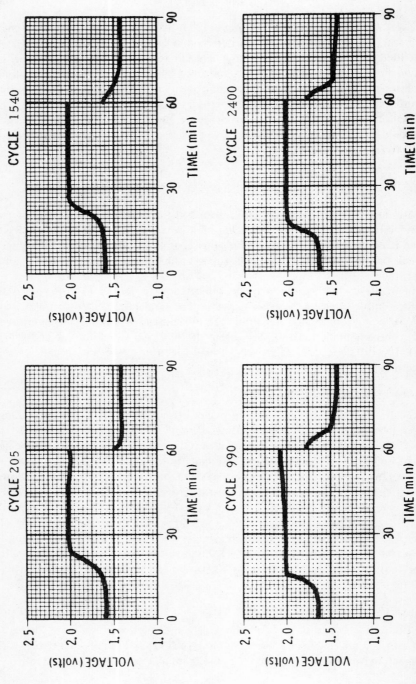

Figure 110 Cycle performance data for 5-Ah cells using rigid inorganic separators at 20 mA/cm² at 25°.

discharge at 25° and 30% depth at 100°. These cells completed 2500 cycles at 20% depth and 2000 cycles at 30% depth at 25°; 500 cycles at 30% depth were obtained at 100°.

Figure 110 shows the cell performance for cycles 205, 990, 1540, and 2400. It is also notable that the capacity retention of these cells was significantly good. Figure 111 shows the capacity retention up to 2500 cycles for a group of 15 cells tested on this regime. All cells in this test group delivered in excess of 4 Ah and the average for the group was 5 Ah or higher (the rated capacity). Separator integrity of these cells is further demonstrated by the constancy of the open-circuit voltage for these cells up to about 1800 cycles. There was no degradation of nominal (1.86 V) OCV voltage.

In the tests run at 100° to evaluate the effect of high operating temperature on cycle life, one cell in this group completed 591 cycles, and most of the cells operated satisfactorily for 500 cycles, which was the average for the ten cells' runs. When these cells were disassembled after testing, it was found that failure was caused by carbonation and to erosion of the silver electrode lead wires. All other cell components, including the separators, were in satisfactory condition.

Rigid inorganic separators were also evaluated in total discharge cycle tests at C/2. In these tests on 5 Ah cells using Type 3420-09 separators, 125 total discharge cycles were completed at C/2 prior to separator failure.

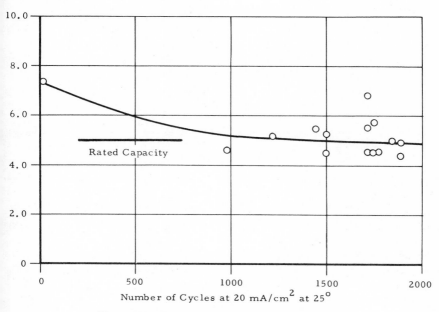

**Figure 111** Capacity retention of groups A and D.

Another use of silver-zinc cells is in applications where the cells are totally discharged to 1.0 V at rates ranging from C/10 to 3 C or higher and then recharged in 15 to 30 hours at intermediate current densities. Cells fabricated with rigid inorganic separators have been cycled at 100% depth of discharge for more than 130 times without evidence of separator degradation. Table 46 is a summary of these results. Separator integrity was verified after 100 cycles by placing the cells on charged stand for 72 hours, after which they exhibited 1.86 V.

TABLE 46.   TOTAL DISCHARGE CYCLE PERFORMANCE OF RIGID
TYPE 3420-09 INORGANIC SEPARATORS IN MULTIPLATE
SILVER-ZINC CELLS

| Cell No. | Original Capacity Ah | Final Capacity Ah | Total Cycles No. | Rate A |
|---|---|---|---|---|
| Group I: | | | | |
| 1 | 9.5 | 8.7 | 138[a] | 3 |
| 2 | 9.5 | 9.0 | 57 | |
| Group II: | | | | |
| 1 | 6.3 | 3.5 | 135[a] | 3 |
| 2 | 6.4 | 4.5 | 76 | |
| 3 | 6.3 | 4.5 | 78 | |
| 4 | 6.3 | 4.5 | 51 | |
| 5 | 6.2 | 4.1 | 63 | |
| Group III: | | | | |
| 1 | 6.0 | 4.8 | 47[a] | 2 |
| 2 | 6.2 | 5.2 | 47[a] | |
| 3 | 6.0 | 4.6 | 47[a] | |
| 4 | 6.4 | 4.4 | 47[a] | |
| Group IV: | 105 | 82 | 120 | 20 |

[a] Tests in progress

In other tests, the integrity of the rigid inorganic separators after sterilization was established by charged stand tests. In these tests, 1 Ah primary silver-zinc cells maintained nominal OCV and retained 100% of capacity for 5 months after being sterilized at 145° for 108 hours. These results demonstrate that cells using inorganic separators are capable of long, continuous orbital cycling and long charged stand life after heat sterilization. The results obtained are also consistent with the results of specific KOH compatibility tests reported in the preceding section which, it was found, had both rigid and flexible inorganic separators substantially unaffected by KOH ranging in composition from less than 30% to over 50% at temperatures as high as 145°.

## SILVER-ZINC BATTERY TESTS

In addition to these cell tests, preliminary tests have been made in order to evaluate silver-zinc batteries comprised of a number of series-connected cells to determine performance characteristics.

When 5-cell groups of silver-zinc cells fabricated with Type 3420-09 rigid separators were tested, they completed 700 cycles at 25% depth of discharge when 45% KOH was used as the electrolyte. With 30% KOH, 550 cycles were completed prior to separator shorting. At 50% depth of discharge, 440 cycles were completed with batteries employing 45% KOH and 285 cycles when 30% KOH was used. At 100% depth of discharge (based on rated capacity), the cells containing 45% KOH completed 85 cycles, and the cells using 30% KOH were cycled 95 times before failure occurred. In these battery tests, failure was defined as the cycle where the first deviation from specification in any one cell occurred.

## SUMMARY

The development of inorganic separators has made it possible to fabricate silver-zinc batteries having substantially improved performance characteristics. Silver-zinc cells constructed with both rigid and flexible inorganic separators are capable of:

1. Orbital regime long cycle life. Experimental cells have been cycled for as many as 2500 times.

2. Extensive deep discharge and total discharge cycle life. Cells rated 5 to 100 Ah have been totally discharged more than 100 times.

3. Operation at elevated temperature. Cells rated 5 Ah have been cycled more than 500 times at 100° in an orbital regime.

4. Withstanding heat sterilization at 145°. Cell characteristics and performance are substantially unaffected by heat sterilization.

5. Long charge stand. Test cells have maintained rated capacity and separator integrity for more than 5 months of charged stand.

# PART VI

## CELL AND BATTERY
## DESIGN FEATURES

# 22.

## Cell and Battery Case Materials, Cell Sealing Techniques for Sealed Silver–Zinc Batteries

### A. M. Chreitzberg

Sealed Zn/KOH/AgO batteries developed in the period 1960 to 1968 for Ranger, Mariner, and Surveyor space missions were designed to minimize space and weight. Batteries for these missions were sealed by the manufacturer before shipment. Active lives date from time of activation until failure or last contact in the space mission and range from three months to four years, including cold storage before flight.

Cell case materials were high impact styrene-acrylonitrile copolymer sealed with catalyzed polystyrene cement; acrylonitrile-butadiene-styrene terpolymer sealed with catalyzed ABS cement; or epoxy cast as an integral structure around duplex electrodes in pile battery array. Table 47 gives typical material characteristics. Jars, covers, and sealing components were injection molded, annealed 5 to 10 degrees below heat distortion temperature, and inspected for dimensions, crazing, and flow lines. Cell jars were individually pressure tested one minute at 50% of the unsupported jar-burst pressure in test fixtures simulating a battery chassis. Table 48 summarizes cell jar material, capacity size, and burst pressures typical of flight cells.

A primary concern was surface crazing of high-impact polystyrene induced by machining operations and aggravated by the presence of styrene monomer in cements. Development of new catalyzed polystyrene cements decreased the incidence of crazing significantly (R604).

Replacement of high-impact polystyrene with ABS reduced surface crazing but uncovered an acute problem of stress cracking induced by cleaning solvents and reactive diluents used in potting operations. Clamping pressures on cemented parts had to be reduced to 1-2 psi. Surface cleaning solvents had to be carefully selected or replaced by detergent and water. Some solvent and catalyzed cements bonded well but reduced the tensile strength of ABS by 50%. Best bond strengths (2400-2600 psi butt tensile) were achieved with new catalyzed ABS cements and by some epoxies. Figure 112 shows the effect of temperature (R95) on the tensile strength of catalyzed polystyrene cement. Cell

## TABLE 47. TYPICAL MATERIAL CHARACTERISTICS FOR SEALED CELL CASE-COVER MATERIALS

| Physical Property | Test Method D– | Unit | Temp °F | High Impact | | High Heat | |
|---|---|---|---|---|---|---|---|
| | | | | SAN | ABS | PSO | PPO |
| Strength tensile | 638 | k | 73 | 11.0 | 5.9 | 10.2 | 11.6 |
| | | psi | 160 | 7.5 | 3.0 | 8.5 | 8.5 |
| | | | 300 | – | – | 5.9 | 6.0 |
| Flexural yield | 790 | k | 73 | 20.0 | 9.6 | 15.5 | 16.5 |
| | | psi | 160 | 13.0 | 5.5 | 14.0 | 14.0 |
| | | | 300 | – | – | 10.0 | 11.5 |
| Flexural modulus | 790 | k | 73 | 590 | 320 | 390 | 375 |
| | | psi | 160 | 570 | 230 | 380 | 365 |
| | | | 300 | – | – | 320 | 340 |
| Impact zod | 256 | ft-lb/in. | 73 | .55 | 5.5 | 1.3 | 1.1 |
| | | | −20 | .50 | 2.3 | 1.2 | 1.0 |
| Maximum use temp. | | °F | | 175 | 175 | 340 | 300 |
| Specific gravity | 792 | | | 1.08 | 1.04 | 1.24 | 1.06 |
| Linear expansion coefficient | 696 | $10^{-6}/°F$ | | 36 | 55 | 31 | 29 |

## TABLE 48. CELL CASE BURST PRESSURES

| Mission | Battery Type | Case-Cover Material | Test Temp. °C | Burst Pressure psi | |
|---|---|---|---|---|---|
| | | | | U [a] | S [b] |
| Mariner | 9SS60 | SAN | 49 | | 95-140 |
| | | | 10 | | 90-95 |
| | 18SS50 | ABS | 25 | 75 | 145-175 |
| Surveyor | 14SS160 | SAN | 25 | | E 70-100 [c] |
| | | ABS | 25 | | E 120-165 [c] |
| | | ABS | −18 | | E 70-195 [c] |
| | | SAN | 51 | | 115-130 |
| | | | 25 | | 115-150 |
| | | | −18 | | 80-135 |
| | | ABS | −18 | | 70-90 |
| Other | SS140 | ABS | 25 | 130-150 | |
| | | SAN | 25 | 45-75 | |
| | SS60 | ABS | 25 | 100-150 | |
| | | SAN | 25 | 70-90 | |
| | SS30 | ABS | 25 | 70-90 | |
| | | SAN | 25 | 45-65 | |

[a] U, Unsupported.
[b] S, Supported by battery chassis.
[c] E, End walls only.

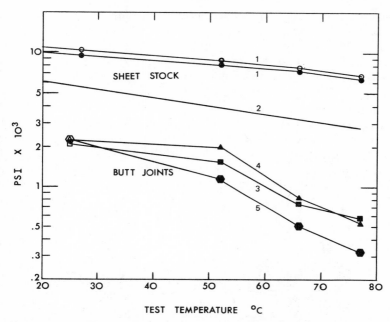

**Figure 112** Effect of temperature on tensile strength catalyzed PS cement. (1) Stock 0.125″ SAN sheet, sources A and B; (2) stock 0.125″ ABS sheet; (3) butt joints, ABS to catalyzed SAN cement to ABS; (4) butt joints, ABS to catalyzed ABS cement to ABS; (5) butt joints, SAN to catalyzed SAN cement to SAN.

seals were designed for operating temperatures as high as 70° by increasing bonding area at least tenfold.

Recent NASA programs require heat sterilizable cell case materials and sealants. Polysulfone and polyphenylene oxide passed sterilization requirements of 120 hours at 135° submerged in 35-45% potassium hydroxide electrolyte. Tests of hermetically sealed cells containing the same electrolyte eliminated polysulfone because of crazing initiated by steam under pressure (R50A). Several polyphenylene oxides passed this test in a nitrogen atmosphere while only one PPO passed the test in air.

Bond strength at 25° for novalac epoxy to PPO test specimens (R51) decreased from 2000 psi to 1300 psi after 3 cycles of 64 hours each at 135° (R50B). Chloroform solvent seals yielded 2100 psi before and after heat sterilization but failed in cell-jar tests when cracks developed in areas of solvent entrapment. Cases sealed with novalac epoxies passed 120 hours at 135° under an imposed pressure of 65 psig and gave helium leak rates of $10^{-5}$ to $10^{-8}$ cm$^3$/sec at 25°.

Cell sealing techniques vary with the environmental requirements of the sealed battery. Early Ranger missions dictated survival of a lunar landing shock estimated to be 2850 gravities for 3 msec (R223). An additional requirement for chemical sterility directed that epoxies be selected on the basis of maximum sporicidal and bacteriocidal strength. The entire structure of the battery and the seal on each cell was cast from epoxy and cured at room temperature in successive steps: first as an epoxy frame on a silver sheet duplex electrode; then as an assembly of electrodes in series array; and finally, after activation and seal of each cell cavity, as the outer shell of the battery.

Sealing techniques of prismatic Ag-Zn cells evolved from encapsulation with epoxy to simpler, more producible, and reliable cell-jar-to-cover and terminal-to-cover seals. Figure 113 shows a top platelock seal used in Ranger missions.

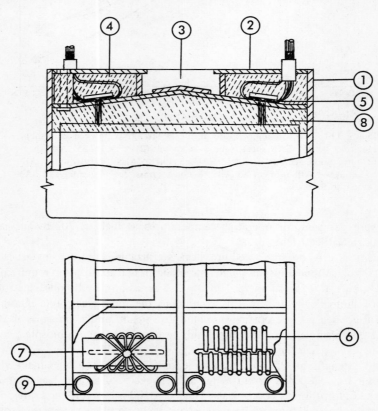

Figure 113. Ranger sealed Ag-Zn cell design. (1,2,3) SAN jar, cover; (4) epoxy sealant; (5) split-wire seal; (6) before cementing; (7) after cementing seal cover; (8) epoxy top platelock; (9) vent plug after seal. U.S. Patent 3,223,558 (R502).

Epoxy sealed the cell and restricted plate movement to minimize damage during shock and vibration. Encapsulation of plate and U folds of separator above the plates did not prevent the growth of zinc moss up along negative lead wires and over the separator to adjacent positive plates.

Figure 114 shows the Mariner '64 and '67 battery cell seal design. The top platelock was omitted and a threaded activation port added. Each plate lead wire was split out from the bundle and sealed individually for increased reliability (R502). Figure 115 shows the Surveyor battery manifold seal (R646). The

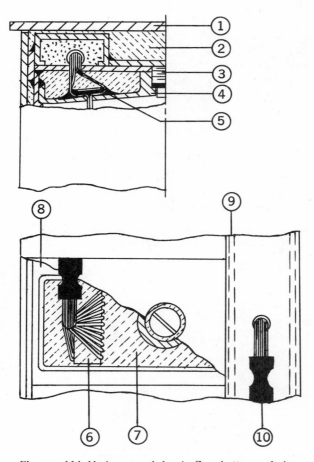

**Figure 114** Mariner sealed Ag-Zn battery design. (1) Magnesium chassis; (2) polyurethane encapsulant; (3) activation port plug; (4) ABS subcover; (5,6) split-wire seal; (7) epoxy sealant; (8) ABS jar; (9) SAN channel seal; (10) battery intercell connectors.

threaded activation port was replaced by a vent valve leading into a common manifold hermetically sealed by assembly of a pressure transducer onto the manifold. Figure 116 shows the seal used in Minuteman instrumented package flights. Here the split-wire seal was replaced by a terminal–O-ring–cover epoxy seal. The reliability of this seal is dependent upon the epoxy-to-plastic, epoxy-to-metal, and O-ring-to-cover seal design.

Overall battery seal reliability was increased by making at least three seals in series at the sites of potential electrolyte leakage. Figures 114 and 115 show a

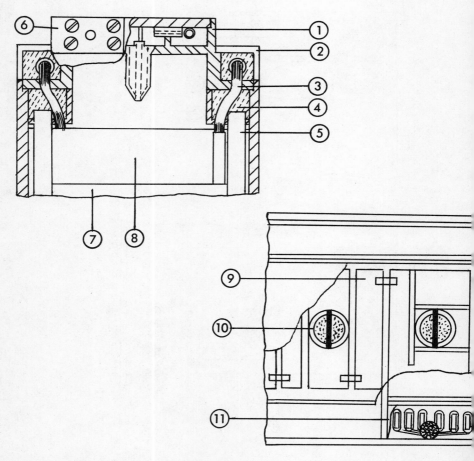

**Figure 115** Surveyor sealed Ag-Zn monoblock design. (1) SAN top manifold; (2) SAN channel seal; (3) plate leads; (4) epoxy sealant; (5) plate strut; (6) pressure transducer mount; (7,8) plate and separators; (9) manifold maze; (10) cell vent; (11) split-wire seal. U.S. Patent 3,282,740 (R646).

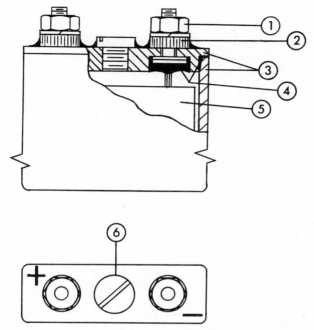

**Figure 116** Sealed Ag-Zn cell design. (1,2) terminal hardware; (3) SAN or ABS cover and jar; (4) epoxy seal; (5) plates; (6) activation plug seal.

combination channel—epoxy seal over cell terminals, a common design feature for Ranger, Mariner, and Surveyor batteries.

Failure modes for seals during the 1960-1968 period have been: (1) cover-to-jar seal failing in peel at high cell pressures, and (2) creepage of electrolyte along negative plate and voltage leads.

Both failure modes are accelerated by high temperature exposure. A combination of good structural design of battery chassis to support seals and prevent peel, selection of epoxy or sealing cements or both, skilled workmanship, and step-by-step inspection decreased the incidence of premature battery failure by electrolyte leakage from one battery in ten in 1960 to near zero in units delivered during recent years.

# 23.

# Theoretical Design of Primary and Secondary Cells

## J. J. Lander

The maximum energy yield of the silver oxide-zinc cell on a weight and volume basis based on active materials only is readily calculated at 25° for the reaction

$$Zn + AgO = ZnO + Ag$$

from the free energy, $\Delta G°$, of $-78.65$ kcal/mol, the sum of the molecular weights of the reactants, 189.3 g/mol, and the sum of the molecular volumes of the reactants, (123.9/7.44) or 16.7 cm$^3$/mol for AgO and (65.4/7.14) or 9.2 cm$^3$ for Zn, equal to a total reactant volume of 25.5 cm$^3$. The energy densities calculated from these basic data are 483 Wh/kg (218 Wh/lb) and 3.6 Wh/cm$^3$ (58 Wh/in.$^3$)

Theoretical values, based on active materials only, do not give a very realistic idea of practical values to be achieved in well-designed cells and batteries because of the need for grids, tabs, electrolyte, separators, cell case and cover, and terminals, all of which add to the weight and volume. They are useful, mostly, for comparing different couples for potential capability. Inefficiency of utilization of active materials, internal resistance, and polarization also contribute to reduction of energy yields from theoretical values. The silver oxide-zinc cell is capable of very good energy yields on a specific weight and volume basis, nevertheless, because of high efficiency of utilization of active materials, 80-90% being readily achieved at the silver oxide plate and 60-80% at the zinc electrode (R251).

The cell reaction as written above does not, however, convey any idea of the part the electrolyte plays in a working cell, and this aspect of behavior has been the subject of many investigations over the years, particularly in relation to the zinc electrode. Considerably less attention has been given to the silver oxide electrode's capability to perform, probably because achieving good performance with it has not been much of a problem. On this basis, and because of a lack of data for the silver oxide electrode, cell performance will be evaluated in terms of zinc electrode behavior and its electrolyte requirement, taking the silver oxide electrode for granted.

## EFFECT OF THE ELECTROLYTE

For a given amount of zinc active material and rate of discharge, how much electrolyte is required? It has been shown that the capacity of sheet zinc electrodes (R558) is a linear function of the volume of the electrolyte times its concentration, up to optimum concentrations in the range of 430-560 g of KOH per liter ($\sim$7.7-10$N$) over a temperature range of $-20$ to $60°$. The optimum concentration decreases somewhat as temperature decreases and as current density increases, and the slopes of the straight-line capacity curves decrease as current density increases and as temperature goes down. The relationship breaks down at very high current densities.

On the face of it, because of the high solubility of the zinc discharge product in the electrolyte (R154) the work might seem to indicate that capacity is limited by the solubility of the discharge product in the available electrolyte, after which precipitation of zinc oxide on the electrode surface passivates it, and the reaction stops. However, precipitation (R154) can occur away from the electrode surface in the body of the electrolyte. Furthermore, electrolyte in contact with a discharging zinc electrode can become supersaturated in zinc (R153), and the amount of supersaturation can be as high as twice the saturation value, depending somewhat on KOH concentration (R172). This fact, and the possibility of precipitation away from the electrode can result in capacities much increased over what would be available if equilibrium solubility limited the capacity.

In Figure 117 Shepherd's data are plotted for 8.05$N$ KOH and an available electrolyte content of 0.36 cm$^3$/cm$^2$ (2.35 cm$^3$/in.$^2$) of zinc electrode surface, for a wide range of temperatures. The horizontal lines indicate limiting capacities based on the equilibrium solubility and on supersaturation value at room temperature for this KOH concentration. Over the range 20-60° the curves lie quite close together, except at very low current density values at the higher temperatures, undoubtedly the influence of self-discharge. The curves appear to be tending toward the equilibrium value of solubility as current density increases. This, however, is probably an artifact, and it is quite likely that they are approaching a value where diffusion-convection of reaction product away from the surface of the electrode becomes limiting. Thus diffusion control (R190) has been shown in 6.9$N$ KOH at vertical electrodes at current densities down to about 0.25 A/cm$^2$ (1.6 A/in.$^2$), a capacity of 72 mA-min/cm$^2$. At slightly lower current densities diffusion-convection effects combine to yield

somewhat larger capacities, for example, 1.55 A-min/cm$^2$ (10 A-min/in.$^2$) at a
current density of 0.15 A/cm$^2$ (0.968 A/in.$^2$).

It is apparent, then, from Figure 118 (R190, 558) that the capacity of sheet
zinc electrodes in the lower current density range is controlled by the volume
and concentration of the available electrolyte together with a combination of

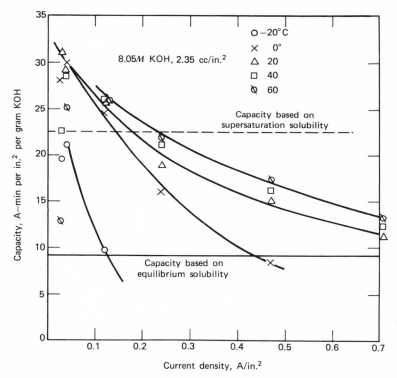

**Figure 117** Effects of current density and temperature on capacity of
sheet-zinc electrodes.

effects depending on degree of supersaturation and rate of precipitation away
from the electrode surface. As current density increases, diffusion-convection
transport phenomena come into play, and, finally, at sufficiently high current
densities the electrode capacity is diffusion limited.

Shepherd (R558) has provided a reasonably complete set of data for negative
plate behavior, from which estimated values of specific energy maxima may be
calculated. The data are, indeed, a very noteworthy contribution to Zn-Ag
battery engineering.

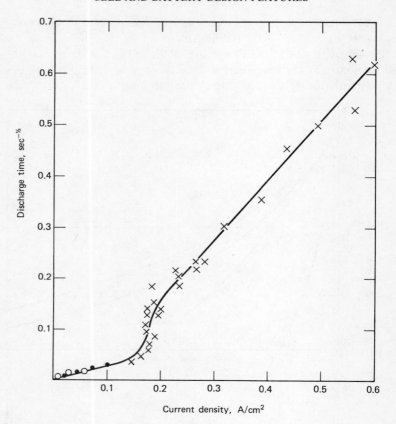

**Figure 118** Capacity limitations on sheet-zinc electrodes in several current density ranges, vertical electrodes. (1) Electrolyte volume x concentration; (2) diffusion-convection; (3) diffusion.

## EVALUATION OF THE ELECTROLYTE REQUIREMENT

The unit capacity expressed as discharge time, min/(g of KOH)(in.$^2$ of Zn electrode area), is plotted as a function of current density at 20° in Figure 119 (R558, Figure 15, has plots from −20 to 60°) for the optimum electrolyte; average discharge voltages are shown.

The curve shows that a 15-minute rate would correspond to a current density of 0.69 A/(in.$^2$)(g KOH) or 0.11 A/(cm$^2$)(g KOH) at 1.33 V and an optimum KOH concentration of 440 grams per liter. Calculations for this discharge rate yield a requirement for electrolyte of 2.27 cm$^3$/in.$^2$ (35 cm$^3$/dm$^2$) equivalent to a weight of electrolyte of 3.0 g/in.$^2$ (46 g/dm$^2$). The current yield is 10.35 A min/in.$^2$ (2.67 Ah/dm$^2$) corresponding to a Zn density of 0.210 g/in.$^2$

(3.26 Ah/dm$^2$) and to an AgO density of 0.398 g/in.$^2$ (6.2 g/dm$^2$). The energy output is 13.78 Wmin/in.$^2$ (3.55 Wh/dm$^2$).

This zinc electrode will discharge at 100% faradaic efficiency. If 10% additional weight is allowed to carry current, the required weight of zinc would be 0.231 g/in.$^2$ If the AgO electrode could discharge at 90% efficiency, the required weight of AgO would be 0.422 g. Total active material weight plus the electrolyte requirement would be 3.0 + 0.231 + 0.442 = 3.673 g/in.$^2$ The gravimetric energy density would be

$$(13.78/60)\,(453/3.673) \qquad \text{or} \qquad 28.3 \text{ Wh/lb (62.4 Wh/kg)}$$

The power density would be

$$(0.69)(1.33)(453/3.673) \qquad \text{or} \qquad 113 \text{ W/lb (249 W/kg)}$$

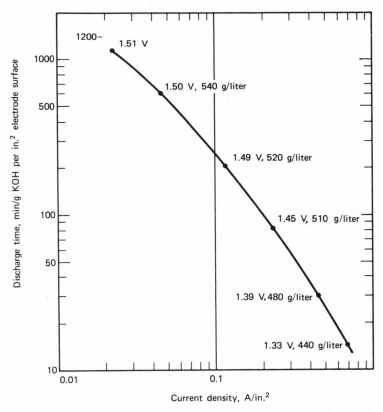

**Figure 119** Minutes discharge time of sheet-zinc electrodes per gram of KOH per square inch of electrode surface versus current density, 20°.

The total volume per unit area of zinc electrode would be

$$2.27 \text{ (electrolyte)} + (0.231/7.14)(Zn) + (0.442/7.4)(AgO)$$

or

$$2.36 \text{ cm}^3/\text{in.}^2 \quad (36.6 \text{ cm}^3/\text{dm}^2)$$

The volumetric energy density would be

$$(13.78/60) \, (16.4/2.36) \quad \text{or} \quad 1.59 \text{ Wh/in.}^3 \, (79.5 \text{ Wh/dm}^3)$$

It is evident that the electrolyte requirement is an overriding factor in determining performance.

Using the data of Figure 119 and repeating the calculations for other discharge times and using the cell voltages and the optimum electrolyte concentrations shown on the basis of 90% efficiency at each electrode, Tables 49 and 50 may well be drawn up for sheet zinc cell construction.

TABLE 49. WEIGHT AND VOLUME REQUIREMENTS FOR 2-PLATE CELLS[a]

| Discharge Time min | Rate A | Electrolyte | | Zinc Plate | | AgO Plate | |
|---|---|---|---|---|---|---|---|
| | | Weight g | Volume cm$^3$ | Weight g | Volume cm$^3$ | Weight g | Volume cm$^3$ |
| 15 | 4C | 3.0 | 2.27 | 0.231 | 0.0323 | 0.442 | 0.0597 |
| 30 | 2C | 2.8 | 2.08 | 0.309 | 0.0434 | 0.585 | 0.079 |
| 60 | 1C | 2.71 | 2.00 | 0.389 | 0.0545 | 0.734 | 0.099 |
| 120 | C/2 | 2.65 | 1.95 | 0.471 | 0.066 | 0.893 | 0.121 |
| 240 | C/4 | 2.62 | 1.91 | 0.552 | 0.078 | 1.05 | 0.142 |
| 600 | C/10 | 2.56 | 1.85 | 0.614 | 0.086 | 1.17 | 0.158 |

[a]Sheet electrodes, geometric area, 1 in.$^2$ (6.45 cm$^2$)

TABLE 50. PERFORMANCE OF TABLE 49 CELLS

| Discharge Rate A | Current Density | | Cell Voltage V | Energy Density | | | Power Density | |
|---|---|---|---|---|---|---|---|---|
| | mA/in.$^2$ | mA/cm$^2$ | | Wh/lb | Wh/kg | Wh/in.$^3$ | W/lb | W/kg |
| 4 C | 690 | 107 | 1.33 | 28.3 | 62.4 | 1.59 | 113 | 249 |
| 2 C | 460 | 71 | 1.39 | 39.2 | 86.4 | 2.38 | 78.5 | 173 |
| 1 C | 290 | 45 | 1.43 | 49.0 | 108.0 | 3.16 | 49.2 | 108 |
| C/2 | 176 | 27 | 1.47 | 58.5 | 129.0 | 3.97 | 29.2 | 64 |
| C/4 | 103 | 16 | 1.49 | 66.0 | 145.5 | 5.31 | 16.5 | 36 |
| C/10 | 46 | 7 | 1.50 | 72.0 | 158.8 | 5.40 | 7.2 | 16 |

## THE EFFECT OF INCREASING AREA

In scanning the data in Figure 119 it seems quite evident that performance is very sensitive to current density (c. d.); not only does decreased c.d. increase the voltage, but higher density electrolytes are usable, and, most of all, the discharge time goes up at something approaching a logarithmic rate. Evidently, substantially better performance could be achieved by increasing area.

If the area is doubled, the current density is halved. For half the c.d. of the 15-minute rate, 0.345 A/in.$^2$ (53 mA/cm$^2$), the capacity will be 47 minutes at 1.42 V and at an optimum electrolyte concentration of 495 grams per liter. This results in an electrolyte volume and weight requirement of 2.02 cm$^3$/in.$^2$ (31.3 cm$^3$/dm$^2$) and 2.73 g/in.$^2$ (42.3 g/dm$^3$), respectively. For a 15-minute rate, however, remembering that capacity is a linear function of volume, only 15/47 or a volume of 0.645 cm$^3$/in.$^2$ (10 cm$^3$/dm$^2$) and a weight of 0.87 g/in.$^2$ (13.2 g/dm$^2$) will be required. Total cell weight will be 0.87 x 2 + 0.231 or 2.313 g/in.$^2$ (13.2 g/dm$^2$).

The gravimetric energy density becomes

$$28.3(1.42/1.33)(3.67/2.313) \qquad \text{or} \qquad 48.0 \text{ Wh/lb (106 Wh/kg)}$$

The power density becomes

$$113(1.42/1.33)(3.67/2.313) \qquad \text{or} \qquad 192 \text{ W/lb (423 W/kg)}$$

The volumetric energy density becomes

$$1.59(1.42/1.33)(2.36/1.38) \qquad \text{or} \qquad 2.9 \text{ Wh/lb (64 Wh/kg)}$$

Doubling the area again (to 4 in.$^2$) gives a current density of 0.173 A/in.$^2$ (26.8 mA/cm$^2$), a capacity of 122 min at 1.47 V and at an optimum electrolyte concentration of 515 grams per liter. Treating the data in the same fashion, for a 15-min discharge the following data are obtained:

| | | |
|---|---|---|
| Total electrolyte volume | 0.955 cm$^3$/in.$^2$ | (14.8 cm$^3$/dm$^2$) |
| Total electrolyte weight | 1.31 g/in.$^2$ | (20.3 g/dm$^2$) |
| Gravimetric energy density | 58 Wh/lb | (128 Wh/kg) |
| Power density | 231 W/lb | (509 W/kg) |
| Volumetric energy density | 3.94 Wh/in.$^3$ | (240 Wh/dm$^3$) |

Continuation of this process yields the data of Table 51. Continuing the process for rates ranging from C/10 to 10C, the curves shown in Figure 120 are obtained. These data are quite instructive. They say that for discharge rates of about C/4 and lower, not much is gained by increasing area. Consequently, for low-rate discharge applications, sheet zinc electrodes might just as well be used. In any case, not much is to be gained by increasing the area ratio beyond 8:1; in fact, electrodes having real area ratios larger than this would probably respond as

TABLE 51.  EFFECT OF AREA INCREASE ON 4 C PERFORMANCE, 20°

| Area Increase | Cell Voltage V | Cell Weight g | Cell Volume cm³ | Energy Density | | | Power Density W/lb |
|---|---|---|---|---|---|---|---|
| | | | | Wh/lb | Wh/kg | Wh/in.³ | |
| 1x | 1.33 | 3.67 | 2.36 | 28.3 | 62.4 | 1.59 | 113 |
| 2x | 1.42 | 2.31 | 1.38 | 46.0 | 101.4 | 2.90 | 192 |
| 4x | 1.47 | 1.98 | 1.05 | 58.0 | 127.9 | 3.94 | 231 |
| 8x | 1.495 | 1.64 | 0.80 | 68.0 | 149.9 | 5.25 | 284 |
| 16x | 1.50 | 1.66 | 0.81 | 70.5 | 155.5 | 5.20 | 282 |

though they had an area ratio of 8-10:1. Increasing the area ratio to about 8:1 pegs the optimum electrolyte concentration in the range 9.3–9.8$N$ (37.5 to 39% by weight), regardless of discharge rate, thus removing the need of varying the electrolyte concentration to fit the discharge rate.

The data also say that about 70-76 Wh/lb (154-168 Wh/kg) is the maximum possible energy density and not much can be done about it because of the electrolyte requirement.

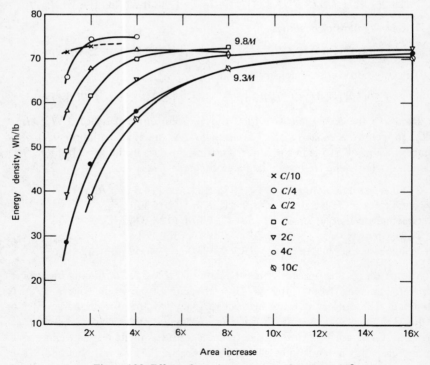

**Figure 120**  Effect of area increase on performance, 20°.

In Figure 121 Shepherd's data covering the temperature range −20 to 60° are reproduced. Similar treatment of the data for the other temperatures yields curves quite like those in Figure 120, that is, the maximum energy yield is attained for an 8x increase in surface area though sometimes a smaller increase in area is indicated. Optimum concentrations of electrolyte range from about 35% KOH by weight at −20° to about 41% by weight for 40°. Maximum energy yield is nearly constant for all temperatures in the range 0-60°. This is evident from the closeness of the curves in Figure 120 at current densities in the range of 0.05-0.15 A/in.² (7.7-23 mA/cm²); at −20° the maximum energy density falls to 50-55 Wh/lb (110-120 Wh/kg).

Proper design is achieved by designing to the discharge rate requirement, if maximum energy and power yields are to be obtained. In Figure 122, for example, the energy yield to be obtained at all rates is shown for cell elements designed for optimum yields at the C/4 rate and the 4 C rate. In a multipurpose cell, that is, one which will be required to perform at a variety of rates, best

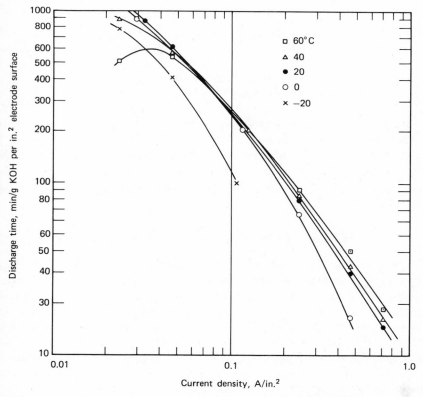

Figure 121 Effect of temperature on discharge time as a function of current density.

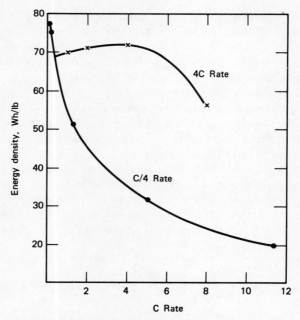

**Figure 122** Energy yields at other rates for cells optimized at the 4C and C/4 rates, elements only.

all-around design will be achieved by designing to the high-rate requirement, because not much performance will be lost at lower rates of discharge. It is noted, however, that the ampere-hour yield will be substantially the same at the lower rates as it is for the high rate. This could put a constraint on the specification writer; that is, his requirements must be self-consistent for low and high rates.

## PRACTICAL ACHIEVABILITY OF HIGH-RATE DESIGNS

It will be stated here that energy yields approaching 70 Wh/lb (154 Wh/kg), elements only, or higher are readily achievable for primary cells at the C/4 rate or lower, provided the ampere-hour requirement is about 10 Ah or larger. This is shown by the upper curve of Figure 123. In calculation of this curve, an allowance of 0.12 g/in.$^2$ (1.8 g/dm$^2$) of electrode area has been made for grid weight, and a plastic cell case and cover of 0.010 in. (0.25 mm) thickness and a density of 2.2 was used. Both electrodes are 50% porous and 90% utilization of active materials is assumed. Calculation of practical yields for the 4 C rate yields the lower curve of Figure 123. For this, a grid weight of 0.24 g/in.$^2$ (3.7 g/dm$^2$) was used, together with the same plastic cell case figures used for the C/4 rate cell. In this case, the zinc electrode is 80% porous and 90% utilization of active materials is assumed. This efficiency is not deemed unreasonable in view of the fact that 70% utilization of zinc electrodes has been demonstrated (R559) at the 7-minute rate. It is noted that a reduction in ampere-hour efficiency to 70% would result in but a 3% reduction in total cell performance because of the relative weight of active materials to the sum of weights of electrolyte, grids and cell case.

The basic design for the 4 C cell is shown in Figure 124, which is a scale drawing of a unit 1-square-inch electrode area, 3-plate cell. This unit cell would provide 0.345 Ah at 1.495 V. Scale-up to higher Ah requirements involving increasing the number of plates per cell would require doubling the thickness of inside positive plates. Note the electrolyte thickness requirement. The cell would have an internal resistance through the electrolyte of 0.013 ohm and at the 15-minute rate (0.069 A/in.$^2$ or 10.7 mA/cm$^2$) the $IR$ loss through the electrolyte would be 0.018 V. Polarization at the negative electrode would be about 0.15 V based on a measured zinc discharge overvoltage curve (R176) and assuming an effective area of 8x for the zinc electrode.

While the 4 C design shown appears to be feasible as far as is known, performance of a porous zinc electrode in combination with adjacent free electrolye has never been checked out. It is noted, however, that attainment of

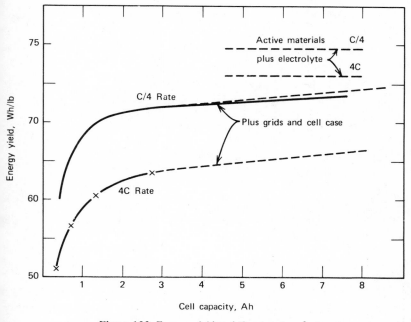

Figure 123 Energy yield and the structure factors.

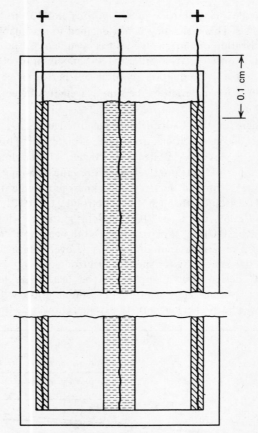

**Figure 124** Design to scale for an optimized cell (0.0345 Ah) for the 4C rate, scale 0.10 in. = 0.01 cm. Plate dimensions, 1 in.² ; electrolyte, 38% KOH, 1.46 cc; 50% porous AgO plates; 80% porous Zn plate. Cell resistance, 0.013 ohm.

maximum energy yield involves an effective plate porosity of about 96%, and fabrication of such a plate to incorporate the electrolyte requirement essentially within its own volume might be difficult to achieve.

## COMPARISON WITH EXPERIENCE

The design, derived from the electrolyte requirement, shown in Figure 124 is by no means the customary design for either high or low rate cells, providing an interplate spacing that accommodates the calculated electrolyte volume require ment. Cells are more likely to have negative plate porosities of the order of 60% and small interplate spacings which would provide a sharply increased Zn : KOH weight ratio compared to that allowed for by the cell of Figure 124 (0.6 : 1). Table 52 shows that high-rate primary battery cells (R108) readily achieve 2-3x the ampere-minutes per gram of KOH yield that would be predicted, at low current densities, by the data of Figure 121, that is, approximately 28 A min/g KOH.

Low-rate primary cells (C/100) have been constructed (R485) which yield between 100 and 221 ampere-minutes per gram of KOH.

The discrepancies between actual data and calculated values will be discussed later. It seems quite possible that the Ah yield in the data of Table 52 is still electrolyte limited, and better yields could be achieved by adding more electrolyte.

TABLE 52. HIGH-RATE[a] PRIMARY CELL DATA

| Ah | Weight KOH g[b] | Weight Ratio Zn : KOH | A-min per g KOH | Wh/lb (Element Only) |
|---|---|---|---|---|
| 0.5 | 1.24 | 1.35:1 | 22 | 38 |
| 1.1 | 1.38 | 2.4:1 | 44.3 | 54 |
| 4.2 | 5.07 | 2.4:1 | 46 | 48.6 |
| 6.3 | 6.75 | 3.0:1 | 51.5 | 49.1 |
| 8.6 | 5.22 | 4.5:1 | 86.8 | 61.3 |
| 11.4 | 10.5 | 4.2:1 | 62.6 | 52.3 |
| 22.3 | 16.7 | 3.9:1 | 74 | 64.7 |

[a]Rates higher than 5 C.
[b]33% of total electrolyte weight.

## SECONDARY CELLS

Secondary cells with initial performance equivalent to primaries can be envisioned. Primary design has been, however, impractical for secondary cells because of the well-known tendency for "shape change" whereby the zinc electrode active material is removed from the top and sides to the center and bottom of the plate as cycling progresses. A picture of this phenomenon is included in Chapter 33, "Failure Modes and Mechanisms" in this book. This process is much accelerated if free electrolyte is available to the zinc electrode, and any reasonable cycle life at all has been achieved only by very tight packing of the element in the case. This has been accomplished by the use of cellulosic separator materials which swell to about double their thickness when wet with the electrolyte. In addition, multiple layers of the cellulosic materials are needed to prevent or slow down the onset of short-circuiting caused by zinc dendritic penetration and metallic silver accumulation in the cellulose. These factors combine to reduce the electrolyte volume available to the zinc electrode. Moreover, because the cellulosic membranes are semipermeable in nature, actual transport of electrolyte out of the zinc electrode compartment can occur if the zinc plates are wrapped with separator material. The outcome is that secondary cell design is not particularly amenable to high energy yield design, where maximum cycle life is a requirement. As a result energy yields in the range of only 20-40 Wh/lb (44-88 Wh/kg) are achieved for cycle life in the range 600-60 cycles. Two major problems exist: one, to achieve cycle life comparable to other secondary battery systems at discharge depths of 60% or more and, two, to reduce the "dead" weight of electrolyte contained in the separator. These are research problems, however, and this paper will be concerned with optimum design as a function of discharge rate and life requirements based on current technology. Recharge time available is a factor which can limit usable capacity in some applications also. As in the case of the primary cell, positive plate performance will be taken for granted.

THE SEPARATOR THICKNESS REQUIREMENT

The thickness of the cellulosic separator required to prevent short-circuiting depends on the type of cycling, that is, depth of discharge, rate of charge and discharge, amount of overcharge, stand time and temperature, and, probably zinc plate design. Moreover, thickness of interseparator is a factor. (Interseparator is a layer of very porous, chemically stable, usually nonwoven material such as Dynel, Pellon, or nylon, 0.001 to 0.004 in. (0.025-0.010 mm) which separates the first layer of cellophane next to the silver oxide plate from that plate. It is used because it decreases, measurably, the rate of attack on the cellulose and the rate of metallic silver deposition in the cellulose).

Single layers of any type of cellulosic separation 0.003-0.006 in. (0.025-0.15 mm), wet thickness, will afford only 15-50 shallow cycles of 2-hours duration, continuous, that is, no stand time. The number of cycles with minimal overcharge permitted before failure by short-circuiting increases out of proportion to the total separator thickness as shown in Table 53.

Charged stand-life of cells made with 3 layers of cellulosic separators (0.46 mm wet thickness) is 10-12 months at room temperature. On the basis of these and some other data (R96) for 24-hour continuous cycling, a rather sketchy idea of the cellulosic separator thickness requirement is obtained, as shown in Figure 125. Even though it is sketchy, it does show the effect of cycling and the disastrous effect of high-rate continuous cycling on cellulosic separator life. This is, no doubt, to be expected, based on Oswin's chapter illustrating the effect of potential on zinc dendrite growth, and the higher temperatures inside cells cycling at high rates. The data also offer a guide to the separator requirement in terms of usage. There is no doubt that high depths of discharge call for relatively thick separation, especially at high rates of discharge and recharge. The data are shown only to give an idea of the separator requirement, because there has been no systematic investigation of the thickness requirement as a function of temperature, amount of overcharge, and zinc plate design.

TABLE 53. CYCLE LIFE OF CELLULOSIC SEPARATORS[a]

| Wet Thickness | | Depth of Discharge Number of cycles | | |
|---|---|---|---|---|
| in. | mm | 25% | 40% | 60% |
| 0.006 | 0.15 | 50 | – | – |
| 0.012 | 0.30 | 200 | – | – |
| 0.018 | 0.46 | 800 | 350 | 150 |
| 0.024 | 0.61 | 1700 | 700 | 200[b] |

[a]Fiber-strengthened sausage casing, Food Products Division, Union Carbide Corporation, Chicago, Illinois.

[b]Continuous two-hour cycles, room temperature.

## ANALYSIS OF INITIAL CAPACITY FOR A PARTICULAR CELL DESIGN

Under a series of Air Force contracts, applied research on secondary cells evolved a design (R372) which, although aimed at achieving maximum cycle life, may be instructive to analyze in terms of the electrolyte volume provided and the initial capacity obtained.

The electrochemical element consisted of 8 negative and 7 positive plates, each 10 in.$^2$ (0.65 dm$^2$) separated by 3 layers of fiber-strengthened sausage casing (swollen thickness 0.006 in. or 0.15 mm per layer). A porous, nonwoven acrylonitrile (Dynel 470, The Kendall Company, Fiber Products Division, Walpole, Mass. 02081) interseparator (0.001 in. or 0.025 mm) was used. The cell (limited electrolyte) contained 85-90 cm$^3$ of 40% KOH (d. 20° = 1.396 g/cm$^3$) and accepted 45-50 Ah on formation charging. At the 10-ampere discharge rate, the cell gave 31.6 ± 1.4 Ah and reference electrode readings showed it to be negative limiting. The cell element weighed 1.11 lb (0.5 kg) and, at an average discharge voltage of 1.45 V, yielded 41.3 Wh/lb (91.1 Wh/kg).

If it is assumed, for the moment, that the zinc electrode area is ~2x the geometric area of the zinc plates—140 in.$^2$ (9.0 dm$^2$), which would correspond to sheet zinc electrodes—the data of Figure 119 can be used to calculate the capacity. The current density would be 10/140 or 0.0715 A/in.$^2$ (11 mA/cm$^2$), and the KOH weight would be 90 x 1.396 x 0.4 = 60.4 g for 0.431 g KOH/in.$^2$ (6.68 g/dm$^2$). From Figure 119, the discharge time would be

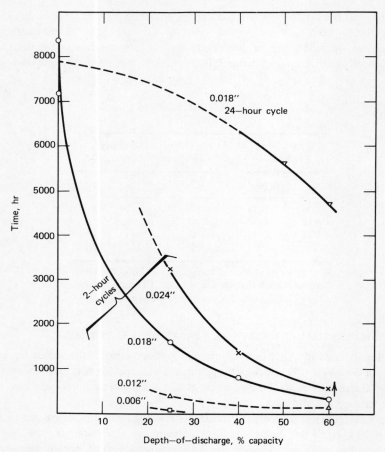

**Figure 125** Life of cellulosic separators, short-circuiting failure, room-temperature ambient.

$375 \times (530/560) \times 0.431$ or 153 min, and the faradaic yield would be $153 \times (10/60)$ or 25.6 Ah.

Assuming an area factor of 2x (280 in.$^2$ or 18 dm$^2$) for a current density of 0.0357 A/in.$^2$ (or 5.5 mA/cm$^2$) the capacity is calculated to be

$$790 \times 550 \times 0.216 \times (10/60) = 28 \text{ Ah}$$

For an area factor of 4x, the yield is calculated to be 29.0 Ah. Higher area factors involve too long an extrapolation of Figure 119 data; however, from the way the calculated values of ampere-hour yield are increasing, a limiting value of about 30 Ah for 8-10x seems reasonable. The latter value is in good agreement with the experimental value, so it is concluded that the zinc plate capacity is electrolyte-limited in this design. The electrolyte limitation is clear, also, because from Figure 119 the capacity could be doubled readily to 60 Ah merely be doubling the electrolyte volume (i.e., disregarding the availability of zinc and silver oxide). It appears that, at the C/3 rate, diffusion of zinc plate discharge product through the separator to the total volume of electrolyte is permitted. The zinc plates are about 55% porous for the half-charged condition and at a thickness of 0.040 in. (0.10 mm) would contain

$$8 \times 0.040 \times 2.54 \times 10 \times 6.45 \times 0.55 = 28.8 \text{ cm}^3$$

of electrolyte. For an area factor of 8-10x, this would yield only

$$30 \times 28.8/90 = 9.6 \text{ Ah}$$

unless diffusion of soluble zincates to electrolyte in the other parts of the cell were allowed.

Removal of one pair of plates with their corresponding layers of separator would provide an additional 8.3 cm$^3$ of electrolyte for the design involved, which should allow

$$30 \times 98.3/90 = 32.7 \text{ Ah}$$

Removal of an additional plate pair-separator combination should allow a yield of

$$30 \times 106.6/90 = 35.5 \text{ Ah}$$

which just might possibly be admitted by the positive plate which would now have a capacity of

$$50 \times 5/7 = 35.7 \text{ Ah}$$

The decrease in weight resulting from trading off plate weight for electrolyte weight would reduce the element weight to 0.95 lb (0.43 kg), and the energy density would be

$$35.7 \times 1.448/0.95 = 54.3 \text{ Wh/lb} \qquad (120 \text{ Wh/kg})$$

If the additional electrolyte were located at the negative plates, the effective porosity would be 64%, which should result in greatly improved high rate performance (Reference 559; see also Figures 117 and 118). Finally, the

ZnO : Ag stoichiometric ratio is about 2 : 1 in the design concerned and if half the zinc oxide weight could be taken out it would amount to 10 g per plate, and the resulting yield would be 63.1 Wh/lb (139 Wh/kg), which approaches the limiting value of 68-70 Wh/lb (150-155 Wh/kg) for primary cells. Energy yields in the neighborhood of 72-78 Wh/lb (159-172 Wh/kg) have been obtained (R343) at rates in the range C/10-C/3 for 60 Ah cell elements with a 1 : 1 ZnO-Ag stoichiometric ratio having 5 layers of fiber-strengthened sausage casing separation.

The foregoing procedures amount to increasing the effective zinc plate porosity to enable a sufficient volume of electrolyte to weight of zinc ratio to obtain very high utilization factors for the available zinc, and are essentially the same as the procedure used for developing the data of Figure 123, except that allowance has been made for separator weight, and a demonstration was required that the zinc plate could, apparently, successfully utilize all the electrolyte in the cell at the C/3 rate and lower with the separation used.

The latterly proposed redesign (i.e., yielding 63.1 Wh/lb or 139 Wh/kg) would require zinc plates of about 87% porosity (or effective porosity) and utilization factors of 90% and 86% of the available zinc and silver, respectively. It would, consequently, represent quite a challenge in zinc plate fabrication techniques, although it has been shown (R559) how zinc plate porosities of 90% can be achieved. Even granting the possibility of achieving such performance with thick separation, however, cycle life of such a design would be questionable.

FACTORS AFFECTING ZINC PLATE CYCLE LIFE

Sufficient separation can always be provided to ensure failure on cycling by modes other than short-circuiting, provided the energy-power requirement will permit. Furthermore, within the author's experience, the silver plate never fails earlier than the zinc plate; consequently only factors affecting zinc plate cycle life need be considered here. The following factors are known to affect zinc plate life.

*Pack Tightness.* Loose packing promotes more rapid redistribution of zinc to the center and bottom of the plate (shape change), resulting in reduced effective geometric area, at least. The fact that cellophane swells in the electrolyte contributing to pack tightness was a substantial factor in the relative success of the silver-zinc secondary cell.

*Depth of Discharge (DOD).* The extent of utilization of zinc per cycle has a very pronounced effect on both numbers of cycles achieved and total ampere-hours yield as shown in Table 54. Cycle life falls off very rapidly, even allowing for the fact that at higher DODs, there is less capacity loss available before failure. This could be caused by an increased rate of shape change and resultant loss in geometric area, but is also unquestionably contributed to by agglomeration and densification of zinc active material (R346, 376), resulting from solution on discharge and plating out of zinc on charge whereby the larger particles grow at the expense of the smaller. This process might very well be accelerated as zinc usage per discharge increases, substantially reducing true surface area, and the capacity.

TABLE 54. CYCLE LIFE vs. PERCENTAGE UTILIZATION OF
ZINC PER CYCLE[a]

| % Utilization of $Zn$[b] | Ah/Cycle | Cycles | Total Ah Yield |
|---|---|---|---|
| 12.5 | 6.25 | 1000-1300 | 7200 |
| 20 | 10 | 300-400 | 3500 |
| 30 | 15 | 60-100 | 1200 |
| 37 | 18.8 | 30-50 | 750 |

[a]Continuous 2-hour cycles, room temperature, 25-Ah cell.
[b]In negatives, 2% polyvinylalcohol.

*Zinc Oxide: Silver, Stoichiometric Ratio.* Increasing the weight ratio of starting zinc oxide to silver to an optimum near 2 : 1 increases cycle life dramatically (R373) probably because of the reservoir of unformed zinc oxide provided. Interestingly enough, the ZnO excess weight did not entail a reduction in specific energy, very likely because of the accompanying increase in electrolyte volume which resulted in an initial capacity increase.

*Electrolyte Concentration.* Cycle life increases roughly linearly, and rapidly, with concentration increase up to about 45% KOH, in spite of an optimum in initial capacity around 30% (R485). The reason for this is not understood.

*Mercury Content.* Mercuric oxide is added to negative plates of secondary cells in concentrations of 1% or more (R648) in order to reduce self-discharge and hydrogen gas production on stand. Cycle life decreases as HgO content increases from 1 to 4%, by approximately 13% per 1% increase in HgO content of the green mix (R372). Fortunately, 1-2% HgO is sufficient to reduce self-discharge to relatively small rates (R567).

*Plate Thickness.* The distribution of a given weight of active material among an increasing number of plates of decreasing thickness results in a steadily decreasing initial specific energy yield, at least where grid design and separator thickness are held constant, at rates in the range C/10-3 C (R343). The steady decrease is caused by the relative increase in grid and separator weight. Unfortunately, information relating cycle life to active material weight distribution among a variable number of plates is too scanty to allow conclusions to be drawn, although some data (R372) indicate the possibility of life increase as the number of plates goes up. Charge acceptance and recharge time available would certainly be factors for design consideration. Further studies are needed.

*Temperature.* The available data (R96) are not self-consistent enough to allow a detailed estimate of temperature effect on cycle life of zinc plates. Life decreases as temperature is reduced from room temperature and probably increases as temperature goes up from room temperature, reaching a maximum around 38-43°.

*Other Parameters.* Almost nothing is known about cycle life as a function of original zinc plate porosity and starting morphology, zinc plate geometry, or amount of electrolyte. Whether or not such factors could be influential in modifying cycle life capability seems questionable anyway, however, unless they exerted some effect on the form of the redeposited zinc.

SUMMARY, SECONDARY CELL DESIGN

The factors bearing on initial capacity of primary cells are effective in secondary cells also; however, in order to achieve cycle life in secondary cells, tight cell packs and appreciable thicknesses of cellulosic separator materials are required so that specific energy yield is reduced.

In order to ensure maximum cycle life allowed by the zinc plate behavior, a minimum cellulosic separator thickness of 0.020-0.024 in. (0.51-0.52 mm), swollen, is required. There seems to be little doubt that excess zinc oxide promotes cycle life and that an optimum excess is near a 2:1 zinc oxide-silver stoichiometric ratio. The mercuric oxide content of zinc plates should be reduced to a minimum consistent with requirements for sealed operation or stand time or both; 1-2% is sufficient.

Electrolyte concentration should be as high as possible, up to 45%, consistent with low temperature requirements. In order to achieve maximum energy density for rates up to about C/3, the active material should be distributed among as few plates as possible; the fewest number of plates would have to be established by additional experimental work.

There appears to be insufficient information relating cycle life to plate thickness, but what is available indicates that reducing plate numbers to a minimum will reduce cycle life. This relationship needs to be established. For cycling at discharge rates higher than about C/3 initial energy yields will be increased by increasing plate numbers and so decreasing the internal resistance; presumably this would increase cycle life also; however, the quantitative relationship is not established.

Optimum plate geometry is uncertain, but plate area would have to be consistent with optimum plate porosities and thicknesses, the ampere-hour requirement, and the discharge rate. Since optimum cell solid geometry for best specific energy is nearly cubic, optimum plate geometry would probably approach a square; it seems likely that considerable leeway is allowable in geometry.

While some information is available relating cycle life to design parameters such as excess zinc or zinc oxide, electrolyte concentration, and distribution of active materials (plate thickness) the major problem is to achieve substantial increases in zinc plate cycle life at good energy yields for the battery-energy yields more in line with the ultimate capability of the zinc-silver oxide battery.

## DISCUSSION

In view of the importance of the electrolyte contribution to cell weight and the apparent lack of agreement between electrolyte weight and volume requirements derived from Shepherd's flat single-plate data and known designs for multiple porous plates, it seems worthwhile to inquire as to the source of the lack of agreement. There seems to be little doubt that his data are affected by self-discharge, which is unaccounted for in his experiments, at the lower rates and higher temperatures. This would have the effect of increasing the electrolyte requirement as calculated from his data, at least where higher rates are concerned. One may also envision the possibility of rapid precipitation of ZnO away from the electrolyte surface, which would effectively continuously renew electrolyte availability for solution of the reaction product. Perhaps a more important factor could exist in the supersaturation phenomenon. In order to achieve a $2^+x$ supersaturation in bulk solution (R153), a considerably higher average supersaturation value would have to exist in the stagnant layer next to the zinc surface. Thus for electrolyte thicknesses of the order of 0.02 cm and using zincate ion diffusion constants recently determined (R440), one estimates supersaturation values of 4-5x at the zinc-electrolyte interface. Such values very likely depend on true current density and temperature, but it is evident that they would allow a considerable increase in the ampere-minute per gram of KOH yield in porous electrodes of large surface area. Extension of supersaturation data to cover a wider range of current densities and temperatures is recommended.

Other factors which should promote better utilization of electrolyte should also be investigated. The precipitation phenomenon should be studied in an effort to speed it up away from the zinc electrode surfaces. Adsorption studies of organics on zinc electrodes in KOH solutions, currently underway (R487), should be extended to determine the effect of adsorption on sheet zinc electrode capacity.

# 24.

## Heat Generation on Discharge for Zinc–Silver Oxide Cells

### Fred Edelstein, Dan Lehrfeld and D. J. Doan

Studies of the heat generated and the electrical energy produced on discharge of zinc-silver oxide cells under controlled conditions have been published only recently (R234, 522). The present study was made primarily to develop a predicting model by multiple regression analysis for the heat generation rate and the voltage while controlling and measuring the discharge current, the operating cell temperature, the activated stand period, activated stand temperature, and the separator type. The cells, descent and ascent types, used for the study were the ones designed for use in the batteries in the corresponding flight stages of the Grumman Lunar Module. Two parallel programs were carried out with differing sets of independent variables. The separator parameter was made necessary because of a design change at about halfway through the plan. A 64-trial fractional factorial design scheduled for two replications was the original plan for each cell. Each cell discharge with a total of 128 per type involved approximately 12 observations so that about 1600 observations constituted the data for the regression analysis for each cell type.

Heat flow calorimeters adapted from that of Calvet (R77-78) were built for three different heat flow rates. These consisted of methyl methacrylate or expanded polyurethane boxes slightly larger than the cells. Uniformly distributed over the bottom and four sides were approximately 500 outside thermocouples in series with 500 matching thermocouples on the inside suface of the box. The polyurethane box had a thin layer of methyl methacrylate cemented inside and outside to support the thermocouples. This also was tight to glycerine, which was used to displace the air interface between cell and calorimeter. The top was insulated with mineral wool and the cell leads exited through it. The leads had thermocouples mounted along their length and an insulated resistance heater was wrapped around each lead about 10 inches from the edge of the calorimeter. A null temperature controller adjusted the heater wattage so that leads would have a zero gradient across the edge of the calorimeter. Each calorimeter was calibrated with the simulated dummy cells

343

containing a main heater (duplicating the cell discharge current) and an auxiliary high resistance heater to adjust total wattage during calibration to duplicate the expected range of heat generation for the cells at each discharge current and ambient temperature. The calorimeters were suspended with wires in an air chamber in which the air was circulated rapidly.

At the highest rates, equilibrium flow was never attained. Also, for the first two hours, the change in temperature, the $\Delta T$, of the calorimeter continued to increase. Therefore, it was necessary to use a procedure for calculating the heat generated during these unsteady state periods. This was done by measuring the decay curve by interrupting the current on some cells and measuring the decay immediately after completion of discharge. A graphical calculation procedure was used to obtain mean heat flow rates for small time increments during the unsteady state between start-up and attainment of equilibrium. Thus, the heat generation over the entire discharge curve was obtained.

The general regression equation was used in treating data:

$$Y = K + A_1 X_1 + A_2 X_2 \cdots A_n X_n$$

where $Y$ is the *response* in appropriate units, $K$ is the *regression constant*, $X_1$ to $X_n$ are the *controlled variables* coded at $-1$ for the lowest level and $+1$ for the highest level, and $A_1$ to $A_n$ are the regression coefficients. The more important terms yielded by the equation are summarized in Table 55. The nominal capacities of the cells were 300 Ah and 400 Ah for the ascent and descent cells, respectively; the descent cell was a lower rate type. Typical voltage curves as a

TABLE 55.   IMPORTANT REGRESSION COEFFICIENTS

| Constant or Coefficient | Unit | Ascent Cell Symbol | Ascent Cell Value | Descent Cell Symbol | Descent Cell Value |
|---|---|---|---|---|---|
| Constant | — | $K$ | 14.07 | $K$ | 6.21 |
| Stand time | day | $A_1$ | 1.05 | $A_1$ | $-0.15$ |
| Stand temperature | °F | $A_2$ | 0.085 | $A_2$ | $-0.035$ |
| Output | Ah | $A_3$ | 0.157 | $A_3$ | 0.021 |
| Current, ampere | A | $A_4$ | 19.69 | $A_4$ | 6.63 |
| Discharge temperature | °F | $A_5$ | $-3.55$ | $A_5$ | $-1.03$ |
| (Output)$^2$ | (Ah)$^2$ | $A_6$ | 3.12 | $A_7$ | 1.52 |
| (Current)$^2$ | (A)$^2$ | $A_7$ | 5.78 | $A_6$ | 0.99 |
| (Current)$^3$ | (A)$^3$ | $A_8$ | $-6.36$ | — | — |
| (Current)$^4$ | (A)$^4$ | $A_9$ | $-5.18$ | — | — |
| (Discharge temp)$^4$ | (°F)$^4$ | — | — | $A_8$ | 1.55 |
| (Current) x (discharge temp) | (A)(°F) | — | — | $A_9$ | $-1.67$ |
| (Current)$^2$ x (discharge temp)$^2$ | (A)$^2$(°F)$^2$ | — | — | $A_{10}$ | $-1.36$ |

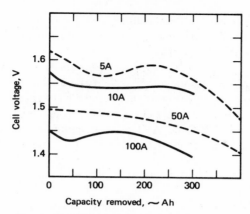

**Figure 126** Voltage curves as a function of output. Dashed lines, descent cell; solid lines, ascent cell; 15 days' stand at 60°F (15.6°); 90°F (32.2°) discharge.

function of the ampere-hours of output for both batteries are shown in Figure 126. The corresponding heat generation curves for two rates of discharge for each type of cell are shown in Figure 127.

At the time of presenting this study, the final regressions have not been fully determined (to yield the smallest error variance). Factor effects of major significance are discharge temperature and discharge current. Minor effects are

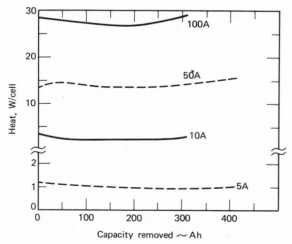

**Figure 127** Heat generation curves. Dashed lines, descent cell; solid lines, ascent cell; 15 days' stand at 60°F (15.6°); 90°F (32.2°) discharge.

due to stand time and stand temperature. The discharge current and the plate design were such that the discharge voltage dropped from an open circuit voltage of 1.85 to below the $Ag_2O$ plateau (1.6 V) very quickly after the start of discharge. The heat generation during discharge is approximately level except for a slight drop-off at the beginning and a slight rise at the end when the voltage decays, confirming the previously found general character (R522). However, the mechanism of discharge is interpreted as a one-step process (*cf.* Wilburn in this volume on chemically prepared AgO plates) combining both positive entities simultaneously and expressed by the equation,

$$AgO + Zn = Ag + ZnO$$

This reaction is corroborated by the qualitative observation of the presence of silver in partially discharged plates and the presence of AgO in nearly fully discharged plates.

# 25.

## The Shelf Life of Unactivated Dry-Charged Zinc–Silver Oxide Cells

### Frederic M. Bowers and Elmer Gubner

"One-shot" or reserve primary batteries have performed at energy densities of 90 Wh/kg and 0.15 Wh/cm$^3$. Such batteries are used where there are requirements for delivery of a single discharge anytime within a period of several years. During this storage time, the temperature may fluctuate between $-50$ and $+71°$. In the reserve type, the cells are assembled with dry and charged elements; the KOH electrolyte is stored in a reservoir until needed. The battery is activated for use by transferring the electrolyte into the cells.

The purpose of this investigation was to study the shelf life of unactivated dry-charged cells by determining the changes in the chemical composition of dry silver oxide cathodes assembled into cells and stored at specific temperatures for specified lengths of time and to relate these changes in composition to capacity.

### EXPERIMENTAL PROCEDURES

The cells were procured from two manufacturers. The methods and processes used in fabrication were, in general, representative of those commonly employed by the various manufacturers of zinc-silver oxide batteries. Pertinent details are described in Table 56.

The cells were stored at 4, 21, 43, 54, and 71°, controlled to ±1.5°. At 54 and 71° the storage times ranged from one month to one year. At 4, 21, and 43° the maximum time was two years. Some cells were stored at 43, 54, and 71° in sealed containers with desiccant. Most of the cells were stored in pairs; however, because of the limited number of samples, some single cells were used to examine the effects of a specific environment. When a pair of cells was removed from programmed storage, one cell was activated with electrolyte and discharged. Its counterpart and singly stored cells were dismantled without activation, and cathodes were analyzed.

347

TABLE 56.    CELLS USED IN TESTING SCHEDULE

| Cells | Manufacturer 1 | Manufacturer 2 |
|---|---|---|
| Cells per manufacturing lot | 50 and 26 | 60 |
| Electrode preparation | | |
|   Anodes | Zn plated from alkaline bath on Cu foil | Zn plated from alkaline bath on 0.04 mm silver f |
|   Cathodes | Sintered Ag powder | Resin-bonded process |
|     Support | Expanded Ag mesh | |
|     Formation | Electrolytic | Electrolytic |
| Element | | |
|   Electrodes | | |
|     Anodes, no. | 20 | 13 |
|     Cathodes, no. | 19 | 12 |
|     Size | | |
|       Height, cm | 10.80 | 8.4 |
|       Width, cm | 8.25 | 7.8 |
|       Thickness | | |
|         Anode, mm | 0.46 | 0.64 |
|         Cathode, mm | 0.38 | 0.40 |
|   Separators | Alpha-cellulose envelopes | Nonwoven polyamide m envelopes surrounded by nonwoven cotton-rayon |
|     Thickness mm | 0.08 | 0.10 + 0.07 |
| Cell Case | Modified polystyrene C-11 | Modified polystyrene |
|   Height, cm | 14.4 | 11.4 |
|   Width, cm | 9.2 | 10.2 |
|   Length, cm | 3.5 | 3.2 |
|   Wall thickness, cm | 0.3 | 0.3 |
| Initial Capacity, Ah[a] | ca. 100 | ca. 50 |
| Activation conditions | | |
|   Reservoir pressure before | | |
|     activation, $kg/cm^2$ | 5.5 | 3.5 |
|   During activation | 2.6 | 2.0 |
|   Electrolyte | Manufacturer's specifications | Manufacturer's specification |
|     KOH concentration % by wt | 45 | 32 |
|     $K_2CO_3$ % by wt | <0.25 | <0.25 |
|     Volume, $cm^3$ | 160 | 100 |

[a] After storage at 21° for 3 months.

Activation was accomplished by placing the cell in a support to prevent the cell case from rupturing and then forcing the electrolyte from a reservoir into the cell with nitrogen gas. Activation conditions are summarized in Table 56.

The cells of both groups were discharged at an apparent current density of 30 $mA/cm^2$ to 1.2 V which, in all cases, was below the knee of the voltage-time curve. This removed about 90% of the capacity of fresh cells in about 40 minutes. The remainder was removed at reduced rates of discharge until the voltage was substantially zero at a current density of 2.4 mA/cm$^2$. Cathodes from some of the discharged cells were analyzed for silver oxides to determine if discharge was complete. No AgO was found and, although some $Ag_2O$ was present, the amount in terms of capacity was less than 0.5% of that removed during discharge.

The chemical composition of the cathodes in cells designated for analysis was determined using the method developed in our laboratory (R623). On dismantling the cells, the cathodes were numbered consecutively. Numbers 5, 10, and 15 were usually taken from Group One cells and 3, 6, and 9 from Group Two. Cathodes which were damaged or irregular in shape were replaced because their content of active material was not representative, and an adjacent cathode was selected. The weight of active material and the chemical composition was determined on each plate selected for analysis. Typical analytical results for the initial standards are given in Table 57.

TABLE 57. COMPOSITION OF THE ACTIVE CATHODE MATERIAL
AND CAPACITY FOR THE INITIAL STANDARDS

|  | Manufacturer 1 | | Manufacturer 2 |
|---|---|---|---|
|  | Lot No. 1 | Lot No. 2 | Lot No. 1 |
| Active material/cathode | | | |
| Average weight, g | 17.1 | 17.1 | 11.8 |
| Analysis | | | |
| AgO, % | 49.8 | 52.4 | 73.8 |
| $Ag_2O$, % | 46.1 | 36.3 | 20.8 |
| Ag, % | 3.6 | 10.1 | 4.6 |
| Capacity | | | |
| By electrical test, Ah | 103 | 103 | 52 |
| By calculation from the analysis, Ah | 103 | 102 | 53 |

The weight of the cathode active material and its chemical composition was used to calculate the total capacity or the residual total capacity of the analyzed cells. For the purpose of this paper the total capacity is the ampere-hour capacity of the initial standards obtained either by electrically discharging the cells to zero volt or by calculation using the chemical composition and weight of cathode active material. The residual total capacity is the total capacity after storage. The total capacities of the initial standards determined electrically and calculated from the analyses are shown in Table 57 and are in excellent agreement.

The total and residual total capacity values obtained from electrical discharges were usually within 5% of the values calculated from the chemical composition of cathodes from counterpart cells and never greater than 10%.

## RESULTS

The analyses and capacities after storage for the two groups at 43, 54, and 71° are shown in Figures 128-133; the data for cells on desiccated storage are included. There was no measurable change in the capacity of Group One cells after a 2-year undesiccated storage at 4 and 21°; the same applies to Group Two cells stored for two years at 21°.

At 43°, the Group One cells showed no change in capacity after one year of storage (Figure 128) and a decrease of about 15% after two years compared to

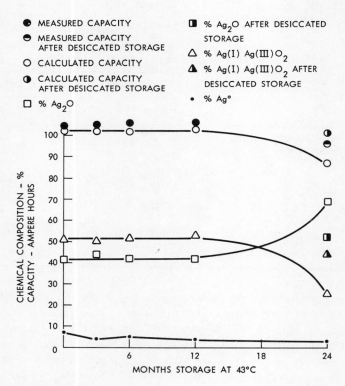

**Figure 128** Capacity of Group One cells and the chemical composition of their cathodes.

no change of desiccated storage for the same time. Group Two cells (Figure 129), however, showed a decrease of about 20% after one year. There was no difference between desiccated and undesiccated storage for 6 and 12 months at this temperature.

At 54°, as shown in Figures 130 and 131, the loss in capacity of the Group One cells is about 10% in one month, and it then changes about linearly to 25% in 20 months. In contrast, the cells on desiccated storage remained uniform for 6 months and then dropped to 25% loss in 20 months, the latter being the same as on undesiccated storage. The Group Two cells showed a lower loss over 12 months with no difference between desiccated and undesiccated storage.

Figure 132 shows the results obtained from the Group One cells stored at 71° without desiccant. Desiccated cells were stored for 1, 2, 3, 6, and 12 months and the data obtained on these were essentially the same. The capacity and the amount of AgO decreased rapidly during the first two months of storage. The capacity tended to stabilize at 70% of that of the initial standard. For the Group Two cells (Figure 133) the rate of AgO decomposition was higher during the first month; the curve flattened to a capacity of 60% of the initial standard. The difference between the two groups is possibly caused by the higher concentration of the unstable AgO initially present in the Group Two cells.

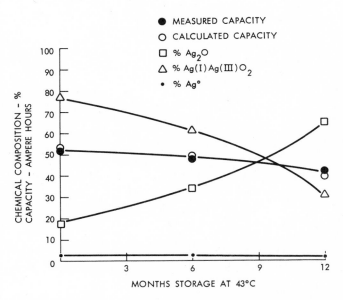

**Figure 129** Capacity of Group Two cells and the chemical composition of their cathodes.

The storage conditions also affected the average voltage during discharge and
the capacity to the 1.2 V endpoint. Figure 134 shows the typical discharge
curves for the initial standards of both groups and for cells after storage at 71°.
The curves for the initial standards from each group are typical of fresh Zn/AgO
cells when discharged at an apparent current density of 30 mA/cm². These
curves show that the voltage is virtually constant at about 1.5 V until the knee
of the curve is reached and that about 90% of the total initial capacity is
delivered before the 1.2 V endpoint. The Group One cells when discharged after
storage for one year at 71° had a slightly lower voltage plateau than the initial
standard. Although there was a large decrease in total capacity, the capacity
delivered to the 1.2 V endpoint remained at about 90% of the respective total
capacities.

The plateau voltage of the Group Two cells, stored for six months at 71°, was
also less than that of their initial standards. These cells delivered only 60% of
their residual total capacity to the 1.2 V endpoint when discharged after six

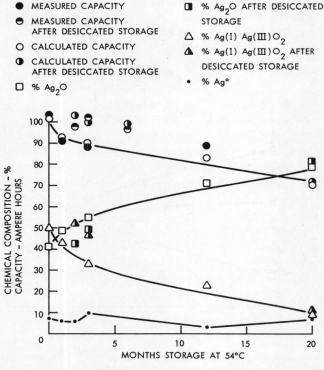

**Figure 130** Capacity of Group One cells and the chemical
composition of their cathodes.

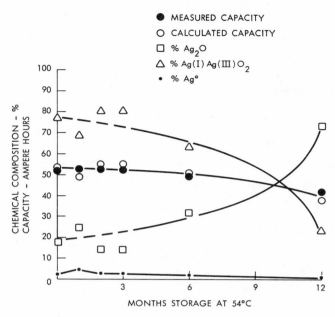

**Figure 131** Capacity of Group Two cells and the chemical composition of their cathodes.

months' storage, whereas the Group One cells delivered 90% of their total residual capacity to the same end point after 12-month storage at the same temperature. Table 58 contains the average voltage values and the percentage of residual total capacity delivered to the 1.2 V endpoint for the discharge of typical cells from the two groups after storage. These data suggest that storage time and temperature adversely affect the percentage of capacity delivered to the 1.2 V endpoint in Group Two cells but had a lesser effect on Group One. Temperature had a greater effect than time.

Storage affected the polyamide mat separators used in the Group Two cells but did not affect either the cotton-rayon mats used in conjunction with the polyamide mats or the alpha cellulose separators used in the Group One cells. The polyamide mats were assembled next to the cathodes. In addition to providing some mechanical separation of electrodes, the polyamide mats, sometimes called absorbers or electrolyte retainers, are supposed to keep the surface of the cathodes irrigated with electrolyte. The first indication of change in the polyamide mats was a slight discoloration from white to pale yellow. This was observed in cells stored for one month at 71°, six months at 54°, and one year at 43°. There was no change in the polyamide mats in the cells stored at 21°. After 90 days at 71° the color was a deeper yellow and there were signs of

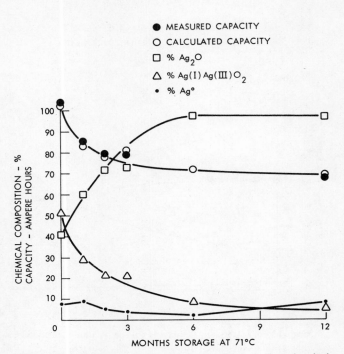

**Figure 132** Capacity of Group One cells and the chemical composition of their cathodes.

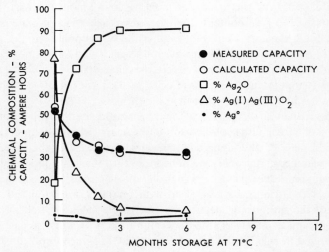

**Figure 133** Capacity of Group Two cells and the chemical composition of their cathodes.

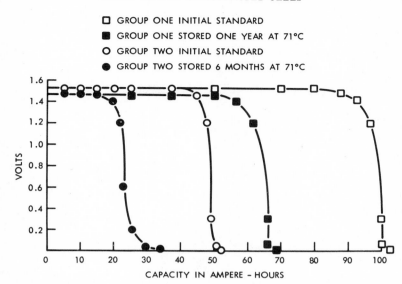

**Figure 134** The effect of storage at 71°C on the discharge characteristics of Group One and Group Two cells (current density, 30 milliamperes/square centimeter).

embrittlement. Figure 135 is a photograph of the separators taken from Group Two cells after storage at 71°. The cathodes from which these separators were removed were located approximately at the center of the plate packs of their respective cells. Those marked with + are the polyamide mats and those with – are cotton-rayon. The 90-day sample shows some chipping along the top edge of the polyamide mat due to embrittlement.

There was also some shrinkage of polyamide fibers that resulted in partial separation of the material. This condition is shown by a series of irregular lines in the lower right-hand section of the 90-day sample in Figure 135. The condition grew worse with time until the type of cracks shown on the 180-day sample were common. In addition, the mats had become so brittle that, in spite of careful handling, the tops crumbled while the cells were being dismantled. Polyamide mats which were removed from the cathodes in the initial standards and stored at 71° for 120 days turned pale yellow but showed no signs of embrittlement or other deterioration. Since the embrittlement and cracking occurred only where the mats were assembled into cells stored at elevated temperatures, the deterioration may be caused by a reaction with the oxygen released from the decomposition of the AgO. The deterioration of the polyamide mats may have also affected their ability to keep the cathodes irrigated with electrolyte and thus may be responsible for the decrease in the percentage of residual total capacity delivered to the 1.2 V endpoint.

TABLE 58.   DISCHARGE DATA OF GROUP ONE CELLS AFTER STORAGE

| Storage Temperature °C | Storage Time, Months | Average Voltage to 1.2 V | | % of Residual Total Capacity Delivered to 1.2 V Endpoint | |
|---|---|---|---|---|---|
| | | Group 1 | Group 2 | Group 1 | Group 2 |
| 21 | 3 | 1.48[a] | 1.50[a] | 95[a] | 94[a] |
| | 12 | 1.49 | 1.48 | 95 | 91 |
| | 24 | 1.48 | 1.47 | 92 | 89 |
| 43 | 3 | 1.48 | – | 93 | – |
| | 6 | – | 1.46 | – | 83 |
| | 12 | 1.48 | 1.45 | 91 | 78 |
| | 24 | 1.46 | – | 92 | – |
| 54 | 1 | 1.49 | 1.49 | 93 | 86 |
| | 2 | 1.48 | 1.47 | 94 | 87 |
| | 3 | 1.45 | 1.48 | 92 | 82 |
| | 6 | 1.48 | 1.47 | 94 | 84 |
| | 12 | 1.45 | 1.46 | 90 | 84 |
| 71 | 1 | 1.47 | 1.47 | 90 | 69 |
| | 2 | 1.45 | 1.45 | 89 | 65 |
| | 3 | 1.45 | 1.44 | 92 | 59 |
| | 6 | – | 1.44 | – | 64 |
| | 12 | 1.45 | – | 91 | – |

[a] Initial standard: Initial standards were cells stored for three months at 21°.

During storage, the surface of some of the cathodes in the Group One cells became coated with a yellow material. X-ray diffraction analysis showed that the material was $Ag_2CO_3$. The amount ranged from no yellow color to almost complete coverage of the surface of the electrode. The condition was first observed in cells stored for one month at 71°. It also appeared after storage of one year at 21°, six months at 43°, and three months at 54°. The color was present only on the surface, which suggests that the surface of the cathode was reacting with the surface of the alpha cellulose separator. However, some cathodes were buckled and the yellow $Ag_2CO_3$ was present on surfaces that were not in contact with the separator at the time the cells were dismantled.

Samples of the cathodes from the cells of both groups were analyzed for $CO_2$ by the combustion method (R265). The results expressed as $Ag_2CO_3$ are contained in Table 59. The table shows that there were large amounts of $Ag_2CO_3$ initially present in the Group One cells indicating that the carbonate was introduced during manufacture. This agrees with others (R606) who found 8.5% $Ag_2CO_3$ in freshly prepared cathodes and who were able to reduce the

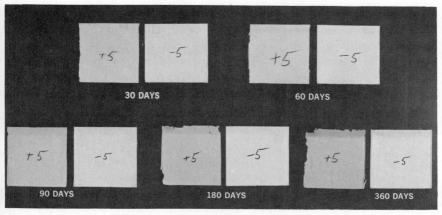

Figure 135 Changes in Group Two cell separators after storage at 71°.

amount to 0.75% by carefully controlling the manufacturing processes. No color caused by $Ag_2CO_3$ was noted in their study. The presence of the yellow color in our study was not significantly related to the amount of $Ag_2CO_3$ found in the bulk of the cathode active material. Several uniformly coated cathodes were divided into sections. The yellow $Ag_2CO_3$ surface was stripped from some sections but not from others. The amount of $Ag_2CO_3$ found in the stripped and unstripped sections was essentially the same. The data on Group Two cathodes show that storage conditions did affect the amount of carbonate: the lowest amounts are found after storage at 21° and the highest at 71°. The measured and calculated residual total capacity data on the cells from both groups indicate that $Ag_2CO_3$ in the amounts found has little effect on the total capacity of the

TABLE 59. PERCENTAGE OF $Ag_2CO_3$ IN CATHODE ACTIVE MATERIAL

| Storage temperature | 21° | | 43° | | 54° | | 71° | |
|---|---|---|---|---|---|---|---|---|
| Time months | Cell 1 | Group 2 | Cell 1 | Group 2 | Cell 1 | Group 2 | Cell 1 | Group 2 |
| 1 | – | – | – | – | 11.3 | 5.6 | 8.8 | 10.0 |
| 2 | – | – | – | – | 15.0 | 5.0 | 11.3 | 11.2 |
| 3 | 10.0[a] | 4.4[a] | – | – | 8.8 | 8.1 | 8.1 | 11.9 |
| 6 | 11.3 | – | 7.5 | 9.4 | 11.3 | 7.5 | 12.5 | 11.3 |
| 12 | 10.0 | 2.5 | 8.1 | 9.4 | 15.7 | 9.4 | 16.9 | 11.9 |
| 24 | 9.4 | 1.2 | 14.4 | – | – | – | – | – |

[a] Initial standard determined on cells stored for 3 months at 21°.

cells. Possibly the following reaction shows the manner in which $Ag_2CO_3$ reacts with 32% KOH to form $Ag_2O$:

$$Ag_2CO_3 + 2KOH = K_2CO_3 + Ag_2O + H_2O$$

If the AgO reacts with the separators to form $Ag_2CO_3$, there would have been a loss in capacity.

There was deterioration of the C-11 plastic cell cases in Group One and the acrylic plastic cement used for sealing cell cases in Group Two at 71°. The C-11 plastic cell cases crazed during the first month so that it was necessary to reinforce them to prevent rupturing during activation. The softening of the acrylic plastic cement used to fasten the fill tube to the case and form a seal between the terminal pole and the case was less serious but might have caused malfunction of a multicell battery. Deterioration of plastic parts was observed only after storage at 71°.

## SUMMARY

The results of this study show that unactivated, dry-charged zinc-silver oxide batteries can be stored at 4° and 21° for at least two years without any change either in the chemical composition of the cathodes or the total capacity. Batteries stored at 43° for two years may experience a decrease in total capacity of approximately 10% but this can be minimized by keeping the internal cell components dry. Storage at 54° results in approximately a 15% loss in total capacity during the first three months. After 20 months the loss is about 30%. During this time about 80% of the AgO decomposes to $Ag_2O$. The data for the storage at 71° show that most of the AgO decomposes during the first three months. This is accompanied by a corresponding decrease in total capacity. After this time the capacity remains relatively stable at 60-70% of its initial value.

The capacity to an endpoint of 1.2 V was about 90% of the residual total capacity irrespective of storage conditions in cells containing alpha-cellulose separators. In cells containing polyamide mat separators the capacity to the 1.2 V endpoint of cells stored six months at 71° dropped to about 60% of the residual total capacity.

The polyamide electrolyte retainer mats suffered severe deterioration at the higher temperatures after only several months of storage. There were also some signs of deterioration at the lower temperatures after approximately one year. The alpha-cellulose separators showed no signs of deterioration. Silver carbonate was dispersed throughout the cathode active material. From 7.5 to 16.9% by weight was found in the cathodes of Group One cells, and in Group Two cells the range was from 1.2 to 11.9%. The presence of carbonate did not affect either the capacity obtained from electrical discharge or that calculated from the chemical composition.

Calculating the capacity available from the chemical composition is a useful and accurate method for predicting the total or residual total capacity that may be obtained from electrical discharge. It also provides a measure of the condition of the cathode. Since other factors may affect the capacity to a preselected end voltage, the calculated value is only accurate for the total or residual total capacity unless the effect of the other factors is known.

# 26.

## Activated Stand Capability of Batteries

### Sidney Gross

One important shortcoming of zinc-silver oxide batteries is their limited life following activation. To learn more about this limitation, secondary cells were tested by studying the effect of activated storage on life. The results of these tests apply to cells that were state of the art in 1963.

The testing conditions are summarized in Table 60. The cells with nominal capacities up to 30 Ah were procured from four vendors in 1963, and the tests were started in 1964. Cells from manufacturers A and B were factory sealed whereas cells from manufacturers C and D were vented through low-pressure relief valves.

TABLE 60.    TEST CONDITIONS TO DETERMINE EFFECT OF
ACTIVATED STORAGE ON LIFE

| | |
|---|---|
| Cell size, nominal capacity | 10, 20, 25, and 30 Ah |
| Charging current | Manufacturer A and C at C/15 |
| | Manufacturer B at C/40 |
| | Manufacturer D at C/60 |
| End-of-charge voltage | Limited to 1.98 V |
| Charge completion | 1/5 of initial current |
| Temperature | 24° |
| Discharge | C/10 rate to end voltage 0.9 V |
| Storage Conditions | |
| Temperature | 10°, 24°, and 38° |
| Cell state | Charged with a few in discharged condition. |
| Capacity test | Charge-discharge cycles at intervals of 2 to 10 months |

## CELL STORAGE AND TESTING

Different cells were stored at temperatures of 10, 24, and 38° in a fully charged state. A few cells were stored discharged but not shorted. At predetermined intervals ranging from 2 to 12 months, the cells were removed from storage, allowed to stabilize at 24°, and discharged. This was followed by a full charge, another discharge-charge cycle, and return to storage. This routine was repeated for each cell until it failed.

The prestorage capacity of the cells, obtained from the second discharge, is shown in Figure 136. Cells from Manufacturer A had closely grouped capacities,

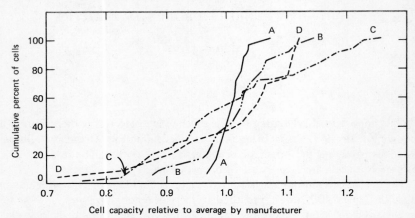

**Figure 136** Initial capacity distribution of cells by Manufacturers A, B, C, and D.

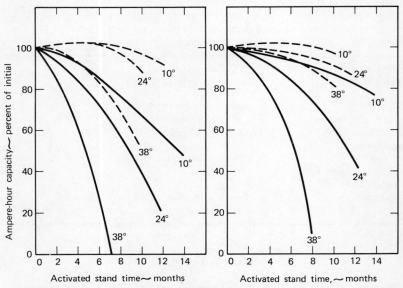

**Figure 137** Capacity of cells of Manufacturer A (dashed lines) and of Manufacturer B (solid lines) after storage without reconditioning. (a) Charge retention during storage; (b) permanent capacity after storage.

whereas cells from Manufacturers C and D had the widest distributions. These initial capacities, based on the second discharge, are the bases for expressing relative performances.

Figure 137 shows typical effects of storage duration on the capacity of cells from two manufacturers stored at three temperatures without interruption. Charge retention (Figure 137a) is the capacity obtained on the first discharge after storage. Permanent capacity (Figure 137b) is the capacity measured on the second discharge after storage. The permanent capacity is only slightly greater than the charge retention; thus most of the capacity loss is permanent.

## CELL RECONDITIONING

Periodic withdrawal of cells from storage followed by two discharge-charge cycles had a beneficial reconditioning effect on many cells. Figure 138 shows this reconditioning effect on cells from Manufacturer B. Cell life is defined as the time for capacity of a cell to fall to one half of its initial value. Reconditioning sometimes more than doubled cell life. The optimum reconditioning cycle duration appears to be temperature related and seems to be from 4 to 6 months for these cells.

The effects of reconditioning on cells from different manufacturers is shown in Figure 139 for one temperature. Reconditioning benefits the cells variously, and the optimum reconditioning cycle is different for each manufacturer.

In some of the tests, large differences were observed in performance among cells from a given manufacturer at a given storage temperature. This random effect at times exceeded the systematic effects of differing conditioning frequencies. However, to establish performance trends, all data from each

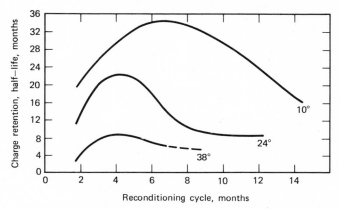

**Figure 138** Effect of reconditioning cycle period on charge retention, Manufacturer B.

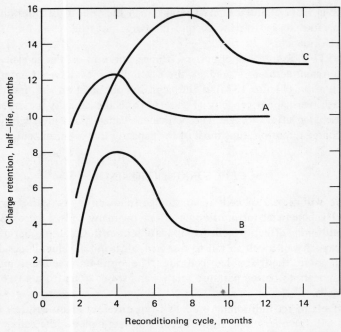

**Figure 139** Effect of reconditioning cycle period on charge retention during 38° storage. (A) Manufacturer A, (B) Manufacturer B, and (C) Manufacturer C.

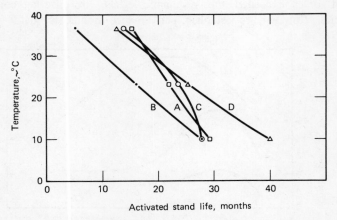

**Figure 140** Activated Stand Life to attain half of initial capacity. Data are for permanent effects of storage on life and for average value. (A) Manufacturer A, (B) Manufacturer B, (C) Manufacturer C, and (D) Manufacturer D.

manufacturer's cells at a given storage temperature were averaged (Figure 140). With the possible exception of Manufacturer C, the temperature life relationships do not follow the Arrhenius equation. It also is evident that the relative performance of a manufacturer's cells at one temperature may not hold at another temperature.

## OBSERVATIONS

A cell user likes to see predictable behavior. Gradual degradation, which gives advance indication of failure, is preferable to a surprise failure. Figure 141 shows two cells tested under identical conditions. One cell failed gradually, and the other failed suddenly. The occurrence of such sudden failures was found to be 5, 12, 19, and 11% for Manufacturers A, B, C, and D, respectively.

A few cells were stored at 10° in a discharged state but not shorted. After 40 months of storage, these cells were charged, and upon discharge had the same capacity they had when new. Cells stored charged are compared with cells stored discharged in Figure 142.

It will be noted in Figure 136 that there was a wide variation in initial

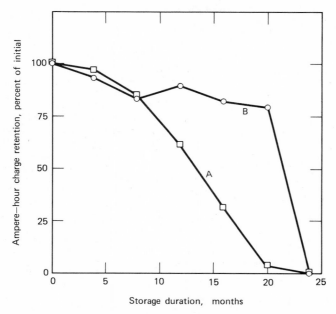

**Figure 141** Sudden failure behavior. Basis: 24° storage, 4 months conditioning cycle. (A) Gradual failure of Cell 5, Manufacturer A; (B) sudden failure of Cell 5, Manufacturer B.

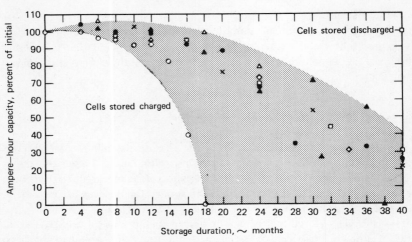

**Figure 142** Effect of storage mode on permanent capacity, Manufacturer A.

capacity of the cells from Manufacturers C and D. No correlation between relative initial capacity and cell life was found for the cells from Manufacturer D. However, with the cells from Manufacturer C, it was found that low capacity cells had a shorter life relative to the average, and high capacity cells had a longer life relative to the average (Table 61).

TABLE 61.   EFFECT OF ABNORMAL INITIAL CAPACITY
ON CELL PERFORMANCE

| Cells | Cell Number | Ratio, Initial Cell Capacity to Average | Life Relative to Average Manufacturer C |
|---|---|---|---|
| | 13 | 0.823 | Bad |
| | 6 | 0.838 | Bad |
| Low capacity | 1 | 0.854 | Poor |
| | 15 | 0.854 | Poor |
| | 8 | 0.875 | Poor |
| | 12 | 1.139 | · Good |
| | 16 | 1.139 | Good |
| High capacity | 3 | 1.150 | Average |
| | 20 | 1.176 | Good |
| | 5 | 1.255 | Good |

## CONCLUSIONS

The major findings of this test are summarized as follows:

1. Life of silver-zinc cells is enhanced by storage at low temperatures.
2. Most of the loss in charge retention during storage was permanent.
3. Degradation rate increases slightly with time.
4. Important differences were observed in the performance of cells from any one manufacturer and from one manufacturer to another.
5. Sometimes conditioning of cells at regular intervals was beneficial to life. In other cases, the random effects overshadowed any systematic effects of differing conditioning frequencies.
6. In limited tests, discharged cells stored over three years did not degrade in capacity.

# PART VII

## APPLICATIONS OF ZINC–SILVER OXIDE BATTERIES

# 27.

## Applications of Zinc–Silver Oxide Batteries
### Paul L. Howard

Since 1940 demands for greater and greater power levels have brought about the development of new battery couples. One of these was the zinc- silver oxide battery which has been responsible for much of the torpedo, missile, and other new equipment developments. Due to its high power and energy density capabilities it has found wide use both as a primary and a secondary battery. Specific applications and design parameters are present in other papers so only a general discussion is given.

During the early 1940s it became evident that advances in electrical and electronic equipment requiring d-c power established an ever increasing demand for more power in less space and weight. During this period, effort was put forth to establish new battery systems which would be able to meet these demands. One such system was the zinc-silver oxide couple which lends itself to either a primary or secondary battery system. Primary systems were developed during the early 1940s; the secondary systems remained to be developed for production during the late 1940s and early 1950s.

The available energy densities were 4 to 5 times that of the conventional battery system (R293, 295, 552) at much higher power levels. Thus it was possible to develop many new products, both military and commercial, to meet the demands for higher power.

Literature and manufacturers' data have been published over the last 20 years. These will probably be cited under specific applications which will be discussed by other authors. Some of the more general ones are given in References (1, 116, 180, 295-299, 368, 375, 398, 635).

Problems of wet life, cycle life, and silver cost have been a major factor in the extent to which this system may be used generally. There has been a great deal of progress in these areas so as time goes on more and more uses will arise.

## DISCUSSION

The particular characteristics of the zinc-silver battery lend themselves to use in four forms: dry charged primary which can be automatically or manually activated; dry-charged or dry-unformed secondary which can be stored dry and activated when needed; the wet secondary stored in the unformed or discharged condition which is ready for use on charging; and most recently, the sealed secondary.

Each of these conditions has been designed in a range of capacities from 0.1 to several thousand ampere-hours. Most of the designs were originally made for specific applications but have found other uses during the years and are still finding new uses as new equipments are developed.

As in the case of other battery systems, no one type of battery is suited for all applications. It is necessary to determine the particular requirements of the application and match the battery type to these. Thus, there are many applications where the zinc-silver will not meet the requirements as well as other systems, as for starting, lighting, and ignition for automobiles, trucks, and industrial trucks ; marine communications; and sealed battery applications.

The primary system at present is limited to those types of application which require high energy densities for short life applications; the secondary to relatively short wet life with a small number of full cycles. It is expected that future developments will overcome some of the problems so that new areas of application will arise.

Because of system limitations and cost of silver, the major applications are governmental, involving batteries made of various size cells in voltages from 1.5-500v depending upon the application. These may be divided into the following areas of use, namely, underwater, space, ground and atmosphere.

The original application of this battery was in the underwater area for use in torpedos, mines, protective devices, and buoys. Later this expanded to special underwater test vehicles, swimmer aids, underwater lighting and cameras, and presently to the field of deep submergence and rescue vehicles as well as various exploratory vehicles such as the Deep Star, Aluminaut, Trieste, and Sealab. The latter field is in its early stage of development so there is a good potential use for specially designed batteries. In these applications the need for maximum power in a minimum weight and volume dictated that the system be developed for them. Many of these might not have been possible otherwise.

With the advent of the space era, the battery in primary and secondary designs has found use where space and weight considerations are critical. Vanguard satellites established a background of experience for the requirements in exploratory unmanned and manned vehicles. Most missile functions, such as guidance and control, telemetering, and destruct, requiring d-c power are quite generally based on the use of zinc-silver oxide batteries.

Since it is difficult to cycle a sealed zinc-silver cell for very long, a great deal of development and testing is being done to improve this condition so that it may be used more extensively in exploratory satellites. Some satellites use special tight containers so that the vented type cell could be used. The Ranger, Mariner, Surveyor, and Voyager series use a specially designed sealed zinc-silver battery which is capable of a few cycles after long charged stand. Whenever charging is involved, this is accomplished from solar panels so as to ensure a controlled current on a modified constant potential regime.

The ground applications are mainly electronic for back packs and walkie-talkie equipment, camera drives, transponders, rocket sleds, TV camera, medical instruments, radar, and night vision. At present there is much study and evaluation going on to use these in electric drive vehicles (R116). At this time only test vehicles are being evaluated. The major drawback for this application seems to be the cycle life and cost. Extensive commercial application has developed for small, 60 and 170mAh, button cells used in electric watches and hearing aids.

Aerial application consists of helicopters, aircraft, target drones, pilotless aircraft, stratosphere balloons both manned and unmanned. For the helicopter and aircraft uses a specially designed medium high-rate long-life cell is used which is comparable in useful life and service to the lead-acid, that is, 6-12 months depending on the severity of the service. Various ground-to-air and air-to-air missiles are controlled by zinc-silver oxide primary or secondary batteries.

## CONCLUSIONS

Although several attempts have been made to establish a standard line of cells to be used in all applications, it has been found that each application requires a specially designed cell, either in case or element size or both. The net result is that all the cells presently designed for specific applications automatically become the cells available for other applications since they are already tested to meet stringent environmental requirements and are essentially production items. Because of specific demands and requirements these cells are not stocked as with other types of batteries for general use. Thus cells are made in specified quantities required for each application.

There still seems to be very little correlation between battery design and initial systems designs. The proper battery could be designed into the overall system rather than the usual procedure of specifying, "Here is the cavity, fit the battery into it." This practice always leads to compromises which may sacrifice some of the overall equipment performance. It is important that the originator of a new equipment requirement and a systems design group should include the power source as a design item of the system rather than to treat it as an afterthought.

# 28.

## Aircraft Zinc–Silver Oxide Batteries

### G. H. Miller and S. F. Schiffer

Rechargeable zinc-silver oxide batteries were first applied to United States military aircraft in July 1954 to fulfill a need for a lightweight, rechargeable, and small-volume battery power source. They range in nominal electrical ratings from 3 Ah, 28 V to 100 Ah, 24 V and are used as emergency power (when generator power fails) and supplemental power (when power demands exceed generator capacity) sources.

## U.S. MILITARY AIRCRAFT USING ZINC-SILVER OXIDE BATTERIES

### AIR FORCE AIRCRAFT

Zinc-silver oxide batteries are used on Air Force fighter type aircraft. These aircraft include: models F-84F, RF-84F, F-105B, F-105D, and F-105F manufactured by Republic Aviation Corporation; and models F-106A and F-106B manufactured by the Convair Division, General Dynamics Corporation. The F-105 series aircraft has one battery and the other aircraft have two batteries.

### NAVY AIRCRAFT

The Navy uses zinc-silver oxide batteries on helicopter type aircraft only (R128). These aircraft include: models H-2A, H-2B, H-2C, and H-43C manufactured by Kaman; and models H-13N and H-13P manufactured by Bell Helicopter. A single battery is used on each of these aircraft except the model H-2C which also has a 0.75 Ah 24 V nickel-cadmium battery to operate electric throttles during emergency situations.

## AIRCRAFT CHARACTERISTICS

The characteristics of the Air Force model F-105 aircraft, particularly the direct-current electrical power subsystem, are given in this section. The electrical system characteristics of the other aircraft are presented in Table 62.

## TABLE 62. AIRCRAFT ELECTRICAL POWER SYSTEMS

| Aircraft Model | A-C Electrical Power Source | A-C Electrical Power Rating | D-C Electrical Power Source | D-C Electrical Power Rating | Battery Type |
|---|---|---|---|---|---|
| Air Force F-84F and RF-84F (See R129-130) | Two inverters | 2500 VA, 1-phase 250 VA, 3-phase | Engine-driven generator: One on F-84F Two on RF-84F | 400 A, 30 V 400 A, 30 V 200 A, 30 V | One lead-acid 36 Ah, 24 V One zinc-silver oxide 6 Ah, 26.25 V |
| Air Force F-105B, F-105D and F-105F (see R131-133) | Air turbine-driven generator | 30 kVA, 400 Hz 120/208 V, 3-phase | Engine-driven generator | 400 A, 30 V | One zinc-silver oxide 100 Ah, 21 V |
| Air Force[a] F-106A and F-106B (see R134) | Two hydraulic motor-driven generators  One air turbine-driven generator | 22 kVA, 400 Hz 120/208 V, 3-phase primary a-c power  3.3 kVa, 400 Hz 120/208 V, 3-phase; 12 kVA, 400 Hz 120/208 V, 3-phase; emergency a-c power | Engine-driven generator and two battery/transformer rectifier units for emergency d-c and canopy power | 100 A, 30 V | One zinc-silver oxide, 15 Ah, 28 V, for emergency d-c power  One zinc-silver oxide, 3 Ah, 28 V, for canopy power |
| Navy H-2A, H-2B, H-2C | | | | | One zinc-silver oxide, 40 Ah, 24 V The H-2C has also a 0.75 Ah, 24 V Ni-Cd |
| H-13N, H-13P | | | | | One zinc-silver oxide, 11Ah, 24 V |
| H-43C | | | | | One zinc-silver oxide, 27 Ah, 24 V |

[a] The models F-106 aircraft have also a combination ac-dc power package for the electronics of the Aircraft and Weapons Control

## AIR FORCE F-105 SERIES AIRCRAFT

The models F-105B and F-105D aircraft are single-place, high performance, swept-wing fighter bombers powered by a dual axial-compressor, turbojet engine equipped with an after burner (R131-132). The model F-105F is a two-place aircraft (R133).

The electrical power system for these aircraft consists of a 3-phase, 120/208 V, 30 kVA, 400 Hz air turbine-driven alternator, and a 400 A, 30 V engine-driven generator supplemented by a 100 Ah, 24 V zinc-silver oxide battery.

The alternator and generator are voltage regulated and have undervoltage and overvoltage protection. The generator has also reverse polarity protection. The generator output voltage is maintained at 28 V by a carbon pile voltage regulator. Whenever generator voltage drops below battery voltage, reverse current from the battery opens a differential (undervoltage) relay and disconnects the generator from the bus system. The overvoltage relay disconnects the generator from the bus whenever generator voltage reaches and holds 32-34 V for ten seconds. The reverse polarity protection consists of a resistor in series between the primary bus, the voltage regulator and the generator, and a rectifier in series with a field-relay reverse polarity trip coil. When the generator has correct output polarity, the high inverse resistance of the rectifier does not allow sufficient current to flow to energize the reverse polarity trip coil. With reversed polarity the low direct resistance of the rectifier allows sufficient current to flow to energize the reverse polarity trip coil and trip the field-relay to de-energize the main contactor and disconnect the generator.

The d-c power is distributed through the battery and primary and secondary buses. The aircraft structure is electrical ground. During normal operation, the generator output is connected to all buses. Switches in the cockpit permit connecting and disconnecting the battery and generator from the bus system. Whenever the generator is disconnected from the bus system, either intentionally or by a system fault, the battery is connected to the primary bus, and the systems operated from the secondary bus are inoperative.

The aircraft is equipped with an indicating system for warning the pilot of adverse battery-charging conditions. This indicating system constantly monitors the magnitude, polarity, and duration of current and voltage in the battery circuit. The current-sensing section of this sensor unit is connected in series with the positive battery lead; therefore, all battery current flows through the sensor. When current in excess of approximately 160 A flows into the battery (charge direction) for approximately one minute's duration, a "battery high charge current" indicator is illuminated. The sensor unit is insensitive to discharge current. The voltage-sensing section of the sensor unit is powered by the primary bus. A "battery high voltage" indicator is illuminated when battery voltage exceeds 28.75 V for one minute.

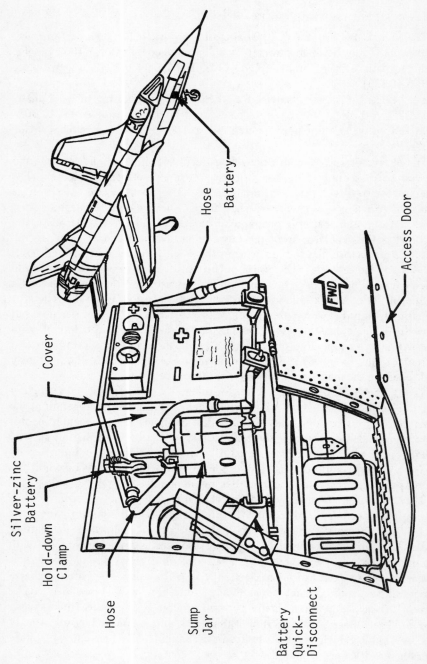

Cover

Silver-zinc Battery

Hold-down Clamp

Hose

Sump Jar

Battery Quick-Disconnect

Hose

Battery

FWD

Access Door

Figure 143  F-105 Battery Installation.

The battery is located in the bottom of the aircraft's nose section. There are no special provisions for controlling battery temperature. The battery enclosure, shown in Figure 143 is vented overboard through a sump jar.

## BATTERY REQUIREMENT–SPECIFICATION–PROCUREMENT CYCLE

### ORIGIN OF BATTERY REQUIREMENTS

With the authorization of the Department of Defense, the requesting service agency contracts usually with a single air frame manufacturer for the development of a particular aircraft. The air frame manufacturer subcontracts with other manufacturing firms for the development of particular systems, subsystems, and components. The battery is a component of the d-c electrical power subsystem. The battery requirements for those United States military aircraft using the zinc-silver oxide battery were originated by the air frame manufacturer. The actual battery developments were accomplished by the battery manufacturing firms best satisfying the air frame manufacturer's defined requirements.

### GENERAL

There are several operational requirements common to most aircraft batteries. These requirements include: operation in all positions; operation without generating radio interference radiation; and unscheduled discharging. Other requirements common to all military aircraft batteries and cells pertain to packaging, environmental testing, marking, and delivery and storage. The procedures of MIL-E-5272 (R135) are used for the environmental tests. The tests include: altitude-temperature, shock, vibration, ventilation, leakage, explosive atmosphere, crash safety, humidity, fungus, salt spray, dielectric, and special tests such as step discharges and discharging in inverted position.

The specific requirements for the battery(ies) used on particular aircraft are dictated by many factors. The chief factors are the functions and characteristics of the aircraft, its environmental conditions, and electrical loads to be battery powered. The intended use of the battery (i.e., emergency and/or supplemental power) and the judgment, previous experience, and prejudices of the aircraft's designers also enter into the formulation of the requirements.

### AIR FORCE F-84F AND RF-84F SERIES AIRCRAFT

The zinc-silver oxide battery aboard these aircraft functions strictly as an emergency power source. It powers the aircraft's radio equipment when primary power fails. The battery receives no charge from the aircraft, and evidence of its use is given by a broken lockwire. The basic requirements for this battery, initially defined by Republic Aviation Corporation, are presented in Table 63

TABLE 63.  BATTERY REQUIREMENTS

| Aircraft Model | Physical Requirements | Electrical Performance at Environmental Requirements | | | | | | |
|---|---|---|---|---|---|---|---|---|
| | | Capacity at Rate | Cycle Life | Capacity Retention | Altitude and Temperature feet and °F | Shock | Vibration | Acceleration |
| F-84F and RF-84F | 9.5 x 4.88 x 3.5" Filled weight 8.7 lb Integral heater | 6 Ah at 20-min rate to 26.25 V | 50 capacity discharge cycles | 5.7 Ah at 20-min rate after 90 days, $\geq 0°$ and $\leq 32°$ | 50,000 at −65°F | 3 mechanical shocks of 3 g for 11 msec | 200 Hz and 0.036" amplitude at 1-4 resonances | 15 g vertical; 6 g horizontal |
| F-105 | 10.75 x 10 x 10.31" Filled weight 69 lb No special feature | 35A for 3 hr to 21 V followed by 800 A for 10 sec to 13.0 V | 10 capacity discharge cycles | 70% of C after 90 days at 70°F | sea level at 160°F 35,000 at 160°F 50,000 at 150°F 70,000 at 120°F Temperature-discharge tests @ −40°F, −20°F, and 0°F | Same as for F-84 and RF-84F | Same as for F-84F and RF-84F | 10 g vertical; 3 g horizontal |
| F-106 Emergency | 9 x 5.5 x 7.69" Filled weight 15 lbs Integral heater | 15Ah at 1-hr rate to 24.5 V | 30 capacity discharge cycles | 70% of C after 30 days 70°F | 65,000 at 100°F and temperature discharge at 160°F | 18 mechanical shocks of 10 g for 11 msec and 3 thermal shocks of 71°F to −40°F | 500 Hz and 0.036" amplitude at 1-4 resonances | |

Electrical Performance at Environmental Requirements

| Aircraft Model | Physical Requirements | Capacity at Rate | Cycle Life | Capacity Retention | Altitude and Temperature feet and °F | Shock | Vibration | Acceleration |
|---|---|---|---|---|---|---|---|---|
| F-106 (contd) Canopy | 6.37 x 6.14 x 7.37 Filled weight 7.5 lbs Integral heater | 15 Ah at 1-hr rate to 24.5 V | 30 capacity discharge cycles | 70% of C after 30 days 70°F | | | | |
| Navy H-2 | 7.68 x 9.93 x 8.75" Filled weight 40 lbs No special feature | 40 Ah at 1-hr rate to 24 V | 50 cycles of 30 capacity discharge cycles and 20 frequency cycles | | 30,000 at −65°F | 18 mechanical shocks of 15 g for 11 msec and 3 thermal shocks of 71°F to −40°F | 500 Hz and 0.036" amplitude for 1-4 resonances | |
| H-13 | 4.93 x 6.37 x 5.12" Filled weight 15 lbs No special feature | 11 Ah at 1-hr rate to 24 V | 45 capacity discharge cycles | | | | | |
| H-43 | 7.75 x 7.67 x 7.75" Filled weight 30 lbs No special feature | 27 Ah at 1-hr rate to 24 V | 45 capacity discharge cycles | | | | | |

381

(R509A). The Air Force uses this specification for competitive bid procurement of batteries from three qualified sources.

### AIR FORCE F-105 SERIES AIRCRAFT

The 100 Ah zinc-silver oxide battery, intended as an emergency power source, furnishes both emergency and supplemental power aboard the F-105 series aircraft. It powers those loads necessary to sustain flight for thirty minutes when no other power is available. The heaviest drawing emergency loads are: inverter (29 A continuous); radio direction finder (27 A continuous); landing lights (29 A continuous, when used); and trailing edge flap motors (260 A total at an average of 130 A for 0.5 minute, average 32.5 A for 2 minutes, and average 2.2 A continuous). Control actuators, valves, solenoids, relays, and indicating lights and transmitters comprise the low-drain emergency loads. The average emergency load is 113 A for 30 minutes for the F-105F model and 77 A for 30 minutes for the F-105B and D models. These are load estimates (actual flight instrument measured values not available) based on a battery voltage of 21.5 V and do not reflect in-rush or surge currents which may be two to ten times greater than normal. On starting, the trailing edge flap motors draw between 800 and 1350 A. The 30 V, 400 A generator is capable of supplying 800 A (100% overload) for short durations. Conceivably, the battery supplements the generator during every takeoff-and-landing operation since the trailing edge flaps combine with the leading edge flaps to provide control and lift during takeoff and landing.

The Air Force procures this battery as a part number item from Yardney Electric Corporation. The battery was granted Industry Developed Equipment (IDE) approval in May 1959 after performing satisfactorily in test aircraft for approximately two years. The leading requirements for this battery (R509B) are presented in Table 63.

### AIR FORCE F-106 SERIES AIRCRAFT

The leading requirements for the batteries aboard the F-106 series aircraft are given in Table 63.

The Air Force uses purchase documents PD SANE-83A/1 (R136) and PD-SANE-77A/1 (R137) to procure the emergency and canopy batteries respectively from qualified sources.

### NAVY HELICOPTERS

The batteries used on the Navy helicopters employ the vented, cell replaceable design. None of these batteries have heaters. The leading requirements for these batteries are listed in Table 63.

The 40 Ah battery on the H-2 models acts also as a vibration absorber. The Navy uses the general specification MIL-B-8565E(AS) (R138) and particular military standards for competitive bid procurement of these batteries.

# BATTERY DESIGN

## GENERAL CONSIDERATIONS

Given the physical and electrical requirements for the battery, the designer proceeds to analyze them with regard to optimization of the design. Starting with the basic parameters, the design will consist of a series of compromises to obtain the most favorable combination of voltage, electrical capacity, and cycle life characteristics within the allowable battery weight and volume.

## F-84F AND RF-84F BATTERY

The F-84 battery is nominally a 26.25 V, 6 Ah unit at the 18-ampere rate. A capability of fifty 20-minute (18 A) discharge cycles is required.

The battery case dimensions are 3.07 in. (7.8 cm) x 4.73 in. (12 cm) x 9.14 in. (23.2 cm) wide. Provided are 21 cells, each being 2.90 in. (7.4 cm) high (2.49 excluding terminals) x 0.79 in. (2 cm) x 2.08 in. (5.3 cm) wide. The battery case material is stainless steel; the cell cases and covers are molded using Bakelite C-11 styrene-acrylonitrile copolymer. Two #10-32 terminals are provided per cell.

Yardney Electric Corp's. HR5V-6 cell, used in this application, consists of 20 positive (Ag) electrodes and 21 negative (ZnO) electrodes with a total active positive electrode area of 19.5 in$^2$. (1.26 dm$^2$). The electrolyte used in 45% KOH.

Four layers of Yardney Electric Corporation's C-19 (silver-treated cellophane, R420), are used for the main separator in these cells.

Yardney Electric reports that the cell provides a plateau voltage of 1.32 V at the 18-ampere (0.925 amps/in.$^2$) discharge rate and 1.44 V at the 4.75-ampere (0.024 A/in.$^2$) discharge rate. It yields 37 Wh/lb (82 Wh/kg) and 2.2 Wh/in.$^3$ at the 60-minute rate, 21.1° (70°F) to 1.10 V.

## F-105 BATTERY

The F-105 battery is nominally a 21 V, 100 Ah battery (at the three-hour, 35 A rate).

As noted previously, the actual electrical use on the aircraft has never been documented. However, the battery must have the capability of providing discharges in the range of 35 A for 3 hours to 1350 A for 30 seconds. The battery can provide a minimum of 40 minutes of discharge at 35 A at −40° and must be able to withstand "indefinite" storage in the temperature range of −54° (−65°F) to +71° (160°F). Fourteen cells are used for this application. The (stainless steel) case and cell dimensions are tabulated in Table 64.

Because of the wide temperature range desired, the cell cases and covers were constructed from nylon. Four $\frac{3}{8}$-24 screw terminals were provided per cell to accommodate the high-rate discharge requirements. At the 1350-ampere rate,

TABLE 64.   F-105 BATTERY AND CELL DIMENSIONS

|         | Height | | Length | | Width | |
|---------|--------|------|--------|------|-------|------|
|         | in.    | cm   | in.    | cm   | in.   | cm   |
| Battery | 10.31  | 26.2 | 9.69   | 24.6 | 9.94  | 25.2 |
| Cell    | 9.44   | 24.0 | 1.81   | 4.60 | 2.81  | 7.14 |
|         | 8.75   | 22.2[a] |     |      |       |      |

[a]Excluding terminals.

each electrode must carry 67.5 A. Each positive electrode is provided with one $\frac{1}{8}$ x 0.010 in. (3.2 x 0.25 mm) thick Ag tab to safely carry this current. Two positive terminals (and two negative terminals) were required to accommodate these tabs. As the electrolyte, 45% KOH was used. The battery has no heater.

The cell, Yardney Electric Corporation's HR85, consists of 20 positive (Ag) electrodes and 21 negative (ZnO) electrodes with a total active positive electrode area of 575 in.$^2$ (37 dm$^2$).

Yardney Electric Corporation's data for the 65 pound F-105 battery (their P/N 2781) rate it as yielding 35 W/lb and 2.1 Wh/in.$^3$ (77 Wh/kg and 128 Wh/dm). The individual HR85 cell provides 49 Wh/lb (108 Wh/kg) and 3.7 Wh/in. at the 60-minute rate, 21.1° (70°F) to 1.10 V.

### F-106 CANOPY BATTERY

The F-106 Canopy Battery is nominally a 28-V, 3-Ah battery at the one-hour (3A) rate.

The battery is required to provide 30 charge-discharge cycles. The output voltage must remain within the limits of 17 to 29 V at any altitude from sea level to 6000 feet and at any temperature within the range of −54 to +71°. The battery must be capable of delivering 85% of its rated ampere-hour capacity after having been stored for 72 hours in a 38° ambient temperature and 70% of its capacity after 30 days in a 21 ± 6° ambient. The battery must also be capable of withstanding 71 to −40° thermal shock. The maximum allowable wet weight is 7.5 lb (3.4 kg).

Seventeen cells were used for this application, resulting in an allowable voltage range of 1.0 to 1.71 V per cell under load. A 75 W (nominal) 115 V a-c heater blanket, controlled by thermal switches, is provided for use during low-temperature operation. All components, cells, heaters, thermostats, and electrical wiring assemblies are encapsulated in epoxy within the glass-reinforced polyester battery case to protect against environmental extremes. (The cell vents are left open to allow gases to vent during operation.) The activated battery weight is approximately 6 lbs (2.7 kg). The battery case dimensions are tabulated in Table 65.

TABLE 65.  F-106 CANOPY BATTERY AND CELL DIMENSIONS

|         | Height | | Length | | Width | |
|---------|------|------|------|------|------|------|
|         | in. | cm | in. | cm | in. | cm |
| Battery | 5.19 | 13.2 | 3.77 | 9.58 | 6.09 | 15.5 |
| Cell    | 3.36 | 8.53 | 0.59 | 1.50 | 1.72 | 4.37 |
|         | 3.0 [a] | 7.62 | | | | |

[a] Excluding terminals.

Yardney Electric Corporation's HR4A-2 cell, used for this application, consists of 6 positive (Ag) and 7 negative (ZnO) electrodes with a total active positive electrode area of 33 in.$^2$ Again, 4 turns of C-19 separator are used and the electrolyte is 45% KOH. The cell cases and covers are molded using Bakelite C-11. The two 10-32 screw terminals provided per cell are more than adequate for the 45 A—0.2 sec—maximum current application. The cell yields 39 Wh/lb (86 Wh/kg) and 2.7 Wh/in.$^3$ (164 Wh/dm$^3$) at the 60-min rate, 21.1° (70°F) to 1.10 V.

## F-106 EMERGENCY BATTERY

This battery is nominally a 28-V, 15-Ah battery at the 1-hour (15-A) rate. The Yardney Electric Corporation's 18HR15V (their P/N 10987) has been tested and qualified to the applicable Air Force Purchase Description.

The environmental requirements are identical to those for the canopy battery. Thus the cell case and cover and battery case material used were essentially the case as those for that battery. The internal battery components are encapsulated up to the cell-case tops in a polyurethane potting compound. An ac-powered heater with controlling thermostats is also provided for low-temperature operation.

Electrically, the battery must supply 15 A for 30 min at a voltage within the limits of 24.5 and 31.0 V followed by 30 A for five minutes at a minimum voltage of 24.5 V (1.36 V per cell). The battery must provide 30 cycles of operation and deliver 15 Ah initially and no less than 13.5 Ah at the end of the cycle life.

The Yardney Electric HR15V cell contains 8 positive (Ag) and 9 negative (ZnO) electrodes, with a total active positive electrode area of 83.4 in.$^2$ As in the other aircraft-application cells, the separator used is 4 turns of Yardney C-19 and the electrolyte is 45% KOH. Each cell (18 per battery) comes equipped with two $\frac{1}{4}$-28 screw terminals. The cell and battery case dimensions are shown in Table 66. The cells are capable of 44 Wh/lb (97 Wh/kg) and 3.0 Wh/in.$^3$ (183 Wh/dm$^3$) at the 60-min rate, 21.1° (70°F) to 1.10 V.

TABLE 66. F-106 EMERGENCY BATTERY AND CELL
DIMENSIONS

|  | Height | | Length | | Width | |
|---|---|---|---|---|---|---|
|  | in. | cm | in. | cm | in. | cm |
| Battery | 5.125 | 13.0 | 5.00 | 12.7 | 8.05 | 20.5 |
| Cell | 4.89 | 12.4 | 0.81 | 2.06 | 2.30 | 5.84 |
|  | 4.22[a] | 10.7[a] | | | | |

[a] Excluding terminals.

## BATTERY MANUFACTURING AND PROCESSING INFORMATION

The basic component of the zinc-silver oxide aircraft battery is, of course, the cell. The function of the remaining battery components is to provide electrical interconnections for the cells, to bring the electrical energy to a usable junction and to protect the assembly of cells from any expected environment or other external stresses.

The manufacturing techniques used by the various battery producers are, for the most part, kept in strictest confidence, the competition in the industry being extreme. Customers will be provided with information on battery components and assembly techniques, upon request down to but *not* including the cell itself. Yardney Electric Corp. manufactured the positive silver electrodes by their patented (R65, 149, 420, 421) rolling mill processing technique (R622).

The basic specification for quality control requirements for the military and their suppliers is MIL-Q-9858A "Quality Program Requirements." This specification "requires the establishment of a quality program by the contractor to assure compliance with the requirements of the contract."

## OPERATION AND MAINTENANCE OF ZINC-SILVER OXIDE
## AIRCRAFT BATTERIES

Military aircraft batteries are packed and shipped according to the requirements of the applicable procurement documents. This section of the paper reports on the operation and maintenance tasks, procedures, and equipment involved from receipt of the batteries until they are removed from aircraft service.

The exact nature and frequency of battery maintenance depends on the functions of the aircraft and the battery. Technical manuals, prepared for each aircraft and each battery, provide instructions for flight-line level and battery-shop level maintenance. Also, the battery manufacturer supplies instructions with each shipment for preparing, operating, and maintaining the batteries. The technical manuals used by the Air Force and the Navy are listed in Table 67.

Most military aircraft batteries are received dry and unformed; thus the initial preparation-for-use step is filling with electrolyte, using the kit supplied with each battery. A soak period of 72 hours minimum to 10 days maximum is required before forming the batteries. The technical manuals recommend constant-current charging and discharging during formation and servicing operations; however, the values for the rates, cutoff voltage, and capacity are provided for both constant-current and modified constant-potential methods. The formation and servicing operations should be accomplished between 15.5° (60°F) and 32.2° (90°F). The Navy manual gives a temperature-capacity correction.

The maintenance routine for aircraft batteries generally consists of: (1) scheduled periodic servicing of batteries in storage; (2) scheduled periodic removal of batteries from aircraft for servicing; and (3) removal of batteries from aircraft for servicings as dictated by the result of preflight and postflight checkouts.

The dry, unformed aircraft batteries are stored at $\leqslant 60°$ (140°F). The filled and formed batteries are stored at $\leqslant 32.2°$ (90°F). They are removed from

TABLE 67. OPERATION AND MAINTENANCE MANUALS

| Aircraft Battery | Aircraft Electrical System | Battery Operation and Maintenance | Overhaul and Parts |
|---|---|---|---|
| F-84F and RF-84F 6Ah | T.O.IF-84F-2-9, 12 March 65 T.O.IF-84(R)F-2-4, 29 Jan 65 | T.O. 8D2-4-11, 1 Nov 60 | T.O. 8D2-4-13, 1 Nov 60 |
| F-105 100 Ah | T.O.IF-105B-2-9, 25 Jan 66 T.O.IF-105D-2-9, 25 Jan 66 T.O.IF-105F-2-9, 20 Feb 68 | T.O. 8D2-2-1, 5 Sept 66 | T.O. 8D2-2-4, 1 Aug 62 |
| F-106 15 Ah | T.O.IF-106A-2-10, 1 Apr 68 | T.O. 8D2-8-1, 20 May 64 | T.O. 8D2-8-3, 1 Apr 64 |
| F-106 3 Ah | T.O.IF-106A-2-10, 1 Apr 68 | T.O. 8D2-7-1, 1 Oct 64 T.O. 8D2-7-15-1, 1 Jun 67 | T.O. 8D2-8-3, 1 Apr 64 |
| Navy All batteries | | "Handbook, Service and Maintenance Instructions for NAVAL Aircraft Storage Batteries", Jan 1965 | |

storage and service discharged and charged every thirty days, and drain discharged (0 to 2 V) before reshipment. Batteries are removed from aircraft for servicing after a specified usage (5 cycles for F-84 series aircraft) or duration (usually 90 days) or whenever flight-line checkouts indicate a weak cell or battery. The flight-line measurements are taken no sooner than two hours after use of the battery to assure the battery is stabilized. Cell(s), after servicing, continuing to show an open circuit voltage outside the range or 1.82 to 1.86 V are replaced with comparable aged cell(s) or a different battery if of a cell-nonreplaceable battery design. The electrolyte level is examined and adjusted as necessary in the Navy batteries only. The Navy batteries are removed from the aircraft every sixty days for examination and adjustment of the electrolyte level. Other items such as battery heaters and sump jars are maintained per the instruction manuals. The final step is to record the vital information associated with each maintenance action. Important information such as dates of formation and servicing are recorded on the service placard affixed to the outside of each battery case. This information and other details are entered on special "battery maintenance record forms" kept on file in the Battery Shops.

## FIELD PERFORMANCE OF ZINC-SILVER OXIDE AIRCRAFT BATTERIES

Inventory management records are kept on each component of the aircraft. For the batteries, these records consist of "battery maintenance record forms." Maintenance personnel are instructed to complete these forms following each maintenance action, either scheduled (i.e., according to the technical manual) or unscheduled (i.e., dictated by the situation). The information contained on the forms includes: dates; nature and duration of maintenance (i.e., number of scheduled and unscheduled manhours); nature of malfunction; and reason(s) for repair or replacement of the component. This information is coded, thus enabling the use of machines for tabulating and retrieving all or selected portions of information such as the number of unscheduled manhours during a particular time period. The inventory management records are entered into an inventory information network established and operated by Logistics Command.

The Logistics Command uses the inventory management records to assure that aircraft repair and replacement parts are available, and overhauls of major components are accomplished. These records also indicate how well a component is performing and its associated maintenance and costs aspects. Recent data on the military aircraft zinc-silver oxide batteries indicate the following average service lifes:

| | |
|---|---|
| Navy batteries | 8 to 9 months |
| Air Force F-84F, RF-84F, and F-106 batteries | 6 to 7 months |
| Air Force F-105 battery | 3 to 4 months |

Service life is the time period after activation during which the battery is capable of powering the required aircraft loads. The F-105 battery performs satisfactorily only slightly longer than the initial, 90-day capacity maintenance interval. Scheduled maintenance on this battery has been averaging about only 10% of the total maintenance, with all maintenance averaging about 5.5 hours/action. The reasons (malfunction) for unscheduled maintenance are given by descriptors which typically include: incorrect voltage; incorrect current; low or no output; cell unbalance; insulation breakdown; defective contact; burned; leaking; shorted; bent; corroded; and dirty. These same type descriptors are used also to describe the reasons why cell(s) and batteries are replaced. Incorrect voltage is the most frequently occurring malfunction. Failure is usually given as loss of capacity or shorting.

Scheduled maintenance exceeds unscheduled maintenance for the other zinc-silver oxide aircraft batteries. The factors contributing to the less-than-expected performance of all the mentioned batteries, particularly the F-105 battery, are discussed in the next section.

## FACTORS AFFECTING BATTERY PERFORMANCE

The major factors affecting the performance and life of the military aircraft cells and batteries are: (1) operating practices, (2) battery conditions aboard aircraft, and (3) battery limitations.

### OPERATING PRACTICES

Most cell/battery malfunctions and failures can be attributed to improper operation, either because of inadequate procedures or nonadherence to correct procedures. The situation of having ill-trained personnel assigned occasionally to battery shops accounts for instances of nonadherence to procedures. Regarding questionable procedures, the following examples are noted. The technical manual instructs a minimum of three or possibly five formation cycles for the F-105 battery and three to five cycles every time that cells are replaced. Also, one to two cycles are completed for every scheduled (90 days) and unscheduled maintenance. These cycle requirements subtract substantially from the total expected life of the battery. The technical manual instructs that no adjustments of the electrolyte of the Air Force batteries are necessary. Examination of cells of three to ten months age removed from F-105 aircraft operated in arid environment showed over half of these cells had no electrolyte, and most of the remaining cells showed only a trace of electrolyte. Regarding nonadherence to procedures, the violations most detrimental to battery performance and life are: missing scheduled maintenance; abuses in the field; and charging at high ambient temperatures ($>32.2°$, $>90°F$) and high battery temperatures ($>65.5°$, $>150°F$). The abuses in the field generally are unauthorized charging of the battery using ground power, and unauthorized discharging of the battery such as for electrical checkout of other systems on the aircraft. Also, use of out-of-calibration instruments during maintenance; improper storage of cells; and failure to keep accurate and up-to-date battery maintenance records detract from the expected battery performance.

## BATTERY CONDITIONS ABOARD AIRCRAFT

Batteries are charged by the constant potential method aboard aircraft. They are connected at all times during aircraft operation to the d-c bus through a voltage regulator, thus are either being charged or discharged. Charging is limited only by the upper voltage, whose setting at the regulator, although adjustable, is kept usually at a selected single value such as 27.75 V for the F-105 battery. This practice does not compensate for the effect of temperature (ambient) ranging from less than 0° (32°F) to above 32.2° (90°F) on battery-charging characteristics. Since individual cells are not monitored, they are vulnerable to excessive overcharging during extended float-charging situations and reverse discharging during high-rate discharges. The F-105 battery, because it supplements the generator system for flap motor operation, is subjected to high-rate charging following discharge in order to be fully charged for emergency situations.

## BATTERY LIMITATIONS

The zinc-silver oxide battery system is sensitive to overcharging and must be designed specifically for high-rate charging. The F-105 battery encounters both these situations. The four-terminal-post design of this battery provides a high current-carrying capability, however, it has been thought to introduce cell imbalance problems. With this design, the cell consists of two cell packs of positive and negative plates contained in a single cell case. Potential differences are frequently measured between any two terminals when the intercell connectors between the common terminals are removed. This potential difference indicates a probable difference in the state-of-charge of the two packs. Tests of these cells that exhibit this phenomena, however, have shown that this potential difference is a result of one of two occurrences: there is an internal short circuiting that develops in one of the cell packs, in which case a two-terminal cell would show the lower voltage more quickly; and the cell has been discharged sufficiently to bring it to the borderline between the first and second silver plateau levels, in which case any small amount of charge brings both packs to a 1.86 V open-circuit-level or any amount of discharge brings both packs to a 1.62 V open-circuit level. Equalization of the potential unbalance can be easily achieved by connecting the common terminal; however, the technical manuals presently do not instruct that activated cells be stored with the common terminals connected.

## OVERALL ASSESSMENT AND FUTURE PROSPECTS OF ZINC-SILVER OXIDE AIRCRAFT BATTERIES

The rechargeable battery system costs more and has a shorter average service life than the lead-acid and nickel-cadmium aircraft battery systems. It will, however, remain competitive with these battery systems for specialized aircraft applications because of its qualities of high specific energy (Wh/lb), high specific volume, (Wh/in$^3$), and ability to deliver high currents. The service performance of existing zinc-silver oxide aircraft batteries must be improved considerably before the rechargeable system becomes a contender battery system for across-the-board aircraft applications. The prospect of such batteries for future military aircraft applications is somewhat dim. This is because of the inherent mismatch between the rather delicate rechargeable battery system and the abusive nature of the aircraft application.

# 29.

## Zinc–Silver Oxide Torpedo Batteries

### Francis G. Murphy

It was approximately twenty years ago that the Navy became interested in the silver-zinc battery for torpedo propulsion. The problem at that time was very simple—provide a replacement for the heavy, troublesome lead-acid batteries. As evidenced by their continued usage silver-zinc secondaries have provided an excellent replacement. The next step—to provide a primary silver-zinc battery for war-shot application—has not been so successful for a variety of reasons, principally torpedo system design and usage.

The torpedo is not designed, built, or used in a one-shot application. Of course in its wartime usage it is self-destructive. But meantime, proofing runs, tactical-data runs, development runs, and fleet training runs add up to a sizeable number. One estimate places the average number of runs per torpedo at 10. Run time varies depending on torpedo type and purpose of the run. We should immediately distinguish between the aircraft-launched (missile, surface-ship) torpedo and the submarine-launched torpedo. The former is small, lightweight and has a short run time. The submarine-launched torpedo is approximately twice the diameter, two to three times the length, and has twice the run time for war-shot use.

While there is considerable variation in the types of motors used, the series motor is best suited and has been used for the most part with the silver-zinc battery. To date there has been no speed regulation and the battery has been directly coupled to the motor load, although through a contactor for secondary batteries.

Specific details of torpedoes cannot be mentioned for security reasons. Broadly speaking, in excess of 50% of the weight and volume of a torpedo is taken up for propulsion. The energy density of the battery is therefore very significant.

Returning again to history, primary batteries were introduced into the system in order to provide long-term operational readiness without maintenance. An increase in energy density was not the goal. As a matter of fact, since the torpedo systems were developed around the secondary battery, it was a mandatory requirement that the primary battery be equal in weight and occupy the same volume as the secondary battery that it replaced. Performance capabilities remained essentially static for electric propulsion systems. Meantime hot gas propulsion systems were showing major performance gains as a result of concentrated system engineering. Ten years ago the future of batteries for torpedo propulsion looked very dim. Today serious consideration is being given once again to electric propulsion. System engineering and a major change in battery design concepts have caused this change in attitude. The silver-zinc battery in a primary pile design has demonstrated on a volume basis the maximum in energy storage capability.

The results of our system studies may be translated into battery requirements as follows:

1. Operational readiness must be maintained without maintenance after months of shipboard environment.

2. High transient electrical and dynamic load during torpedo startup must be maintained while being subjected to dynamic loading as the torpedo is launched

3. Constant voltage output at high current density, 250 mA per cm² for a time duration between 5 and 15 minutes quantifies the range of interest.

4. Ability to supply an auxiliary electronic load is required under the conditions listed above without transmitting to the auxiliary load the electrical noise produced by the main motor load.

5. There must be no emission of explosive gases within the main torpedo volume.

6. The battery must serve as its own heat sink during discharge.

7. The battery must present no safety hazard to fleet personnel while carried aboard ship, during the torpedo launch, recovery and postrun maintenance.

## OPERATIONAL READINESS

There are a number of specifications that define the shipboard environment General familiarity with shipboard environment can be safely assumed for the purpose of this presentation. The major point in the first requirement is no maintenance.

## HIGH TRANSIENT ELECTRICAL LOAD

The second requirement—high transient electric load—is of major importance as the power level increases. The surge current at motor startup can be five to six times steady-state current. While the analysis for a specific system is quite detailed, the following simplification will illustrate the electrical transient. The equation for voltage applied to a motor is

$$E = I_m(R_m + R_b) + K_v N_m$$

where $E$ = emf developed in battery; $I_m$ = current in amperes; $R_m$ = armature circuit resistance; $R_b$ = battery internal resistance; $K_v$ = motor back emf constant volts/rad-sec; and $N_m$ = motor shaft speed, rad/sec.

Since in high power motors $R_m$ is in the order of 0.02 ohm, we are essentially short circuiting the battery when $N_m$ is zero at motor startup. A motor start circuit is not desired becase of the large weight and volume penalty. Ideally the comeup time of a primary battery can be matched to the acceleration time of the motor to provide a safe limit to the motor inrush current. The common practice is to directly couple the motor to the battery; remote activation of the electrolyte filling system starts the torpedo. There is approximately a 15-second transient period with a maximum current value approximately double the steady-state value. Meanwhile the torpedo has exited the torpedo tube in approximately one second with a maximum acceleration of 10 g. For the smaller torpedoes in which there is air travel prior to water entry; the water-entry shock is of major consideration. For aircraft launching a maximum of 250 g has been specified for water-entry shock. Torpedo control is dependent both on torpedo speed and electrical power to make the control surfaces effective. Thus during the launch phase there is a potential problem area in torpedo trajectory. This is reflected back into additional transient dynamic loads on the battery system.

## CONSTANT VOLTAGE OUTPUT

Constant voltage output merits some discussion. As was mentioned earlier, it has been the usual practice to couple the silver-zinc battery with a series motor. Since the effect of heating on motor speed is negligible for a series motor, the torpedo speed is constant as long as the battery voltage is constant and the torpedo is on straight and level path. No speed control system is required. A capability to vary torpedo speed is desirable in some situations, but the basic simplicity of the silver-zinc battery coupled to a series motor is a major advantage. The intercept of a torpedo with a target is a simpler calculation when torpedo speed is constant. Torpedo fire-control systems can handle variable speeds as long as they are known and reproducible. The third requirement may be restated as follows: constant voltage output of the battery is a desirable attribute in keeping with a major advantage of electric propulsion systems— simplicity.

## AUXILIARY LOAD

The auxiliary load of a torpedo consists of a variety of a-c and d-c supplies to control and guidance electronics. It has been the practice to date to supply 28 V dc to a rotary inverter, either as a tap voltage from the main propulsion battery or from a separate set of cells. Solid-state inverter supplied by full propulsion battery voltage appears to be the optimum method for powering electronic systems in future torpedoes. Cross talk between main power circuits and low-level electronic circuits is a problem in any torpedo system design.

## EXPLOSIVE GASES

The problem of explosive gases in an electric torpedo is self-explanatory. The silver-zinc system operating within its limitations has not caused problems in this respect except at the end of the run. Two safety precautions have been taken. The torpedo is purged of air and backfilled with nitrogen; the battery compartment is isolated by a pressure bulkhead. Further safety considerations will be reserved for discussion of the seventh requirement.

## BATTERY TEMPERATURE

Even though the ocean is an ideal heat sink and heat transfer from a metal torpedo shell to the ocean is excellent, the thermal conductivity of the medium between the battery proper and the torpedo shell precludes any sensible heat transfer. This may be shown as follows: the total heat $Q$, produced by the battery inefficiency is equal to $Q_s$, the heat absorbed by the battery plus $Q_c$ the heat conducted to the ocean.

$$Q = Q_s + Q_c$$

$$Q_s = mC_p \frac{\partial T}{\partial t}$$

$$Q_c = KA(T - T_c)$$

$$Q = mC_p \frac{\partial T}{\partial t} + KA(T - T_c)$$

$$\frac{\partial \theta}{\partial t} + \frac{KA\theta}{mC_p} = \frac{Q}{mC_p}$$

where $m$ = weight of battery, $C_p$ = specific heat of battery, $T$ = temperature of battery, $t$ = time in hours, $K$ = conductivity of medium between battery and ocean, $A$ = surface area of battery exposed to ocean, $T_c$ = ocean temperature, $\theta = T - T_c$, and $\partial \theta / \partial t = \partial T / \partial t$. Thus we have

$$\theta = Q \exp(-KAt/mC_p) + Q/KA \tag{1}$$

and when $t = 0$, $\theta = 0$, therefore $Q = -Q/KA$, and

$$\theta = [Q/KA] [1 - \exp(-KAt/mC_p)] \tag{2}$$

Since batteries are rated in useful energy, the energy wasted in heat is determined by the ratio of the inefficiency to the efficiency, $R$, times the rated energy. $P$;

$$Q = (RP/1050) \text{ BTU/sec} \qquad \text{or} \qquad (RP/4200) \text{ kcal/sec.}$$

The curved suface of a right cylinder is related to the volume

$$A = \frac{2Pt}{rD_v} \quad \text{and} \quad m = \frac{Pt}{D_m}$$

where $D_v$ = watt hours/in.$^3$ and $D_m$ = watt hours/lb. Therefore

$$\theta = [rRD_v /2100Kt] [1 - \exp(-2KD_m t/rC_pD_v ] \tag{3}$$

By inserting the appropriate values in Equation 3, it can be shown that the time constant for all systems without dynamic cooling is in the range of hours. For a 10-minute run Equation 3 reduces to

$$\theta = RD_m /1050tC_p \tag{4}$$

That is, all the heat produced by a battery is absorbed by it unless dynamic cooling is provided. In reviewing possible systems for development, temperature limitations must be examined closely.

## SAFETY

The last requirement listed, no safety hazard, is relative and may be met by a variety of means. Suitable instructions, safety equipment, and accepted practices in the handling of high voltages and caustic solutions suffice for the majority of hazards. Because of their size and usage, special consideration has to be given to the primary silver-zinc torpedo batteries. Accidental activation has occurred a number of times and not been noticed until the battery started to consume itself chemically. Since both hydrogen and oxygen are produced, there is an explosive danger. There is also a danger of fire. In most cases there were large emissions of toxic gases resulting from the high-temperature oxidation of plastic materials of construction. The most obvious solution to this problem is incorporation of a bleed resistor across the battery terminals. The power lost during a normal discharge would be only a small fraction of the power delivered to the motor.

In the launch phase there are two considerations. The battery must be constructed to withstand the launching forces and thus prevent internal shorting and resultant fire or explosion or both. In the smaller torpedoes, activation time encompasses both the air- and water-entry phases and hence the batteries have to contend with higher dynamic forces.

The second consideration relative to launching safety has to do with activation time. As mentioned earlier, control of the torpedo is dependent on electrical power. Slow comeup time of the battery could lead to erratic torpedo trajectory and pose a real threat to the launching ship.

During torpedo recovery and postrun maintenance, the accumulation of gases in the battery must be vented to eliminate this source of danger. The battery compartment has a relief valve to vent any pressure built up in the compartment. For this valve to be effective, the battery must have its own venting system into the compartment.

Another item in recovery and postrun maintenance applies to primary batteries. For a number of reasons the torpedo run may be shorter than the battery capability. For the more powerful batteries, the small (on a percentage basis) residual capacity after a full-length run is sufficient to provide a fire hazard. For both these cases, a bleed resistor is required to drain the battery completely. Both in the case of accidental activation and in postrun safety, the wet-stand capability of the primary battery should be sufficient to match the drain time of the bleed resistor.

## PILE DEVELOPMENT

As mentioned earlier, the future of electric propulsion for torpedoes looked very dim ten years ago. The weight, volume, and cost of batteries were prohibitive for the high-powered torpedoes being developed. The concept of a silver-zinc pile integrated with the torpedo shell originated at Naval Underwater Weapons Station (NUWS) eight years ago. Initial estimates were that the energy density of torpedo-propulsion batteries could be doubled. These estimates were based on the following weight and volume savings:

1. Separate cell cases and intercell connectors could be eliminated.
2. Battery case and support structure could be eliminated.
3. Thickness of torpedo shell could be reduced since the pile supports the shell against external sea pressure.
4. The full cross-sectional area of torpedo could be used for maximum plate area.
5. Energy losses in intercell connections would be decreased.
6. Energy losses would be decreased because of more fully wetted electrodes.

To ensure proper filling, the cells are evacuated before pressure injection of electrolyte.

Because of the dim outlook for electric propulsion, essentially all development funds were allocated to thermal propulsion, and the pile development started on an extremely low budget. Fortunately interest in pile development was prevalent in several companies, and as part of a study contract some pile modules were being tested. It was decided to extend the work to manufacture and test three 50-cell piles in accordance with NUWS concept and a Yardney design. Results were quite successful. With each test an improvement was made over the previous one. Since all three batteries were built with the same hardware (simulated shell), which is not representative of the final design, data on these tests are given on a cell basis in Table 68.

Results obtained showed that, at equivalent current densities, cell voltages were consistently higher than those obtained in conventionally constructed batteries. Utilization of active material was also better, particularly when vacuum activation (Battery No. 3) was used. Intercell leakage, previously regarded as a major problem in the pile, was of little consequence. This limited program demonstrated the feasibility of the pile battery for torpedo propulsion.

Based on the success of the initial tests, parallel contracts were let to Yardney Electric Corporation and Electric Storage Battery Company for continued investigation. Ultimate objective of the contracts was the design of a primary battery, complete with activation system and integrated with the torpedo shell.

Much of the state of art was not applicable since pile design is such a radical departure from conventional practices. The size of the electrodes, their area

TABLE 68.  PERFORMANCE OF YARDNEY DESIGN
PILE BATTERIES, EXPERIMENTAL MODELS

| Assembled and Activated Battery | No. 1 | No. 2 | No. 3 |
|---|---|---|---|
| Number of cells | 46 | 50 | 50 |
| Discharge rate, kW (battery) | 19 | 7.5-27 | 15.7 |
| Energy, Wh to 1.0 V/cell | 3280 | 3863 | 5113 |
| Weight, | | | |
| pounds | 90 | 92 | 99 |
| kilograms | 41 | 42 | 45 |
| Volume, | | | |
| in.$^3$ | 680 | 750 | 780 |
| dm$^3$ | 11.1 | 12.3 | 12.8 |
| Energy density, | | | |
| Wh/lb | 36.4 | 42.0 | 51.7 |
| Wh/kg | 80 | 92 | 114 |
| Energy Density, | | | |
| Wh/in$^3$ | 4.8 | 5.2 | 6.6 |
| Wh/dm$^3$ | 295 | 310 | 400 |

compared to their thickness, the close spacing, vacuum activation, temperature rise during discharge—all these factors combined to make design parameters uncertain. Within the limits of the fixed-price contracts, the following items were investigated:

1. Techniques to manufacture reproducible duplex electrodes
2. Chemically formed AgO versus electrolytically formed AgO
3. Materials for duplex barrier foils
4. Electrode spacing
5. Separator material
6. Electrolyte concentration
7. Electrolyte temperature
8. Zinc density and zinc electrode fabrication.

Rather than dwell on the separate investigations at this time, comments will be reserved for the final portion of the discussion. The net results of the two development contracts were two basically similar designs. Electric Storage Battery Company provided a final demonstration of their design with a 10-cell 21 in. (53-cm) diameter pile test whereas Yardney Electric Corporation used a 155-cell 10-in. (25-cm) diameter pile test.

A follow-on contract was let to make a detailed design of a 340-kW, 12-minute battery and manufacture 3 production prototypes. The design was completed after testing 12 additional 10-cell, 21 in. (53 cm) diameter modules. Figure 144 illustrates the design. Assembly of the first battery was plagued with the usual bugs of any new development. Special tooling and assembly fixtures had to be designed and built. Stacking of the duplexes, larger perimeter gaskets, and separators was a tedious painstaking task. Compression of the two parallel piles with the electrolyte tank sandwiched between them in the torpedo shell had to be accomplished with insufficient visual checks. As a result when the battery was discharged, the face seals at the ends of the electrolyte manifold failed. Gross leakage of electrolyte occurred, shorting the piles directly to the shell. Failure analysis disclosed that the individual piles did operate and were essentially intact. Practically the entire battery energy was expended in electromachining a groove in the aluminum shell with the bus plate serving as the tool. The major design deficiencies in the battery were the face seals and the method of application of the Teflon liner on the aluminum shell. Redesign is currently underway with a repeat test scheduled for the end of 1968.

Much has been learned in this program to date. There is every reason to believe that a pile design is the logical choice for silver-zinc battery application to torpedo propulsion. More development is required both in mechanical design and the electrochemistry of high rate—high temperature discharges.

To conclude this presentation, a summary of the data gathered during the pile development will highlight the electrochemical problems encountered. As

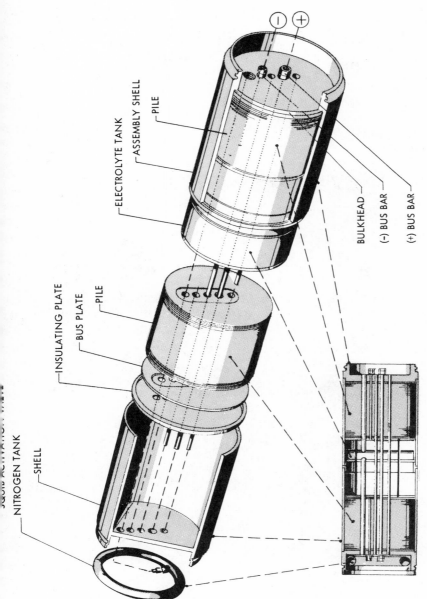

NITROGEN TANK

SHELL

INSULATING PLATE

BUS PLATE

PILE

ELECTROLYTE TANK

ASSEMBLY SHELL

PILE

BULKHEAD

(−) BUS BAR

(+) BUS BAR

Figure 144  Pile battery assembly.

401

mentioned earlier, the battery serves as its own heat sink. Consequently under the conditions of high-rate, short-time discharge, battery design is predicated upon a limiting temperature at the end of the discharge. In the 155-cell discharge an estimated 300°F (149°) was reached at 17 minutes of run time with a current density of 230 mA/cm². Figure 145 presents a picture of the battery, and Figure 146 is a copy of the discharge data. The 155-cell test cell voltage (1.45) was approximately 0.06 volt higher than previous 10-cell tests because of the higher operating temperature. Silver utilization was approximately the same. In an earlier test, a 10-cell battery was activated with 31% electrolyte at 93° (200°F) cell voltage showed the same increase; in addition, silver utilization was increased by more than 6%. Final battery temperature remained unchanged.

**Figure 145** Duplex electrode battery.

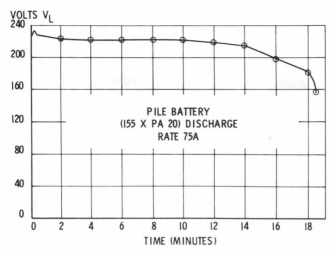

**Figure 146** Pile battery discharge data.

Early tests on the concentration of KOH demonstrated an optimum of 31% when zinc density was 1.4 g/cm$^3$ and separator spacing was 0.009 in. (0.23 mm). Final 10 cell tests of the 21-inch (53 cm) diameter pile demonstrated a silver utilization of 2.9 g/Ah with 40% KOH compared to 3.7 g/Ah with 31% KOH. In this case zinc density was 2.0 g/cm$^3$ and separator spacing was 0.016 in. (0.41 mm).

Many attempts were made to form the electrodes directly on the barrier foil. Final design involved separate formation of both the silver and zinc electrodes and physical contact only between the electrodes and barrier foil. While the pressure of the compressed pack maintained a good physical contact, spots of glue were used to position the electrodes on the foil and to ease the pile assembly problem. The decision to rely on physical contact between barrier foil and electrode has lessened the restrictions on barrier foil material. Presently the use of stainless steel is being investigated. To date, no degradation in performance has been found.

Chemically formed silver oxide is still of interest; however, tests to date have not been encouraging. For the present, charged sintered silver strip for the cathode and charged pasted zinc oxide for the anode appear the best choice. Batch processing of active materials and dry fabrication of electrodes directly on the barrier foil offer economy in manufacture and therefore justify continued research.

# 30.

## Air-to-Air and Air-to-Ground Missile Batteries

### R. L. Kerr

Selection of an electrical power supply for air-to-air and air-to-ground missile applications requires consideration of several factors. The energy densities must be high, storage life must be long prior to use, and activation time must be short once operation is initiated. Consideration must also be given to the environmental constraints such as shock, vibration, and temperature range for both storage and use conditions.

Because of its high gravimetric and volumetric energy densities, the zinc-silver oxide battery has been selected for extensive use in both air-to-air and air-to-ground missile applications. To achieve the long storage life desired in this use, automatically activated dry-charged reserve type cells have been developed with emphasis on methods to provide rapid activation.

Typical requirements for missile applications include storage between $-54°$ ($-65°F$) and $75°$ ($165°F$) for years followed by heating of the battery from $-46°$ ($-50°F$) in approximately 5 minutes. Operation of such batteries is required within less than ½ second following initiation of activation. In use, the battery can experience a shock and acceleration up to 100 g together with vibration and temperatures up to $66°$ ($150°F$).

Storage of the battery in a dry condition provides the long life needed for an operational missile; however, potassium hydroxide electrolyte must then be supplied to the cells rapidly when power is needed. To accomplish this, many designs have been tried and many refinements made. Early designs attempted to use a gravity feed by placing the electrolyte reservoir above the cells and relying on gravity to move the electrolyte when needed. As might be supposed, this was a slow and orientation dependent process which was unsuited for most applications.

To improve on the speed of activation, an evacuated battery compartment was added and this helped to move the electrolyte from the atmospheric pressure reservoir into the cells. The electrolyte in this case was held in a collapsible rubber or plastic bag and released into the hermetically sealed battery compartment by puncturing a separating diaphragm with a movable plunger.

Upon entering the battery, the electrolyte flowed through a manifold into the cells. Gas generated within the battery, especially at high storage temperatures, and possible leaks could destroy the vacuum. This was difficult to detect. In addition, at low temperatures, it was difficult to heat the electrolyte for rapid activation. Because of these problems, this concept was also abandoned although it had shown promise of light weight.

To eliminate the difficulties which had been encountered, positive methods of introducing electrolyte were developed. These designs fall into two classes, one using gas pressure and the second using piston motion. Several gas pressure concepts for forcing the electrolyte into the battery have been examined. One concept opens a gas pressure chamber into a second chamber from which the electrolyte is forced into the battery manifold and then into the cells. In this design, the activation chambers can be an integral part of the battery or can be removed after use. Another method uses a coil type reservoir surrounding the battery. A mechanically or electrically fired gas generator at one end of the coil forces the electrolyte through the tube, into a manifold, and into the cells. Frangible diaphragms are generally used at both ends of the coil to isolate the electrolyte, and these are burst when the generator is fired. These diaphragms and the metal tube allow a metal-to-metal seal to contain the potassium hydroxide electrolyte until the battery is activated. The metal tube also provides a large and efficient surface for supplying heat to the battery.

The first battery designed with a coil reservoir, the BA472/U, was developed against extreme environmental requirements. This battery can withstand a 240-g shock and acceleration, 40-g sinusoidal vibration from 5 to 2000 Hz, and temperature variation from $-62°(-80°F)$ to $100°$ $(212°F)$. Although the specification for this battery required only 5 Wh/lb (11Wh/kg), the actual output was about 9 Wh/lb (20 Wh/kg). Later designs using the coil reservoir have produced 15 Wh/lb (33 Wh/kg).

The second design for automatic activation which is currently in use employs a gas-driven piston to force the electrolyte from a reservoir as opposed to the gas generator in which a gas bubble is actually a form of piston. This concept offers weight advantages for large size batteries over the coil design.

In the case of the piston activator, an electrically or mechanically ignited gas generator moves a piston which pushes the electrolyte through a rupture plate and into the cells. The piston delivers a precise amount of electrolyte to the cells by moving until it strikes the end of the tank in which it is located.

Design of these activators must consider good seals with long life, materials and welds which can withstand prolonged exposure to the electrolyte, and ignitors which provide a smooth but rapid pressure rise together with an evenly sustained pressure throughout the piston travel time.

Battery designs are of either the plate or pile types, one employing a manifold and individual cells and a second using a bipolar type construction. The manifold concept allows uniform distribution of the electrolyte to the cells based on the pressure and is often combined with an overboard vent to release excess gas or electrolyte which might swell the battery. The vents are usually protected by a sump or orifice to eliminate the possibility of electrolyte overflow.

In the case of bipolar cells, the electrolyte moves into the cells, wetting the separators. By proper porting, the cells can be activated uniformly but the rate of activation for large batteries can become a problem.

The environmental requirements for vibration, shock, and acceleration are generally met and indeed exceeded by potting the already compact battery designs. Foamed compounds and epoxy resins are both used depending upon weight and environment requirements.

In order to provide for operation at low temperatures, batteries must be heated, and the easiest method of accomplishing this is by supplying heat to the electrolyte. When aircraft power is available, the heating is accomplished through blanket electrical heaters wrapped around the electrolyte reservoir. Electrical heating provides the capability to maintain the required operating temperature for long periods; however, chemical heating has also been employed in some cases.

Heat can also be supplied to missile batteries by the use of chemicals. In this case, the electrolyte is heated on its way to the battery compartment by a heat powder which is ignited by electric matches. The matches can be wired in series with a thermostat and this in parallel with the activating squibs allows the matches to fire when the battery is activated. If the thermostat is closed in response to the need for heat, the heat powder is ignited. If no heat is required, the thermostat will be open and the powder will not be used. Unlike the electrical method of heating the battery, the chemical system can be used only once and even then it places additional constraints on the battery. The electrical heaters can heat the battery and maintain the proper temperature ahead of time while the chemical heater must supply all of the necessary heat within a short span of time. The missile must also be used following that time or the battery must be replaced.

The use of chemical heat requires consideration of rate of electrolyte flow into the battery, maximizing the surface area for heat transfer and materials of construction for good heat transfer. The type and amount of heat powder must also be controlled in order to allow for sufficient heat transfer without jeopardizing the activation time, temperature limits, or voltage regulation of the battery.

A typical air-to-air missile battery (Figure 147), which uses the coil design is the GAP4007. Although this battery delivers only 10 Wh/lb (22 Wh/kg), it represented a large weight reduction and increase in discharge rate when it was introduced.

The GAP4007 battery, which has the requirements listed in Table 69, has

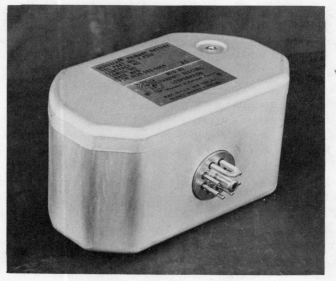

**Figure 147** Typical air-to-air missile battery.

recently been tested to establish its performance after approximately 5 years of storage. These tests reflect the high reliability and long unactivated stand capability of silver-zinc missile batteries. Virtually all of the batteries tested have performed within specification and, in fact, only minor problems have been encountered since this battery was placed in service.

A detailed look at this battery (shown in Figure 148) provides a representa-

TABLE 69.   REQUIREMENTS FOR GAP4007 AIR-TO-AIR
MISSILE BATTERY

| | |
|---|---|
| Output voltage | 13.2-14.6 V |
| Required capacity | 0.85 Ah |
| Discharge rate | 34 A |
| Activation time | 0.5 sec under load |
| Wet life | 2 hours |
| Temperature limits | $-46°(-50°F)$ to $66°$ $(150°F)$ |
| Vibration | 10-500 cps |
| Shock | 100 g (base to cover) |
| | 20 g (cover to base) |
| | 45 g (either perpendicular axes) |
| Acceleration | Same magnitudes as shock for at least 3 seconds each |
| Heater power | 300 W at 108-115 V ac for 5 min prior to activation |
| Weight | 1.22 kg ( 2.7 lb ±0.1 lb ) |
| Dimensions | 3.3 x 5.6 x 3.2 in. (84 x 144 x 81 mm) |

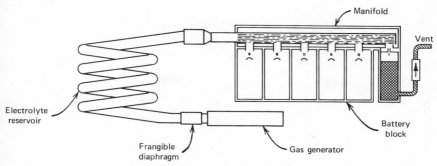

**Figure 148** Coil reservoir battery.

tive sampling of the components and characteristics of coil type missile batteries. The acrylic battery block consists of 10 cell chambers, the cell manifold, and a sump chamber located within the copper electrolyte coil reservoir. At the end of the 5-coil reservoir, away from the battery manifold, there are two gas generators. Each of these generators, which supply 200-250 cm$^3$ of gas, is capable of activating the battery by forcing the approximately 85 cm$^3$ of KOH solution (specific gravity 1.30) into the cells. The generators are designed to fire at 200 mA for 10 msec at 4.4° (40°F) but not to fire at 50 mA for 60 sec at 74.4° (165°F). Actual activation is by 2.7-2.9 V for 100 msec with a minimum of 1 A under load. Pins 4 and 5, which are used to fire the battery, are connected by a shorting spring until used.

The coil reservoir has two frangible pure copper foil diaphragms located at each end to isolate the KOH, and these are ruptured on activation. As the electrolyte leaves the coil, it enters an acrylic manifold with 20 holes for metering the flow into each of the 10-cell chambers. Design of the cells differs, but they contain 3-5 AgO plates and 4-6 Zn plates with pure silver wire or expanded metal grids interleaved with separator material. In some designs the zinc plates are slightly larger than the AgO plates, and a coined groove runs down the fill hole to aid in activation. Excess electrolyte and gas can be swept from the manifold after each cell is approximately equally filled by passing through a torturous cotton packed sump and out a pressure vent set at 16-26 psi (1.1-1.8 kg/cm).

Figure 149 shows the circuit diagram, which terminates in an 8-pin connector. The thermostatically controlled resistance wire heater which is wrapped around the coil reservoir is designed to bring the battery up to operating temperature from a −46° (−50°F) environment in 5 minutes using about 300 W of 108-125 V ac (60 to 1000 Hz) across pins 5 and 6. Although the temperature environment of the battery can vary from −46° (−50°F) to 66° (150°F), the maximum capacity is achieved if storage is at less than 46° (115°F).

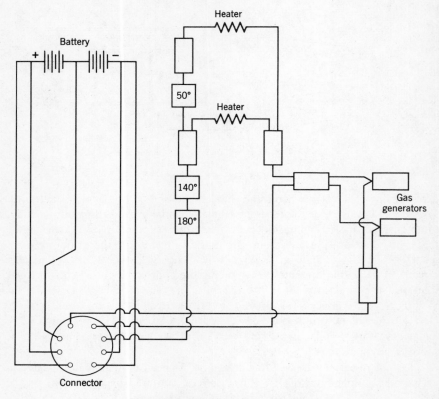

**Figure 149** Circuit for coil type battery.

The coil and gas generators, battery block, heaters and thermostats, and wiring are contained within an 84 x 144 x 81 mm (3.3 x 5.6 x3.2 in.) hermetically sealed, soldered steel case with the octal heater plug mounted on the front. The 1.22 kg (2.7 lb) battery is rated at 13.2-14.6 V under a load of 0.411-0.415 ohm for 90 sec. The capacity of the battery may be used intermittently following activation; however, it must be used within 2 hours after activation. Maximum shelf life is designed to be 42 months, although test data indicate that the units are still usable after 60 months.

After activation, the overall dimensions of the battery are permitted to increase by 0.25 in. (1 mm) in the cover-to-base direction and 5% in the other dimensions. Corrosive by-products up to 0.5 cm$^3$ are allowed to be released during activation and discharge since minor problems have been encountered caused by connector pin's shorting by released electrolyte.

During vibration tests of some of these batteries, ripple has been noted; however, this did not exceed specifications. Irrecoverable spikes of about

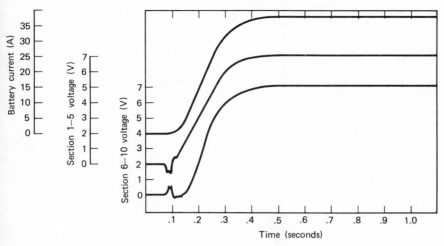

**Figure 150** Performance of coil type battery.

0.13–0.15 V and 0.4–0.5 A were experienced during the 100 g and 40 g shocks, respectively. Again these were within specification and no cause for alarm.

With a 0.413 ohm (± 0.5%) resistive load connected at activation, the battery terminal voltage should be 13.2-14.6 V after 0.5 second. As indicated on Figure 149, this is divided into two voltages by a center tap. Recent tests of 5-year-old batteries produced rise times to the minimum 6.6 V of between 45-90 msec at −50°F and 30-80 msec at 66° (150°F). These same batteries remained above 6.6 V for more than 200 seconds although the requirement is only 90 seconds. Figure 150 shows the performance of this battery at 20-25°.

Figure 151 shows an example of a piston-activator type battery which

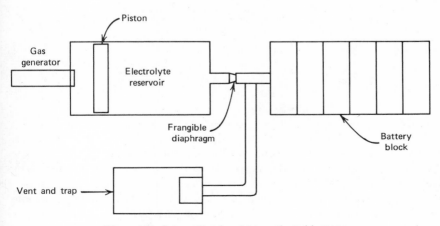

**Figure 151** Schematic of a piston-activated battery.

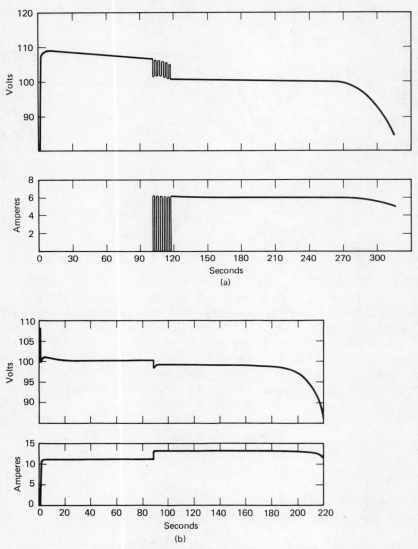

**Figure 152** (a) "A" section battery, 100 volts; (b) "B" section battery, 100 volts.

consists of two 100–V and one 28–V sections. This battery weighs 8.1 kg (17.9 lb) and delivers approximately 290 Wh when operated between −54° (−65°F) and 71° (160°F).

The activator for this battery consists of a stainless steel cylinder containing the KOH electrolyte, two 1.6 liter capacity squibs, a tetrafluoroethylene piston, and tubes to carry the electrolyte to the cell blocks. The piston is retained by a detent inside the cylinder prior to activation, and stainless steel diaphragms protect the squibs and cells from the electrolyte. These diaphragms are broken on activation by the pressure generated from the squibs, which are rated at 1 A/1 W no-fire for 5 minutes and 3.5 A sure-fire with an ignition time of 0.4 msec. Activation time of the battery is generally less than 0.5 msec and is indicated by a monitoring circuit. This circuit consists of a fuse wire connected to 3 cells of one 100–V section which releases a spring-loaded guillotine to cut the monitoring circuit.

Cells for this battery are assembled into groups (74 for the 100–V sections and 22 for the 28–V section) with a manifold for each group. Each group is presealed with epoxy, wrapped with glass cloth, and potted with glass-filled epoxy. Cells for the 100–V sections consist of 3 positive and 4 negative plates while the 28–V section cells have 14 and 15, respectively. The positives are electrolytically formed from silver powder with an expanded silver sheet support material. The negative plates are formed by plating zinc onto a silver grid from a zinc oxide-potassium hydroxide bath. This yields a porous sponge zinc electrode.

The positive plates used in these cells are wrapped in layers of cellulosic separator materials to achieve rapid wettability. The electrodes are then stacked alternately and placed in a nylon case before assembly of the groups.

A heater blanket is wrapped around the activator cylinder to maintain the cell

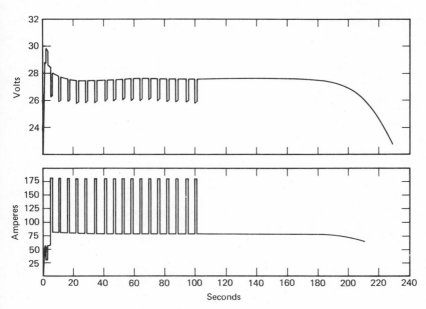

Figure 153  Battery section, 28 volts.

blocks above 27° at the time of activation. Thermostats using 115 V ac and 350 W for a maximum of 30 minutes control the heater by closing at 27° (28–V section temperature) and opening at 82° (180°F) (electrolyte temperature).

A trap and pressure vent are used to contain electrolyte and steam within the battery canister and limit swelling for 153 seconds after activation. After that time, if pressure buildup is sufficient, the canister is vented through a nylon cylinder filled with cotton. The pressure vent between the cylinder and the outside of the canister is isolated from the ambient by a diaphragm which is ruptured by a detonator connected to 3 cells of a 100–V section.

The canister which contains the all elements of the battery is stainless steel 302 with reinforcing depressions. Polyurethane foam is used to pot the unit after assembly to increase its environmental resistance.

The discharge characteristics of this battery are shown in Figures 152 and 153 for the 100–V and 28–V sections. Activation times have varied from 210 to 520 msec and shelf life has been demonstrated for as long as 52 months with no significant change in performance.

The two batteries discussed above are representative of those in use and little fundamental change is expected in future designs employing the highly successful and reliable zinc-silver oxide couple. When future air-to-air and air-to-ground missiles are brought into being, it seems certain this battery will be the major contender as the electrical power source.

# 31.

## Use of Silver Oxide Batteries on Explorers XVII and XXXII

### Thomas J. Hennigan and Charles F. Palandati

### EXPLORER XVII MISSION

The purpose of the Explorer XVII (S-6) mission was to measure the physical and chemical structure of the environment above the earth's atmosphere for a sixty to ninety day period. The spacecraft was to be turned on by ground control with the capability of energizing any or all of the experiments during the five-minute operational periods. The number of turn-ons could vary from zero to seventy-two per day. With this type of control, on-board power could be conserved over the mission life and therefore it was possible to use primary batteries as the main power supply.

The spacecraft was a sealed sphere approximately three feet in diameter. Solar power was not used since outgassing of the array materials could have resulted in erroneous readings of the chemical composition of the environment of the spacecraft. A photograph of the spacecraft, with cover removed is shown in Figure 154. The power supply consisted of zinc-silver oxide primary batteries capable of supplying power for several months without appreciable capacity degradation. The high energy densities of these batteries were attractive in order to minimize the weight of the spacecraft. Total battery weight was 150 lb (68 kg), and the average energy density was 65 Wh/lb (143 Wh/kg).

All major experiments and electronic instruments operated directly from the batteries. Voltage taps of the batteries supplied nominal voltages between 3.1 and 20.2 and between $-3.1$ and $-27.9$. As a result, conversion and regulating devices were eliminated which further resulted in conservation of spacecraft power and a weight decrease. A complete battery layout is shown in Figure 155. Initial planning called for all battery voltages and spacecraft pressures resulting from hydrogen outgassing to be monitored during each turn-on. No data channels were available to monitor current drains from the batteries. A computer program was devised whereby a record of the amount of depleted

**Figure 154** Explorer XVII satellite with cover removed.

capacity for each battery was periodically calculated. This program was based on the number of spacecraft operations and the current drains measured during flight-simulated environmental testing at the Goddard Space Flight Center.

The dry-charged, unsealed cells that were used were deperoxidized, that is, capable of operating at the monovalent voltage level of approximately 1.5 V

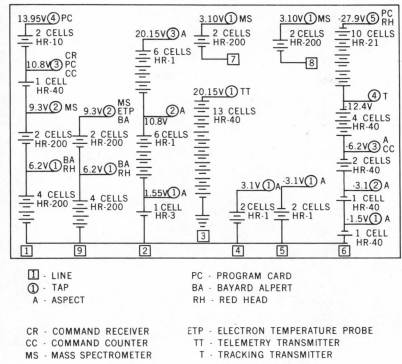

□ - LINE                          PC - PROGRAM CARD
① - TAP                           BA - BAYARD ALPERT
A - ASPECT                        RH - RED HEAD

CR - COMMAND RECEIVER             ETP - ELECTRON TEMPERATURE PROBE
CC - COMMAND COUNTER              TT - TELEMETRY TRANSMITTER
MS - MASS SPECTROMETER            T - TRACKING TRANSMITTER

**Figure 155** Explorer XVII battery layout.

during the entire three-month mission. The power supply consisted of nine batteries, comprising sixty-seven cells, ranging in capacity from 1 to 200 Ah based on the nominal ratings. A photograph of the various sized cells used is shown in Figure 156. Design spacecraft temperatures could vary between $-10°$ and $40°$ depending upon the length of sunlight periods during orbital life.

## TEST PROGRAM

Since the length of the mission was dependent upon the capacity of the various primary batteries it was imperative that simulated tests be performed to determine the actual capacities of the batteries and the amount of capacity degradation that would occur during a 90-day mission. In addition, it was necessary to investigate the composition and amount of gas evolved from the vented batteries in order to determine if the batteries could be placed in individual sealed containers. Initially, an alternate design proposed was to allow the gases to vent into the large, internal free volume of the sphere. As was anticipated and shown during the test programs, the gas evolved was mainly

**Figure 156** Sizes of HR-type used on Explorer XVII.

hydrogen. As a result, either design required purging the free volume of the containers or spacecraft or both with an inert gas such as nitrogen.

Two types of test programs were performed: an accelerated test to determine maximum cell capacities for the mission and a test simulating actual flight load profiles to determine cell capacity, capacity degradation with time, and gassing characteristics.

## ACCELERATED CAPACITY TESTS

In order to determine the actual capacities of the various sized cells, tests were performed whereby the cells were discharged at continuous current drains greater than anticipated, that is, the discharge time was decreased from several months to several days. Although these test results showed that all the various sized cells were capable of delivering more than rated capacity, the HR21, HR40, and HR200 cells did not have sufficient capacity for a 90-day mission. These type cells were modified by increasing the plate thickness and porosity and utilizing the same amount of active material in each cell. No significant increase in weight or size resulted. Accelerated tests on the modified cells showed that capacity would be available for the 90-day mission. The results of the accelerated testing which included the cell modifications are shown in Table 70.

TABLE 70.    AMPERE HOUR CAPACITY

| Cell Type | Rated Ah Capacity | Discharge Rates | | Actual Ah Capacity (24°) | |
|---|---|---|---|---|---|
| | | Accelerated Rate | Simulated Rate | Accelerated Discharge | Simulated Discharge |
| HR-1 | 1.0 | C/5 | C/2000 | 1.75 | 1.5 |
| HR-3 | 3.0 | C/5 | C/2000 | 5.0 | 4.6 |
| HR-10 | 10 | C/5 | C/3000 | 15.0 | 13.0 |
| HR-21 | 21 | C/5 | C/50[a] | 36.0 | 32.0 |
| HR-40 | 40 | C/6 | C/100[a] | 56.0 | 50.0 |
| HR-200 | 200 | C/40 | C/50[a] | 355.0 | 330.0 |

[a]Cells discharged according to test procedure.

## SIMULATED CAPACITY AND OUTGASSING TESTS

Flight simulated capacity tests were performed on the modified cells. In order to determine the amount of capacity degradation and outgassing for a 90-day mission, the various sized cells were tested at the anticipated electrical load profiles. The 1-, 3-, and 10-Ah cells were discharged at continuous C/2000 and C/3000 current rates, C being the rated capacity. The 21-, 40-, and 200-Ah cells were discharged at the C/50 and C/100 rates but simulating the anticipated number of turn-ons during the 90-day period. In other words, these tests were pulse discharges. Also these latter tests were based on the premise that 90% of the spacecraft operations were to be performed on days 1 through 11, 30 through 41, and 60 through 71. Capacities delivered on the simulated test programs are shown in Table 70.

The program for pulsed discharge testing is shown in Table 71. The typical operation of the battery test as indicated in the load timing sequence in the figure is as follows. During day one, item one, the battery was subjected to a total of 72 simulated 5-minute spacecraft turn-ons at 20-minute intervals. During day one, the battery would be discharged for a total of six hours. During the next 48 hours, item two, days two and three, the battery was subjected to a total of 116 simulated 5-minute turn-ons at 25-minute intervals. During this period, the battery was discharged for 9.6 hours. During days one through three, the total time of discharge would be 15.6 hours. The batteries were tested for 90 days, simulating a total of 909 spacecraft turn-ons which resulted in 75.3 hours of discharge involving 18 different turn-on sequences.

TABLE 71.   ELECTRICAL LOAD TIMING SEQUENCE[a]

| | Time | | | Load Time | | | |
|---|---|---|---|---|---|---|---|
| Item | Days | Hours | Elapsed Days | On Minutes | Off hr | min | on Elapsed Hours | Command Number per Item |
| 1 | 1 | 24 | 1 | 5 | | 15 | 6.0 | 72 |
| 2 | 2 | 48 | 3 | 5 | | 20 | 15.6 | 116 |
| 3 | 2 | 48 | 5 | 0 | 48 | | 15.6 | 0 |
| 4 | 6 | 144 | 11 | 5 | | 95 | 22.8 | 87 |
| 5 | 14 | 336 | 25 | 5 | 11 and 55 | | 25.1 | 28 |
| 6 | 5 | 120 | 30 | 0 | 120 | | 25.1 | 0 |
| 7 | 1 | 24 | 31 | 5 | | 15 | 31.1 | 72 |
| 8 | 2 | 48 | 33 | 5 | | 20 | 40.7 | 116 |
| 9 | 2 | 48 | 35 | 0 | 48 | | 40.7 | 0 |
| 10 | 6 | 144 | 41 | 5 | | 95 | 47.9 | 87 |
| 11 | 14 | 336 | 55 | 5 | 11 and 55 | | 50.2 | 28 |
| 12 | 5 | 120 | 60 | 0 | 120 | | 50.2 | 0 |
| 13 | 1 | 24 | 61 | 5 | | 15 | 56.2 | 72 |
| 14 | 2 | 48 | 63 | 5 | | 20 | 65.8 | 116 |
| 15 | 2 | 48 | 65 | 0 | 48 | | 65.8 | 0 |
| 16 | 6 | 144 | 71 | 5 | | 95 | 73.0 | 87 |
| 17 | 14 | 336 | 85 | 5 | 11 and 55 | | 75.3 | 28 |
| 18 | 5 | 120 | 90 | 0 | 120 | | 75.3 | 0 |

[a]Total number of commands = 909; time duration of each command = 5 minutes.

During these tests, the 21-, 40-, and 200-Ah cells were placed in sealed chambers containing pressure gauges. The ratio of the free space volume of the empty container to the cell volume was 1.3 to 1.0 for each cell size. This ratio was equivalent to the original design requirement where each battery was to be contained in a sealed container. Early testing proved that placing the batteries in sealed containers was not feasible because of the large amount of hydrogen evolution observed.

If it had been possible to cycle the cells at least once previous to flight use, hydrogen evolution would have been decreased considerably as shown in Figure 157. However, using recharged cells was not possible because of the higher voltage output of the bivalent silver formed during charging. Limited testing had shown that a 15% decrease of capacity would result from a recharge. Preloading of the recharged cells to assure operation on the monovalent level caused a capacity decrease of approximately 40%. Therefore, in order to obtain maximum capacity from the batteries, it was decided to use dry-charged cells without any preflight cycling and to vent the hydrogen to the interior of the spherical spacecraft.

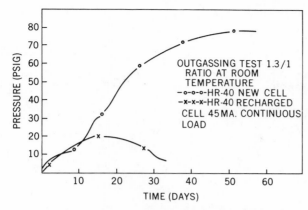

**Figure 157** Outgassing characteristics of new and recharged HR40 cells.

Methods were investigated to reduce the amount of hydrogen accumulation. Outgassing tests of the cells in sealed chambers showed that palladium monoxide powder, contained in a small polystyrene container on the vent of the cells, could retard the amount of hydrogen accumulation in the sphere. The assembly is shown in Figure 158. Three grams of PdO were used for the HR3, HR10, and HR21 cells and 5 grams for the HR40 and HR 200 cells. Test results at 25° are shown in Figures 159 and 160. It can be seen that, for the first 45 days, the pressure rise was decreased approximately by one-half with the PdO. The reduction in hydrogen accumulation in the chamber resulted from the reaction of hydrogen with PdO to form Pd and water and the absorption of hydrogen by the Pd product. Based on these tests, it was decided to use the PdO powder to reduce hydrogen accumulation in the sphere as much as possible.

Actually, the ratio of internal free volume of the sphere to the battery volume was of the order of eleven to one. Therefore tests were run on the 40-Ah cells to determine the pressure increase with the large free volume and also to evaluate the use of PdO. As shown in Figure 161, pressure increase in the eleven-to-one ratio test is negligible. Tests using this large ratio could not be performed readily on the 200-Ah cells because the large outgassing chambers were not available. These cells were tested at volume ratios of 2 : 1 and 6 : 1 with the PdO powder. The results of this test are shown in Figure 162. The results of these tests showed that the amount of hydrogen evolution could be controlled with the PdO powder and that the internal pressure rise in the spacecraft could be within safe limits at 25°. Also, during the 90-day flight simulated tests, the amount of capacity degradation was about 10 to 15% of the amount of capacity realized in the accelerated testing. This comparison is shown in Table 70.

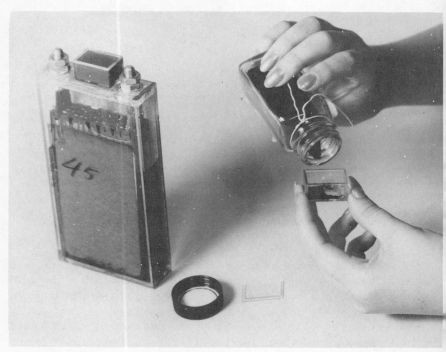

**Figure 158** Palladium monoxide capsule with the HR40 cell.

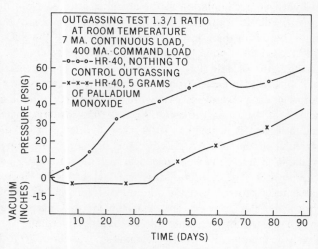

**Figure 159** Outgassing characteristics of HR40 cells with palladium monoxide.

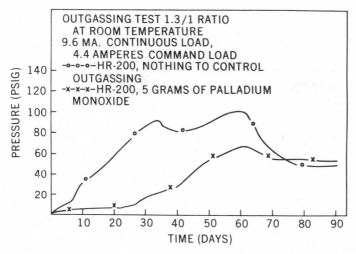

**Figure 160** Outgassing characteristics of HR200 cells with palladium monoxide.

Since the HR40 and HR200 cells were to supply 90% of the required spacecraft power, these cells were tested at 40° to determine capacity degradation. This temperature was the upper design limit of the spacecraft batteries and would be reached if the spacecraft was subjected to prolonged sunlight periods. Capacity degradation of the HR40 cells increased from 10% at 25° to 20% at 40°. The HR200 cells showed an increase in capacity degradation from 8% at 25° to 33% at 40°. These decreases in capacity are based on the

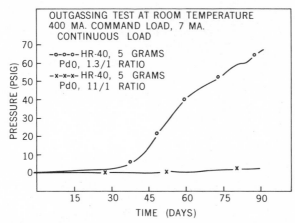

**Figure 161** Effects of free space volume on the outgassing of HR40 cells.

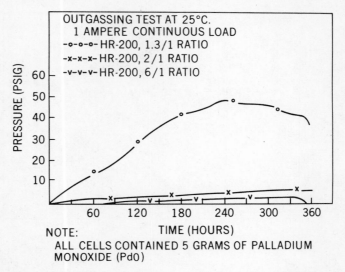

**Figure 162** Effects of free space volume on the outgassing of HR200 cells.

flight-simulated tests at the two temperatures as compared to the accelerated tests. Hydrogen evolution increased severely for both the HR40 and the HR200 cells at the higher temperature. The pressure chambers with a 1.3-to-1.0 volume ratio had to be vented periodically, that is, when the pressures reached 150 psig, the chambers were vented until the internal pressure decreased to 50 psig. The

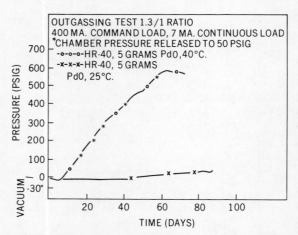

**Figure 163** Outgassing characteristics of HR40 cells at 25° and 40°.

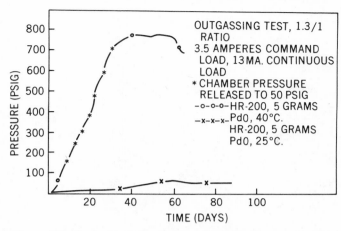

Figure 164 Outgassing characteristics of HR 200 cells at 25° and 40°.

pressure rise caused by hydrogen evolution from these cells is shown in Figures 163 and 164. Several ventings were necessary. However, calculations based on the spacecraft free space ratio of 11 : 1 showed that the hydrogen evolution would not cause a pressure rise exceeding the burst pressure of the sphere. Prolonged periods in continuous sunlight were not expected during the orbital life of the spacecraft.

## EXPLORER XXXII MISSION

The Atmospheric Explorer XXXII (AE-B) consisted of duplicate transmitters, receivers, decoders, programmers, and experiments which were powered from individual batteries or combination of batteries connected in series. On this mission, dry-charged batteries were used as the power supply. However, a body-mounted solar array was included in the design for the purpose of recharging the batteries periodically. Ground control was capable of selecting the equipment to be operated during each turn-on and selecting the battery to be recharged. The 7.5-W solar array was coupled to a constant wattage charger for battery recharge. Charge currents of C/100 to C/300 were available. A photograph of the satellite, mounted on a spin-balance machine, is shown in Figure 165. The addition of the solar array with the battery charger was expected to increase the life of the spacecraft to six-months minimum. This battery power supply used HR40, HR58, and HR200 size cells. The battery layout is shown in Figure 166. Because of the different battery voltage taps, the three different ampere hour capacities, and the limited solar array power, six of the eight batteries were recharged individually while the remaining two batteries were recharged in series with the accompanying battery. In this latter case, when

**Figure 165** Explorer XXXII satellite mounted on spin-balance machine.

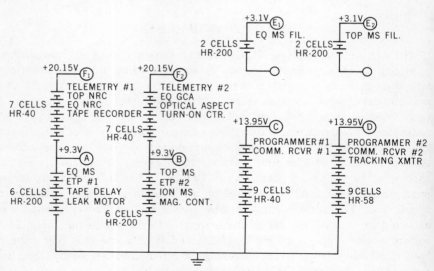

**Figure 166** Explorer XXXII battery layout.

required, F1 battery would be charged with A battery and F2 battery would be charged with B battery. Charge currents were inversely proportional to battery voltage since the battery charger output was constant wattage.

During each turn-on, all battery voltages, battery charge current, battery designation, spacecraft ambient temperature, and internal pressure were monitored during the 4-minute operating period. Actual battery temperatures could be obtained at the discretion of the tracking stations by a series of ground commands. Depletion of battery capacity was determined by calculations based on prelaunch environmental tests and the number of four-minute turn-ons affecting the various instruments and experiments. In-flight recharge capacity calculations were based on ground-monitored charge currents and total hours of available sunlight.

Internal pressure rise within the sphere was minimized with the use of recombination fuel cells. Water vapor diffusion from the batteries was prevented by the placement of thin polyethylene films on each vent port. These films did not prevent the diffusion of hydrogen or oxygen at the evolved rates. The spacecraft was also equipped with a pressure relief valve which could be operated by ground command in the event of an abnormal internal pressure rise.

## BATTERY UTILIZATION PROCEDURES AND TEST PROGRAMS

Because of the limited size of the solar array and the variation in sunlight period from 65% in shadow periods to 100% in continuous sunlight periods, depth of discharge of the batteries was limited to 70% of rated capacity. That is, this is the maximum depth of discharge allowed on any one battery before charging would commence. Although the charge rates of C/100 and C/300 are considered to be extremely low, battery overcharging was recognized as a problem since the spacecraft battery charger was a constant-wattage device. Excessive overcharging could produce oxygen and hydrogen in addition to the amount of hydrogen normally evolved as a result of the zinc/electrolyte reaction. The following three methods were considered in order to minimize the overcharge problem:

1. Termination of the first recharge of each battery when 70% of the calculated discharge capacity was replaced. Subsequent recharging would be terminated when 100% of the previous discharge was replaced. Therefore the batteries would be maintained at some level below 80% of rated capacity.

2. Recharging would be terminated when the charge voltage on the battery being charged reached 1.975 V per cell over the design temperature range.

3. Use of recombination type fuel cells consisting of a hydrogen anode coupled to a cupric oxide cathode with an aqueous KOH electrolyte. In addition

to suppressing hydrogen accumulation, this device was also capable of removing oxygen, caused by limited overcharge, by reacting with hydrogen on the noble metal catalyst in the fuel cell electrode.

Utilization of the vented, dry-charged cells in a secondary battery system presented several problems not encountered on Explorer XVII. The operating life of the spacecraft would be governed by the charged wet stand and limited cycle life of these batteries. It was imperative that accelerated battery tests be carried out to determine the cycling characteristics of the batteries under various load conditions and as a function of temperature. Flight-simulated tests to determine if the batteries would deliver at least three cycles in one year were also necessary. Herein, capacity degradation would be a major factor. In addition, it was necessary to determine the gassing characteristics of the batteries resulting from the cycling regimes.

## ACCELERATED BATTERY TESTS

Batteries comprised of four or six cells representative of the three different ampere hour sizes were periodically cycled at 0, 24, and 40°. The batteries were subjected to a continuous discharge at the spacecraft current drains and recharged at the maximum anticipated charge current under 100% sunlight conditions. Initial testing of the sample batteries was performed to determine the charge/discharge voltage characteristics at the three temperature levels in order to be able to set a prescribed voltage limit for each battery. Test results showed that 1.975 V per cell would allow the batteries to accept a satisfactory percentage recharge without producing oxygen. Ampere-hour charge efficiencies of 90% at 0 and 40° and 97% at 24° were observed. Figure 167 depicts the

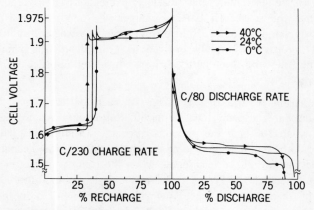

**Figure 167** Typical voltage characteristics at various temperatures.

typical voltage characteristics observed for the three cell sizes at the C/230 charge rate and the C/80 discharge rate. Previous to the cycle shown, the batteries had been discharged to approximately 1.4 V per cell at the C/80 rate. Because of the number of batteries, different capacities and various charge/discharge rates, it was not possible to subject all batteries to tests at all the various parameters.

Following the above test, the sample batteries were subjected to several continuous cycles simulating the first method whereby percentage recharge was limited approximately to 80%. Negligible amounts of hydrogen evolution were observed during the first discharge. No oxygen evolution was detected during the limited recharge. On the same sample cells, the second method was investigated with the batteries recharged to a voltage limit of 1.975 V per cell. Also, during these tests the retention of capacity during charged stand was investigated. The batteries were periodically discharged to 100% depth, 1.4 V per cell, and then recharged to the 1.975 voltage limit. Figures 168-170 show the effects of temperature and wet stand on capacity during the accelerated testing of the HR40, HR58, and HR200 batteries. Gassing characteristics are discussed later in the report.

The data obtained from the accelerated battery tests led to the following four conclusions:

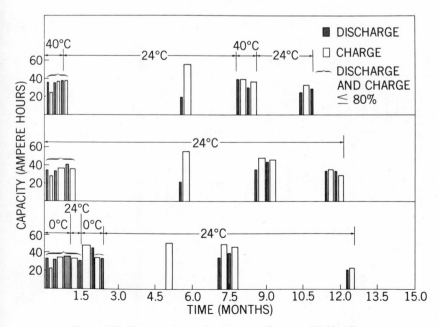

**Figure 168** Temperature and wet-stand effects on HR40 cells.

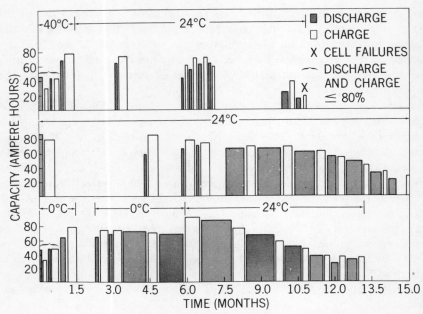

**Figure 169** Temperature and wet-stand effects on HR58 cells.

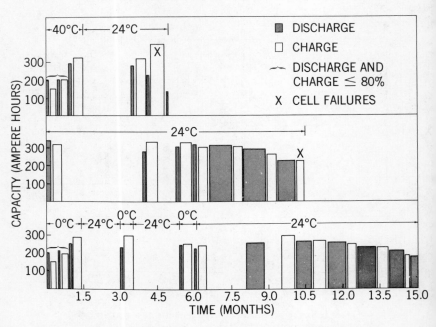

**Figure 170** Temperature and wet-stand effects on HR200 cells.

1. Subjecting charged cells to the long wet stands shown in the figures resulted in severe capacity degradation on the discharge after stand. Dissection of discharged cells, which had shown this capacity degradation, disclosed incomplete discharging of the silver cathodes. Discharging these cathodes against new, charged, zinc anodes showed that the silver cathodes were capable of delivering the full capacity. It was therefore assumed that the cells were negative limited on discharge. However, when the discharged silver electrodes were again cycled against the original zinc electrodes from the test cell, the original capacity of the electrodes was obtained. Although no proof is offered, the removal of the original separator material and the electrode products within the separator structure could have accounted for the capacity improvement.

2. Low rate, continuous cycling was more beneficial for capacity maintenance than prolonged wet-stand periods in the charged state. Cells dissected at the end of test program showed evidence of severe loss of the negative active materials.

3. A decrease in cycle life was observed at $40°$. This was especially true of the HR200 cells. Visual examination of failed cells showed that this was due to a build-up of spongy zinc from the anodes which penetrated the outer two wraps of cellophane separator material. It was suspected that this material had contacted the remaining two layers of cellophane, which was impregnated with silver that had migrated from the positive electrode. This condition could have produced an internal short which prevented the cells from accepting low rate recharges.

4. The maximum cycle life and best capacity maintenance were observed at zero degrees centigrade.

## SIMULATED BATTERY TEST

Before receiving the new HR40 and HR200 dry-charged cells for the Explorer XXXII (AE-B) program, tests at $24°$ were initiated with two batteries, one had three HR40 cells and the other had HR200 cells. These cells had been manufactured for the Explorer XVII (S-6) program. This was done in order to compare the voltage, capacity, and outgassing characteristics of the S-6 and AE-B cells. These cells, placed in sealed chambers with air at 1 atm pressure, were tested simulating 6 spacecraft turns-on per day, equivalent to 20 minutes of current drain per day, for a total of 180 days. The batteries were then removed from the chambers and the remaining capacity discharged to less than 1.4 V per cell. Similar discharges were initiated on 3-cell batteries utilizing the new HR40 and HR200 cells for the AE-B program. The current drains were slightly lower because of modifications in the electronics design for the AE-B. Comparative results of these tests, shown in Figures 171 and 172, indicated that the cells for the AE-B program were capable of delivering approximately 20% more capacity than the cells used on the Explorer XVII (S-6) program. Hydrogen evolution

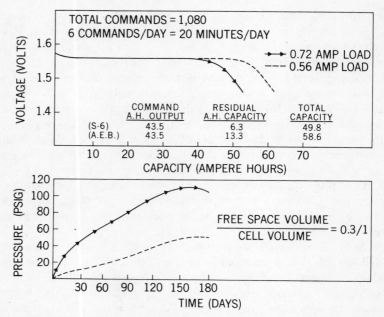

**Figure 171** Capacity and outgassing characteristics of HR40(S-6) versus HR40(AEB) at 25°.

from the AE-B cells was significantly less than from the S-6 cells. Personnel from the battery manufacturer attributed this decrease in hydrogen evolution to improvements in the control of the process to amalgamate the negative electrodes.

Upon completion of the initial 180-day discharge, the AE-B HR40 and HR200 batteries were recharged to 1.975 V per cell limit at the flight-simulated charge currents. The continuing cycle regimes were then somewhat accelerated by decreasing the discharge time from 180 days to 90 days by increasing the number of turn-ons from 6 to 12 per day. This was done in order to be able to determine the effects of wet life and cycle life in regard to capacity degradation. Should severe degradation result from these effects, the number of spacecraft turn-ons per day would be increased in order to obtain as much data as possible during the operational life of the batteries. The continuing cycles were the same as the first whereby the batteries were charged to 1.975 V per cell. The tests were continued until the batteries did not deliver 70% of rated capacity, that is, 28 and 140 Ah, respectively, for the HR40 and HR200 batteries. Recharging the batteries to the 1.975 V per cell limit was sufficient to recharge the batteries with negligible oxygen evolution. Data from these tests are plotted in Figures 173 and 174 for the HR40 and HR200 cells, respectively. The tests also proved that the cells were capable of performing the mission for six months or more on a limited cycle regime.

Since the cells were initially sealed in one atmosphere of air and no leaks had been detected in the outgassing chambers, it was of interest that a decrease in the oxygen content was observed. For example, in Figure 173, on the initial discharge the amount of oxygen should have been 6.6%. However, analysis showed only 1%, meaning that the oxygen was recombining with the negative materials at a very low rate. Similar observations were made during the simulated 3-month discharge test on Explorer XVII program. After each recharge, a significant decrease in pressure was noted during the discharge. This was the result of a decrease in hydrogen partial pressure. The nitrogen in the sealed chambers remained fairly constant with respect to the amount initially sealed in the chamber. The results for the HR200 cell (Figure 174) were similar. The voltage characteristics shown in Figures 173 and 174 indicate that the time to reach the voltage limit at the end of charge decreased as degradation increased. Examination of the discharged cells at the conclusion of the tests showed that the silver electrodes were partially charged while the zinc electrodes had lost 50

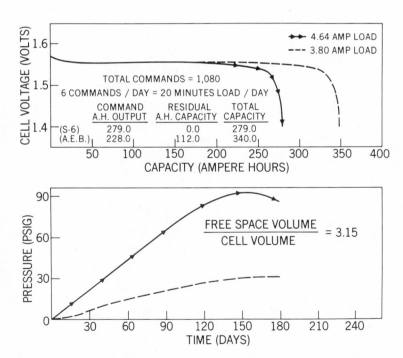

**Figure 172** Capacity and outgassing characteristics of HR200(S-6) versus HR200(AEB) at 25°.

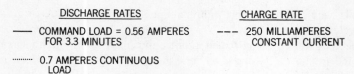

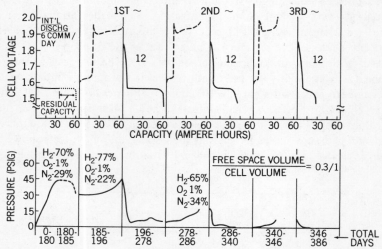

**Figure 173** Simulated charge-discharge characteristics of HR40 at room temperature.

to 90% of active material. The outside zinc electrodes were essentially stripped of active material.

The HR58 cells were cycled in sealed outgassing chambers; the data are presented in Figure 175. They were initially discharged at the flight-simulated continuous current drain of 12 mA for 310 days at which time they were then discharged at 700 mA to less than 1.4 V per cell. The battery was then recharged at the flight-simulated charge current. Although the current drain was not affected by the number of turn-ons, the continuing cycling regimes were also increased, that is, the discharge current was increased from 12 to 25 mA. The test was terminated on day 518 since the battery was capable of accepting only approximately 70% of rated capacity. Because of the larger free space volume, a pressure rise of only 0.34 atm occurred during the initial discharge. The percentage of oxygen was again significantly lower than the amount initially sealed in the chamber with the major decrease occurring during the first discharge. By the end of the second cycle discharge at 511 days, the pressure was 0.88 atm, all of which was essentially nitrogen. Since a negative pressure was

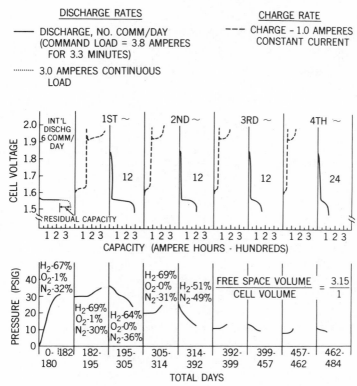

**Figure 174** Simulated charge-discharge characteristics of HR 200 at room temperature.

measured, it is assumed that the sealed chamber did not leak; therefore the decrease in pressure from 1.36 to 0.88 atm was caused by a decrease in hydrogen gas. It has been reported elsewhere that the recombination of hydrogen can occur on the oxides of silver. However, though the outgassing chamber did not show signs of leakage, it cannot be overlooked that the containment of hydrogen is difficult over long periods of time.

As a result of the severe capacity degradation of the HR200 cells at 40° observed during the Explorer XVII tests and accelerated tests for Explorer XXXII, a simulated test on these cells was performed at 40°. The cells were sealed in chambers with a 3.5:1 volume ratio. The cells were initially discharged at a 45 mA continuous drain, approximating the daily capacity requirement of six turn-ons per day. After the first recharge the current drain was increased to 90 mA. This test was terminated at the end of the second cycle discharge because of severe capacity loss, as shown in Figure 176. The battery was capable of 2½ cycles at 40° in 350 days as compared to 4½ cycles at 24° in 484 days.

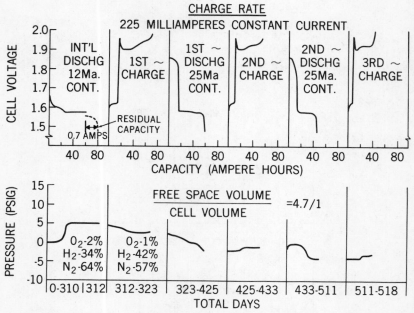

**Figure 175** Simulated charge-discharge characteristics of HR58 at room temperature.

High pressure rises shown in Figure 176 were caused by hydrogen evolution. Simulated cycling tests were not performed at 0° since the accelerated battery test data had shown that wet life and cycle life would increase at the cold temperature.

## FLIGHT BATTERY DESIGN

The final flight battery (Figure 177) consisted of 8 batteries of HR40, HR58, and HR200 cells. The complete battery, including the lightweight cans, weighed 178 lb (80 kg). The average energy density was 45 Wh/lb (99 Wh/kg). On-board equipment was sensitive to water; therefore, the vent ports of the 48 cells were covered with about 1 cm² of polyethylene film, 1 mil (0.025 mm) thick. The films would permit hydrogen and oxygen diffusion while limiting water vapor diffusion. At very small differential pressures of the order of 0.01 atm at 25°, gas diffusion rates were 100 cm³(NTP)/h for hydrogen and 30 cm³(NTP)/h for oxygen. Water vapor diffusion at 40°, considered to be the worst condition, was about 0.4 g/day for the entire battery package.

Three hybrid fuel cell units were used to remove hydrogen and oxygen from the interior of the spacecraft. Each unit was capable of removing approximately 280 liters of hydrogen. Ground testing proved that one of the units, similar to

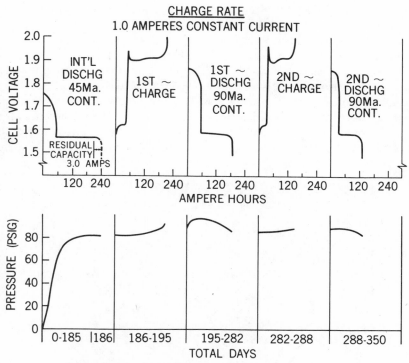

Figure 176 Simulated charge-discharge characteristics of HR200 at 40°.

Figure 177 Explorer XXXII battery pack.

437

the one shown in Figure 178, could remove the anticipated amount of hydrogen evolved in a 6-month period. The three units were redundant. The units were flown with the leads shorted since no telemetry was available for monitoring.

Prior to launch, the proposed 12-month flight schedule shown in Table 72 was formulated. All capacity drains and outputs were related to five turn-ons per day, at $25°$. Capacity inputs were based on 7.5 W of solar power for an 18-hour sunlight period per day. Capacities, based on test results, were used to calculate depth of discharge, which varied from 39 to 65% for the eight batteries for the first 6-month period of the proposed schedule. As a precaution against overcharging, battery charging was to be terminated whenever the maximum charge voltage was observed by telemetry.

Figure 178  General Electric gas recombination fuel cell unit.

## PERFORMANCE OF BATTERY IN FLIGHT

Explorer XXXII was launched in May 1966. Within the first few weeks the main spectrometer filaments failed thereby decreasing the power requirements by approximately 60%. This resulted in a large decrease in current drains for the A and B batteries and completely eliminated the need for the $E_1$ and $E_2$ batteries

## TABLE 72  EXPLORER XXXII PROPOSED BATTERY SCHEDULE

| BATTERY | F1 | A | F2 | B | C | D | E1 | E2 | UNITS |
|---|---|---|---|---|---|---|---|---|---|
| CELL TYPE HR- | 40 | 200 | 40 | 200 | 40 | 58 | 200 | 200 | - |
| RATED CAPACITY | 50 | 300 | 50 | 300 | 50 | 75 | 300 | 300 | A.H. |
| NUMBER OF CELLS | 7 | 6 | 7 | 6 | 9 | 9 | 2 | 2 | # |
| CHARGE VOLTAGE[1] | 26 | 12 | 26 | 12 | 18 | 18 | 4 | 4 | VOLTS |
| LOAD VOLTAGE | 20.15 | 9.3 | 20.15 | 9.3 | 13.95 | 13.95 | 3.1 | 3.1 | VOLTS |
| AVG. COMMAND LOAD[2] | 0.325 | 2.33 | 0.255 | 2.45 | 0.014 | 0.026 | 3.0 | 3.0 | AMPS |
| CONTINUOUS LOAD | | | | | 7.0 | 7.0 | | | M.A. |
| AVG. DRAIN PER DAY | 0.108 | 0.77 | 0.085 | 0.810 | 0.178 | 0.185 | 0.99 | 0.99 | A.H. |
| DRAIN PER MONTH | 3.27 | 23.5 | 2.56 | 24.75 | 5.42 | 5.65 | 30.25 | 30.25 | A.H. |
| 6 MONTH OUTPUT | 19.62 | 141.00 | 15.36 | 148.50 | 32.52 | 33.90 | 181.50 | 181.50 | A.H. |
| RESIDUE CAPACITY | 30.38 | 159.0 | 34.64 | 151.50 | 17.48 | 41.10 | 118.50 | 118.50 | A.H. |
| % RATED CAPACITY | 61.0 | 53.0 | 69.0 | 50.0 | 35.0 | 55.0 | 39.0 | 39.0 | % |
| MAX. CHARGE CURRENT | 0.310 | 0.600 | 0.310 | 0.600 | 0.440 | 0.440 | 1.60 | 1.60 | AMPS |
| MAX. A.H. INPUT PER DAY[3] | 5.72 | 10.80 | 5.72 | 10.80 | 7.92 | 7.92 | 26.8 | 26.8 | A.H. |
| TOTAL CHARGE[3] | 14.80 | 106.0 | 11.5 | 111.3 | 24.3 | 25.4 | 126.2 | 126.2 | A.H. |
| RESTORED CAPACITY | 45.2 | 265.0 | 46.1 | 262.8 | 41.8 | 66.5 | 244.7 | 244.7 | A.H. |
| 12 MONTH OUTPUT | 19.6 | 141.0 | 15.4 | 148.5 | 32.5 | 33.9 | 181.5 | 181.5 | A.H. |
| RESIDUE CAPACITY | 25.6 | 124.0 | 30.7 | 114.3 | 9.3 | 32.4 | 63.2 | 63.2 | A.H. |
| % RATED CAPACITY | 51.0 | 41.0 | 61.0 | 38.0 | 19.0 | 43.0 | 21.0 | 21.0 | % |
| CHARGE CURRENT | 0.310 | 0.600 | 0.310 | 0.600 | 0.440 | 0.440 | 1.60 | 1.60 | A.H. |
| MAX. A.H. INPUT PER DAY[3] | 5.72 | 10.80 | 5.72 | 10.80 | 7.92 | 7.92 | 26.8 | 26.8 | A.H. |
| TOTAL CHARGE[3] | 14.80 | 106.0 | 11.5 | 111.3 | 24.3 | 25.4 | 126.2 | 126.2 | A.H. |
| RESTORED CAPACITY | 40.40 | 230.0 | 42.2 | 125.6 | 33.6 | 57.8 | 189.4 | 189.4 | A.H. |

ALL DATA IS RELATED TO OPERATION OF BATTERIES AT ROOM TEMPERATURE.

[1] CHARGE VOLTAGE IS A MAXIMUM VOLTAGE, NOT TO BE EXCEEDED DURING PAYLOAD OPERATION.

[2] AVG. COMMAND LOAD IS CALCULATED ON EACH TRANSMITTER AND PROGRAMMER OPERATING FOR 50% OF TOTAL COMMANDS.

[3] CAPACITY INPUTS CALCULATED FOR 18 HOURS CHARGE/DAY.

(see Figure 166). Because of the large decrease in power requirements the number of turn-ons were preselected by scheduling the available tracking stations and by predicting atmospheric conditions favorable to experiment operation. This was done to obtain as much information from the mission as possible. Instead of 5 turn-ons per day the operation of the satellite was increased to an average of 27 turn-ons per day.

In September 1966, four months after launch, the first cell failure occurred in the B battery comprised of the HR200 cells. The failure mode was a high impedance internal short, similar to the cell failures experienced during the battery tests. Other cell failures occurred throughout the mission as shown in Table 73. In order to overcome the decrease in the battery voltages caused by cell failures, the batteries were maintained at 80 to 100% recharge, that is, the good cells were operated at the bivalent level to compensate for the low voltage cells. This required continual operation of the battery charger during sunlight hours and complete coverage by all available tracking stations in order to monitor the battery charge voltages.

On January 17, 1967, internal pressure in the spacecraft reduced from 2 to 0 atm within a 2.5-hour period. The cause of the pressure decrease was a leak in the spacecraft shell. By the end of January, the D battery, which was required to operate one of the programmers and command receivers, was suspected of having dry spots within the cells, caused by the leak. This was indicated by extremely low discharge voltages and very high charge voltages. For example, the

TABLE 73　EXPLORER XXXII CELL FAILURE

| BATTERY | MAR | MAY | JUN | JUL | AUG | SEP | OCT | NOV | DEC | JAN | FEB | MAR | TOTAL FAILURES |
|---|---|---|---|---|---|---|---|---|---|---|---|---|---|
| A 6 HR-200 | | | | | | | 1 | | | | DRY | | 1 |
| B 6 HR-200 | ALL CELLS ACTIVATED | | | | | 1 | | | | | | | 1 |
| C 9 HR-40 | | | | | | | | | 1 | 1 | | DRY | 2 |
| D 9 HR-58 | | | | | | | 1 | 1 | 1 | 1DRY | | | 4 |
| E$_1$ 2 HR-200 | | | | | | | | | | 1 | | | 1 |
| E$_2$ 2 HR-200 | | | | | | | | | | | 1 | | 1 |
| F$_1$ 7 HR-40 | | | | | | | | 1 | 1 | | | | 2 |
| F$_2$ 7 HR-40 | | | | | | | 1 | 2 | | | | | 3 |

highest voltage observed for the 9-cell D battery before the leak was approximately 18 V. After the leak occurred, the charge voltage increased to 30 V, the maximum voltage output of the charger. On March 22, 1967, the cells in C battery which operated the other programmer and command receiver developed dry spots thus terminating the mission.

## POSTFLIGHT BATTERY TESTS

A study was made to determine the cause of the premature cell failures which had occurred in the eight batteries. Since the $E_1$ and $E_2$ batteries were not used and had not been charged, it was deduced that the duty cycle could not have caused the cell failures. No monitored spacecraft and battery temperatures had exceeded 33° while internal pressures were normal during the entire mission therefore giving no indication as to the cause of the cell failures. The only anomaly observed during the flight was the amount of radiation that the batteries had received. Because of an extended second-stage burn of the Delta rocket, the apogee had been increased from 1200 to 2700 kilometers thereby subjecting the spacecraft to more severe radiation from the Van Allen belts than had been expected. At the time the first cell failure had occurred, the flight batteries had been subjected to $10^{14}$ electrons per cm² .

Five-ampere-hour HR5 cells were selected for the study of radiation effects at

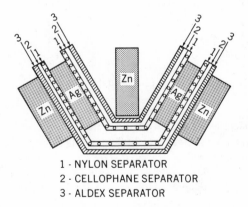

1 - NYLON SEPARATOR
2 - CELLOPHANE SEPARATOR
3 - ALDEX SEPARATOR

| HR-40 dc (10 NM) | HR-58 dc (2 NM) | HR-200 dc (4 NM) | HR-5 dc (11 NM) |
|---|---|---|---|
| Ag { 1 - ONE WRAP / 2 - FOUR WRAPS / 3 - ONE WRAP | Ag { 1 - ONE WRAP / 2 - FOUR WRAPS / 3 - ONE WRAP | Ag { 1 - ONE WRAP / 2 - FOUR WRAPS / 3 - ONE WRAP | Ag { 1 - ONE WRAP / 2 - FOUR WRAPS / 3 - ONE WRAP |
| Zn   NO SEPARATOR | Zn   NO SEPARATOR | Zn   NO SEPARATOR | Zn   NO SEPARATOR |

Figure 179  Cross-sectional view of silver and zinc plates and separators.

$10^{13}$, $10^{14}$, and $10^{15}$ electrons per cm$^2$ at room temperature. A set of cells was not radiated and was used as controls. These 5-Ah cells were constructed in the same manner as the HR40, HR58, and HR200 cells utilized in tests and flight. Figure 179 shows a view of a set of positive and negative electrodes with the separator system and U-fold wrap. The smaller cells were selected because of their compatibility with the opening of the 1.5 MeV electron source. A total of eight cells, comprised of four fully charged and four 70% discharged cells, plus samples of electrolyte wetted Cellophane and dry Cellophane, were radiated at each of the three different levels. The test set-up is shown in Figure 180.

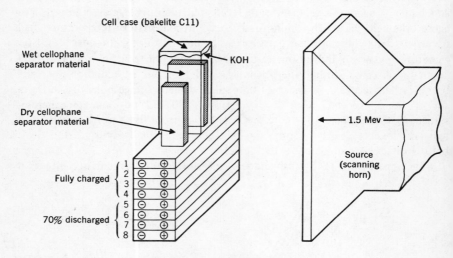

**Figure 180** Cell placement during irradiation.

Visual and physical examination of two cells and infrared studies of the dry cellophane from each radiation level indicated that the cellophane radiated at the $10^{15}$ level had been severely degraded. This caused complete deterioration at the edges of the four wraps of cellophane located nearest to the 1.5 MeV source. The remaining six cells from each radiation level and the control cells were cycled to a 70% depth of discharge per day at 25°. Catastrophic shorts developed in the cells radiated at $10^{15}$ electrons per cm$^2$ within a few cycles. Three of the six cells radiated at $10^{14}$ electrons per cm$^2$ developed high-impedance shorts during cycles 59 and 68. At the end of 72 cycles, the test was terminated. No cell failures occurred in the control group of cells radiated at the lowest level. The results of these tests showed that the premature failures of the flight cells could have been caused by the increased amount of electron radiation encountered during the mission.

## CONCLUSION

The use of dry-charged, primary silver-zinc batteries as the power supply for a scientific spacecraft was demonstrated by the success of the Explorer XVII mission. Telemetered data received during the 780 commands over a period of 100 days showed that all the batteries were capable of delivering the required electrical power. The internal pressure of the spacecraft did not exceed 4.23 atm. This high pressure occurred while the spacecraft was subjected to total sunlight for ten days during the last thirty days of the mission. Internal temperatures monitored at positions remote from the battery indicated that the ambient temperature of the spacecraft may have exceeded the design limit of 40°. This may have accounted for the high internal pressures observed during the extended sunlight period.

During the 10-month period from May 1966 to March 1967, the Explorer XXXII spacecraft completed 8,408 turn-ons during which time 13,000 commands were performed. The utilization of the zinc-silver oxide batteries as secondaries increased the mission life by a factor of three as compared to the Explorer XVII. Although no telemetry data were available, the fuel cell units appeared to be capable of recombining the hydrogen evolved from the batteries since the pressure relief valve was used only twice to reduce the internal pressure from 2.38 to 2.04 atm. This occurred in December and January only after extensive battery charging was required to maintain the batteries at the bivalent voltage level. It was suspected that overcharging had occurred during this period.

# 32.

## Auxiliary Electrodes for Sealed Silver Cells

### W. N. Carson, Jr.

The key problem in the design and use of sealed secondary cells is how to provide for the recombination of the hydrogen and oxygen generated during operation of the cell. These gases are generated by side reactions of the electrodes and electrolytes; it is virtually impossible to operate a secondary cell with an aqueous electrolyte without some gas generation. If a means for recombination of the gases is not provided, continued cycling of the cell will result in rupture of the cell case. The stronger the cell case, the more violent is the rupture when it occurs, a fact sometimes not remembered by cell designers.

If the hydrogen and oxygen were evolved in stoichiometric amounts over the charge-discharge cycle, the recombination problem would be readily solved by using some type of igniter, such as a hot wire activated during charging, to initiate the reaction between the gases. This approach was proposed (R188) but fails in practice because the gases are seldom if ever evolved in stoichiometric amounts. A complicated system (R589) was proposed to overcome this nonstoichiometric generation of gases in which the complementary gas was generated at an auxiliary electrode to provide the stoichiometric ratio that would then allow removal of the gas by burning. This arrangement has not been practical. In addition to the use of hot wires, the use of catalysts of various forms has been proposed to promote recombination; these methods have failed because of the lack of stoichiometry of gas evolution and because the catalysts themselves fail when they are wetted by the electrolyte or condensed water vapor.

Of the two gases generated in a cell, oxygen offers fewer problems in recombination than does hydrogen, largely because oxygen reacts readily with anode metals that are wetted with electrolyte. Hydrogen, in the absence of catalysts, does not readily react with cathode materials. These basic facts have led to the development of what is called the oxygen cycle in sealed cells. In this cycle, the evolution of hydrogen on charging is suppressed by using a large excess of uncharged negative electrode material; the oxygen evolved from the positive material combines directly with anode metal. To promote useful

445

reaction rates, part of the anode is left exposed to the gas space in the cell; this is usually done by limiting the amount of electrolyte used. The oxygen cycle was proposed (R525) for use in a $PbO_2/H_2SO_4$, $CuSO_4/Cu$ cell. In this cell, an excess of copper ions is used in the electrolyte; hydrogen evolution is suppressed during charging since the copper plates out at a lower potential. Oxygen generated at the positive electrode reacts at the exposed Cu electrode in the presence of the electrolyte. The oxygen cycle is used in sealed nickel-cadmium cells (R461).

The success of a cell design based on the use of the oxygen cycle in meeting the requirements for sealed operation depends in large part upon in limiting the evolution of hydrogen during charging, stand, and discharging. The only unqualified successes have been cells with cadmium anodes; sealed-cell designs based solely on the use of the oxygen cycle and using zinc, iron, or lead anodes have not been successful because of the difficulty in suppressing hydrogen evolution.

Because of the slowness of the uncatalyzed reaction, direct recombination of hydrogen with the cathode material cannot be used as the basis of cell design. The use of catalysts such as palladium in the cathode has not been successful because the catalyst is rapidly deactivated on cycling. Small amounts of hydrogen can be recombined in the presence of larger amounts of oxygen by the use of waterproofed catalysts (R87); this approach suffices for zinc-silver cells using very heavily amalgamated zinc electrodes; these cells operate otherwise on the oxygen cycle.

The most useful method for recombination of hydrogen is by use of an auxiliary electrode or the allied small fuel cells attached to the basic cell. Sealed cells with auxiliary electrodes that are used to recombine oxygen or hydrogen have been discussed in general (R528). The use of auxiliary electrodes in silver cells has been studied (R81–82). Recombination auxiliary electrodes functionally are hydrogen or oxygen gas electrodes which have been designed to give relatively high current densities at useful voltage levels.

The most successful recombination auxiliary electrodes have been the so-called fuel cell types, which were originally developed for use in fuel cells. A typical configuration, suitable for oxygen or hydrogen recombination is made as follows: Five parts by weight of carbonyl nickel powder, grade B, or silver flake is mixed with two parts platinum black. This mix is made into a stiff paste by adding Teflon emulsion (DuPont T–30) diluted 1:10 with water. Variations in the powders cause variation in amount of diluted emulsion needed to make the proper consistency paste, which is one that is relatively stiff. The paste is made into an electrode by use of a mold and press. The electrode is built up using 2 layers of Al foil on the bottom of the mold, an 0.005-to-0.008-in. (0.13-0.2 mm) layer of paste, a silver or nickel mesh collector, a second layer of 0.010-to-0.015-in. (0.25-0.38 mm) paste. The final layer is an aluminum sheet on whose surface a layer of porous Teflon of 1 mg/cm$^2$ has been laid down by spraying dilute T–30 emulsion and air drying. This layer of Teflon is placed adjacent to the paste. The mold is closed, heated to 325°, and pressed at 10,000-12,000 psi (700-850 kg/cm$^2$) for 10 minutes. The electrode is removed hot from the mold (the aluminum foils prevent sticking of the electrode to the mold) and air cooled. The aluminum is removed by immersing the electrode in 25% sodium hydroxide; the electrode is then washed free of alkali and air dried. The side with the porous Teflon layer is the "gas" side; the Teflon waterproofs this surface. This electrode typically has 6-12 mg/cm$^2$ of platinum black and is 12-15 mils (0.3-0.4 mm) thick. Silver substrates are used in space cells where low residual magnetism is desired; nickel is used where magnetic properties are not critical.

Figure 181 shows the basic schematic of an auxiliary electrode cell. More than one electrode can be used if desired. In use, the auxiliary electrode is paired with one of the plate groups to form either an oxygen-negative group or a hydrogen-positive group subcell. This cell is electrically connected to some impedance load; for recombination this is usually simply a resistor or resistor plus diode. In alkaline electrolyte cells, the gas electrode reactions are:

$$O_2 + 2H_2O + 4e \rightarrow 4OH^-$$

$$H_2 + 2OH^- \rightarrow 2H_2O + 2e$$

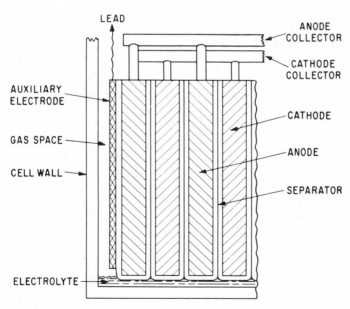

**Figure 181**  Schematic auxiliary electrode.

The principal advantage of using auxiliary electrodes for recombination is the marked increase in recombination rate that can be obtained. Cells capable of recombining the oxygen generated on continuous overcharge at the C/2 to 2C rate have been constructed. Fuel cell type auxiliary electrodes made as above can operate at 20-50 mA/cm$^2$ continuously on oxygen and 10-30 mA/cm$^2$ on hydrogen; peak loads of 100-150 mA/cm$^2$ can be sustained for several minutes. Although most of the work with recombination auxiliary electrodes has been done on nickel-cadmium cells, the major interest with these cells is the use of auxiliary electrodes for charge control; a variety of work has been reported (R79, 80, 83, 84, 85, 88, 91, 121, 264, 313, 409, 526, 555).

For silver cells, the rapid rise in cell voltage at end of charge (unlike that of the Ni-Cd cell) generally provides an adequate charge control signal, and the use of auxiliary electrodes for charge control is restricted to cases where the end-of-charge voltage signal is inadequate. For cells charged at high rates, it is imperative that the charging rate be reduced to a safe value at the onset of overcharge; the voltage signal cutoff is usually not adequate and an auxiliary electrode should be used.

Figure 182 gives the basic auxiliary electrode circuits for recombination and

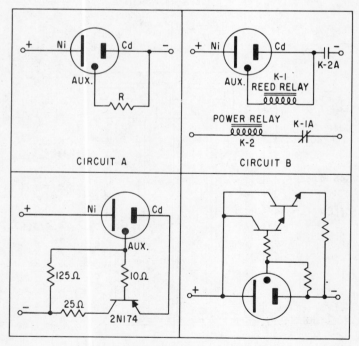

Figure 182   Auxiliary electrode circuits.

charge control. All of the circuits are shown for a cell operating on the oxygen cycle, that is, the auxiliary electrode is connected through the control circuit input impedance to the negative electrode. Similar circuits for cells operating on the hydrogen cycle, that is, the auxiliary electrode returned to the positive, can be set up. Circuit A of the figure is the basic recombination circuit, the resistance $R$ is selected to limit the auxiliary electrode current to a safe value; typical values of resistance are one to ten ohms. In some applications a diode is used, with or without a series resistance. Circuit B is a pilot relay circuit; the reed relay coil is adjusted to close the contacts at a preset value of auxiliary electrode current. The reed relay contacts may be used to open a power relay, as shown in the figure, that shuts off the charging current or the contacts may be used in logic networks when multiple-cell batteries are being controlled. Circuit C is a transistor cutoff circuit; the auxiliary electrode current biases the transistor network to reduce the charging current to the value needed to sustain the cutoff. The circuit can thus be used to provide a suitable trickle charge rate on prolonged overcharge yet allow a full charging rate when the cell is discharged. Circuit D is a bypass control; the transistor network shunts the charging current when the auxiliary electrode current increases to the cutoff value. This circuit can be used with each cell of a multiple-cell battery and in effect converts the charging from series controlled to individual cell control. It is particularly useful for very high charge rates (e.g., 4 to 6C) where cell balance is a problem.

The use of auxiliary electrodes for recombination in sealed silver-cadmium cells is straightforward. The use of the platinum-catalyzed type is recommended, since it will promote the removal of the small amounts of hydrogen generated in the cell by catalytic combustion with the excess oxygen. Modified 5-Ah test cells have been run continuously on overcharge at 300 mA (0.06C) for over 1.5 years without deterioration of the auxiliary electrode performance and minimal degradation of the main cell electrode pack. Gas pressure, after the first 72 hours, was always below atmospheric. In these cells, the auxiliary electrodes were of the platinum-catalyzed type described above, with an area equal to the cross section of the main cell pack ($\frac{1}{8}$ of the total area of the positive electrode).

Constant current charging is seldom used with silver-cadmium cells; the use of auxiliary electrodes in cells charged at constant potentials was tested. Tests were run on six-cell batteries, with the charge control set at $9.12 \pm 0.6$ V ($1.52 \pm 0.01$ V/cell). The cells used in the six-cell batteries were either modified 5-Ah cells like those described above, or the original unmodified cells for the comparison tests. Without auxiliary electrodes, batteries subjected to this charging regime become badly unbalanced with respect to cell voltage, and internal pressures soon reach dangerous levels. With auxiliary electrodes, the cells soon become balanced and pressures are kept low. Figure 183 shows the cell voltage, charge current, and pressure performance of one of the cells with

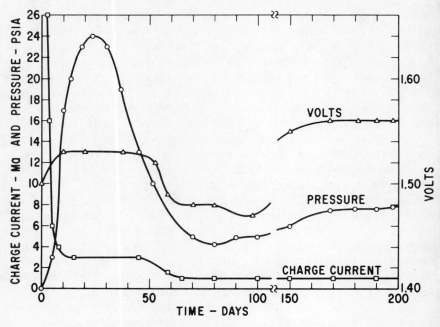

**Figure 183**  Silver-cadmium cell tests.

auxiliary electrodes of the battery on continuous discharge. The data for the other cells are similar. In the test, the auxiliary electrodes were connected to the cadmium electrode through a 5-ohm resistance. After the prolonged charging test, the cells were discharged, all giving over 6.7 Ah to a 1-V cutoff. The cells were then subjected to ten cycles of charge at 300 mA for 20 hours followed by discharge at 2.5 A. All cells gave over 5 Ah capacity above the 1.0 V/cell cutoff.

During the test on continuous constant potential overcharge, it was noticed that hydrogen was sometimes present in excess in the cell, as shown by the voltage of the auxiliary electrode with respect to the cadmium electrode. This hydrogen disappeared later as the charge continued. This suggested that the catalytic behavior of the auxiliary electrode might be its useful property. A test run was made using a battery pack with auxiliary electrode cells, but with the auxiliary electrodes disconnected. The results were similar to the tests with auxiliaries connected with respect to cell voltage balance, but the pressure excursions were greater. The cells showed a pressure rise in the first two weeks to about 2 atm followed by a rapid drop to a low value, then a slow rise to a low pressure maximum. This behavior can be explained by noting that the initial pressure is caused by oxygen (as shown by the auxiliary electrode potential) followed by hydrogen evolution at the cadmium electrode; it combines with the oxygen on the auxiliary electrode catalyst. The final period is for a balanced hydrogen-oxygen evolution.

Since Zn/Ag cells normally show an excess of hydrogen evolution over oxygen evolution, cell designs based on the use of the oxygen cycle require modification of the anode and operation to prevent gross evolution of hydrogen. This is usually done by relatively heavy amalgamation of the zinc electrode. Studies on sealed silver-zinc cell designs not using auxiliary electrodes and based on the oxygen cycle have been reported (R80, 112, 113, 114, 374, 424, 646B).

Palladium catalyst in the auxiliary electrodes (R424) has been found unsatisfactory for recombination because of its ready oxidation by oxygen, even in the presence of hydrogen, and subsequent migration. The test cells using this type electrode failed in a relatively few cycles. Similar tests by Delco-Remy workers on cells with platinum-catalyzed fuel cell type electrodes made in a similar fashion as those described previously were reported (R424) to give satisfactory results on cycling; the tests are continuing.

Of primary concern in using platinum-catalyzed electrodes as auxiliary electrodes is poisoning of the zinc electrode by platinum from the auxiliary electrode which catalyzes the self-discharge reaction of the zinc electrode:

$$Zn + 2OH^- \rightarrow H_2 + ZnO_2^{2-}$$

This migration was not observed in our tests and is believed to be of minimal danger since the Teflon binder is very effective in holding the platinum black, nickel, and silver powders in place with no degradation showing in 12-18 months use and the platinum is used under mainly reducing conditions and is in a highly insoluble form and has shown no tendency to go into solution when the electrode current density is kept below $50 \, mA/cm^2$; the barriers used in silver cells for blocking silver migration are even more effective for platinum. The use of substitutes for platinum is unsatisfactory. Palladium oxidizes readily, and, although it can be reduced by hydrogen, it tends to migrate extensively and is very effective in promoting hydrogen evolution from zinc electrodes. Raney nickel catalyst is satisfactory only if oxygen is rigorously excluded; in the presence of oxygen it undergoes an irreversible oxidation. Some nickel oxide surfaces show a low rate ($1 \, \mu A/cm^2$) for hydrogen recombination; this is satisfactory in cells such as nickel-cadmium in which very small amounts of hydrogen are evolved during normal operation of the cell.

The converse problem, the poisoning of the auxiliary electrode catalyst by the zincate in the electrolyte, does not seem to occur for platinum catalysts.

Even if migration of platinum is prevented, the use of platinum-catalyzed auxiliary electrodes for oxygen recombination in zinc cells is not feasible, since the connection of the auxiliary electrode to the zinc electrode sets up an "inert" cathode cell, with a vigorous evolution of hydrogen. In one test, a $14 \, cm^2$

auxiliary electrode completely discharged a 3-Ah zinc electrode of the same area in fewer than 8 hours when connected to the zinc electrode through a one-ohm resistance. The use of a diode, as recommended (R529B), did not prevent this rapid discharge because of the high voltage found for the $Zn/KOH/H_2$ (Pt) cell.

The use of auxiliary electrodes with silver and mixed oxide (spinel) catalysts also gave high discharge rates when the auxiliaries were connected through a resistor to the zinc electrode. The gassing rate shown for a $14 \text{ cm}^2$ spinel-catalyzed electrode connected to a 3-Ah electrode through one ohm was $39 \text{ cm}^3$ hydrogen in two hours. The use of a diode in the circuit stops the discharge with this type of auxiliary electrode; thus, with the use of a 1N198 diode, the spinel-zinc cell showed zero hydrogen evolution in two hours. Tests run on 3-Ah cells with silver and spinel auxiliaries gave satisfactory performance with respect to oxygen recombination; on cycling, the pressures in the cells increased continuously because of hydrogen evolution and accumulation.

The use of an oxygen auxiliary electrode in sealed zinc cells appears to be unnecessary, since there is normally an excess of hydrogen generated in the cell. As a platinum-catalyzed auxiliary electrode is needed to recombine hydrogen, the catalytic recombination of oxygen will remove the oxygen generated on cell cycling. The direct connection of the auxiliary electrode to the silver oxide cathode does not promote oxygen evolution on stand or discharge, and hydrogen is recombined at a relatively high rate ($15\text{-}25 \text{ mA/cm}^2$ for 5 psia or 0.35 atm). However, on charging and particularly on overcharging, the auxiliary electrode gasses oxygen, which tends to oxidize the silver or nickel substrate at current densities above $2\text{-}3 \text{ mA/cm}^2$. The use of a diode in the circuit prevents this gassing action. Three-ampere-hour cells with $14 \text{ cm}^2$ auxiliary electrodes connected through diodes to the silver electrodes could be overcharged continuously at 50 mA with cell pressures of 150 to 200 mm. The auxiliary electrode potentials showed these cells were operating on the hydrogen cycle; this was confirmed by gas analysis.

The use of small fuel cells to recombine hydrogen evolved from cells is an extension of the use of auxiliary electrodes. The first use of such cells were units built for Explorer XII–B, using copper oxide as the cathode (R86). These cells were placed in the satellite to remove hydrogen vented from the primary zinc-silver cells used. Miniature regenerative cells (Figure 184) capable of being attached directly to cells have been developed (R42). These cells are based on the use of silver oxide cathodes and fuel cell type hydrogen anodes. In well-behaved cells, the amount of recombination needed is small, and cells with $1\text{-}2 \text{ cm}^2$ area can be used. Periodic regeneration of the silver electrode is necessary; this is done by special charging. The test results are favorable, and the units show good long-term stability Figure 184 shows the schematic.

However, it would appear that the use of a cell with a small area auxiliary

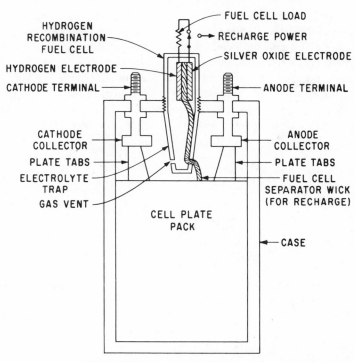

**Figure 184** Schematic small fuel system.

electrode connected internally through the small diode to the silver electrode would accomplish the same objectives without need for providing extra size or large weight to the battery pack (auxiliary electrode plus diode would cost zero volume and 5-10 grams per cell weight) or for providing the regenerative circuit and control. Internal connection is feasible, since no control function is called for from the auxiliary electrode and the circuit components are small. This internal connection eliminates the need for an extra cell terminal. The argument that the fuel cells permit use of standard zinc-silver oxide cells is not particularly valid for space cells at least, inasmuch as these are purchased on special order to specified design anyway.

An alternative to auxiliary electrode and other types of battery charge control is the use of stabistors (R13, 14, 99). These are diode networks that are placed across the cell terminals. They function by shunting charge current when the cell voltage passes a preset value. Although the end-of-charge voltage characteristics of silver cells are, in principle at least, favorable to stabistor use (the sharp rise in cell voltage makes selection of stabistor turn-on voltage level much less critical than for use with nickel-cadmium cells), the use of stabistors in place of recombination auxiliary electrodes cannot be recommended. After all, charge control of sealed silver cells with respect to minimizing overcharge rates is not the problem. Rather, the problem is removal of hydrogen, whose generation in the cell cannot be prevented by any of the present charge control methods, including stabistors. The use of stabistors with auxiliary electrodes with silver cells cannot be recommended either, since their use calls for constant current charging, and the use of constant potential charging is preferred (R281).

# PART VIII

## BATTERY USE, PROCUREMENT, QUALITY CONTROL, AND RELIABILITY

# 33.

## Failure Modes and Mechanisms

### J. J. Lander

The determination of failure modes and mechanisms is one of the most important aspects of battery testing because it discloses where the emphasis in research and development should be placed in order to gain performance improvement. Failures can be caused by a variety of causes such as mechanical damage, subnormal construction, internal short-circuiting, loss in faradaic capacity on cycling or open-circuit stand, leakage of electrolyte, pressure build-up in the case of sealed cells, and so on. In many instances the direct cause(s) of failure are evident, for example, mechanical damage, or the failure of a cell to give adequate capacity. In cycling cells and batteries, however, the remote causes, the factors which combine to degenerate performance gradually, are not always so evident, and special analytical techniques, data treatment, or even specially designed diagnostic experiments may be required to achieve an understanding of remote causes.

## DEFINITIONS FOR CLARITY

Before considering modes and mechanisms of failure it is beneficial to define terms for purposes of clarity. The following definitions (R441) will be used or paraphrased:

*Failure.* The inability of a cell or battery to deliver, on discharge, a quantity of electrical energy or power, or both, commensurate with mission requirements.

*Failure Mode.* The actual dimension of voltage, current, or time, or combinations of these that define mission requirements, which go out of tolerance on discharge. The mode will be evident given adequate electrical monitoring of discharge performance.

*Failure Determinant.* The immediate cause of the failure mode. For example, a mode of low voltage might be caused by too high an internal resistance; a mode of low capacity (current x time) might be caused by loss of material from the plates. The failure determinant will frequently be evident in visual inspection, although some auxiliary measurements such as internal impedance or reference electrode readings to individual plate groups might be necessary.

*Failure Mechanism.* The fundamental chemical or physical processes, or both, which contribute to producing failure determinants. It is these which are frequently difficult to discern and define as a result of testing and failure examination.

While the user of batteries will certainly define failure in terms of electrical performance, he may also wish to impose temperature stresses and mechanical stresses on requirements for meeting performance specifications. The effects of mechanical and other environmental stresses (except temperature) will not be considered here, as being more intimately the concern of engineering design. Temperature, on the other hand, will be considered because of its direct effects on chemical and electromechanical characteristics.

Because of faradaic inefficiencies, cells and batteries are overdesigned in active material contents so that failures from this source are not expected on initial or early cycles.

## REMOTELY ACTIVATED PRIMARY BATTERIES

These battery types have been brought to a high degree of sophistication and reliability through excellent design development work, materials development programs, production quality control, and reliability programs to ensure lot acceptance of component parts, subassemblies, and total batteries. Combining, as they do, problems of design and development of primary cells and batteries with problems arising from storage of electrolyte outside the cell and delivery of electrolyte, uniformly, on demand, to cells in very short times, development of the reserve primary batteries must be reckoned among the most successful battery programs ever undertaken. Involved in this, was the successful development of gas-producing squibs with very long storage life, uniformity of gas quantity produced and its rate of delivery, and resistance to accidental firing.

For controlled room temperature storage conditions, the state-of-art can now provide batteries with $6^+$ years of storage life at a reliability approaching 100%. In fact, storage life and failure determinants have yet to be determined for some designs. For uncontrolled temperature of storage and for controlled elevated storage temperatures failure modes fall into two types: one, an increase in time for voltage to come up to specification and, two, degradation of the plateau voltage on discharge. Life and reliability under such conditions vary quite markedly depending on "tightness" of the specifications, temperature, and mode of temperature control (R639B). Sufficient analytical work has not been done to establish failure determinants and mechanisms.

## SECONDARY BATTERIES

The failure mode of cycling secondary batteries is lack of ability to deliver a required run-time to a minimum voltage under a given discharge load profile. This will be accompanied through life by a more or less gradual diminishing of average discharge voltage and an increase in on-charge and end-of-charge voltages.

Two general failure determinants are evident: one, short-circuiting through or around the separator or, two, loss in negative plate capacity. Two somewhat specialized determinants have been observed: one, a more rapid loss in faradaic capacity accompanying low-temperature cycling and, two, a precipitate loss of negative plate capacity in cell designs which are somewhat electrolyte starved and which have multiple layers of separation where rapid, relatively high depth-of-discharge (DOD) cycling is involved.

Short-circuiting through or around the separator results when insufficient separator thickness is used or when the separator is not extended sufficiently above the tops of the plates. A combination of failure mechanisms can be involved for cellulosic separator materials. Chemical attack by the environment results in physical degradation of the separator material. Reaction of soluble silver oxides with the cellulose or precipitation of colloidal silver, or both, results in accumulation of metallic silver in the texture of the cellulose, progressing toward the negative from the positive plate. When sufficient metallic silver is accumulated from one side to the other of a separator layer, an electronically conducting path is provided. The third contributing factor is growth, through the separator layers, of metallic zinc dendrites from the zinc plate toward the positive plate. Short-circuiting is evidenced by a more or less abrupt loss in capacity and by low end-of-charge voltage. If the short circuit is massive, heat from the short-circuit current can burn holes through separators and even destroy cell cases.

Loss in negative plate capacity as cycling progresses is indicated by a loss in the average discharge voltage or a shortening of the discharge time or both. Higher average on-charge voltages, higher end-of-charge voltages, and earlier voltage rise to the overcharge level are exhibited. These are accompanied by a loss in charging efficiency. Obvious contributing determinants are an increase in internal ohmic resistance, increased polarization, and an earlier onset of the final passivation or concentration polarization, or both, which drives down the zinc electrode voltage, and, consequently, cell voltage goes out of tolerance. Reference electrode readings have established failure to be at the negative plate; moreover, negative plate responsibility is obvious on cell tear-down for failure examination.

Several mechanisms contribute to failure determinants. Increased internal ohmic resistance results from a loss in geometric area of the zinc active material (shape change) thus effectively reducing plate area. This is graphically illustrated in Figure 185, which shows an enlarged photograph of a zinc electrode extracted from a cell which had gone somewhat over 100 cycles at 60% DOD. The darker area in the center of the plate is nodular metallic zinc which has accumulated in the center after transfer from the top and side edges. Nodular zinc can also be seen along the bottom edge. This plate experienced less than 10% loss of active material while the cell was losing 50% of its capacity. The remote causes of shape change are uncertain, it is probably a complex phenomenon. Gravity

Figure 185  Zinc electrode from a failed cell.

effects have been ruled out in tightly packed elements, and the existence of concentration cells on the electrode has been proposed (R439). A contributing factor is undoubtedly access of oxygen to the top and sides of the plate as cycling progresses. It has been observed, when hydrogen gas (substantially only) is in the head space, that shape change is not confined toward the center but that the nodular masses grow randomly over the electrode area.

The major contributing mechanism to zinc plate failure is the decrease in true surface area of the zinc active material which occurs on cycling because of the change from a very porous, finely divided structure to a massive, dense, nodular structure, resulting in greatly increased true current density to the zinc active material. This would, of course, increase polarization (R176) and hasten the end of discharge. This phenomenon is illustrated by the photomicrographs of Figures 186 and 187, which compare cross sections of a new plate and one which had undergone 231 cycles to failure. The agglomeration and densification of the zinc active material is quite apparent. In view of this result, it seems likely that the process is one of growth of the larger zinc particles at the expense of the smaller, a process which is, probably, relatively efficient because of zinc oxide solubility. Thus, during discharge more smaller particles of metallic zinc would be used up

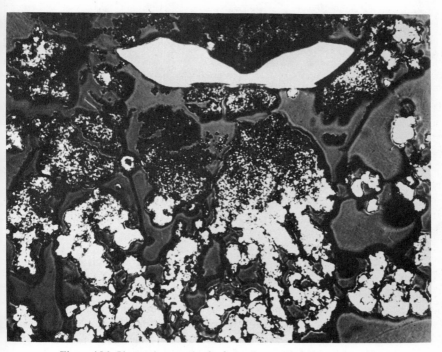

**Figure 186** Photomicrograph of a formed, uncycled zinc plate, ×40.

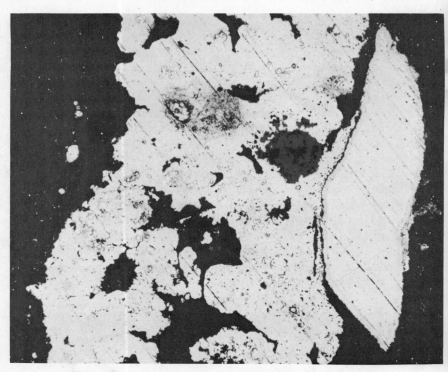

**Figure 187** Photomicrograph of a zinc plate after 231 cycles, x40.

and go into solution and, on subsequent recharge, the larger particles would provide existing nuclei for replating of metallic zinc. Some substance is lent to this view because it has been found that an organic material—Emulphogene BC-610, a nonionic surface-active agent of the tridecyloxypoly (ethyleneoxy) ethanol type—which is strongly adsorbed on zinc from KOH solution (R486), extends cycle life of zinc plates appreciably (R166, 346, 376). It functions, evidently, as a dispersing agent for the metallic zinc, probably encouraging formation of more zinc nuclei on charge. This effect is illustrated in Figure 188, which shows a photomicrograph of a zinc plate after 324 cycles on the same regimen as the plate shown in Figure 187.

In the past eight years or so, a good deal of applied research has been done, seeking to extend the life of zinc plates. This is reviewed in Dalin's earlier paper in this book.

The more rapid loss in capacity at low temperatures is due to enhanced loss of charge acceptance of the cell; possibly it is limited by the positive plate, which is known to have poorer charge acceptance at lower temperatures (R252). Cells will scarcely cycle at all below about 0°. In addition the capacity is reduced as temperature goes down.

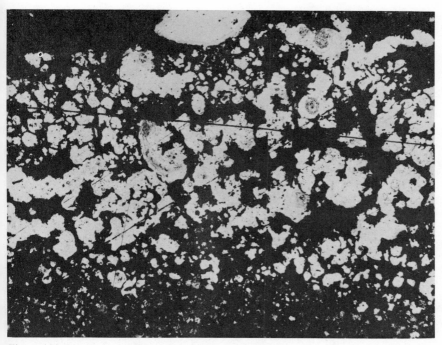

**Figure 188** Photomicrograph of a zinc plate containing 0.1% Emulphogene BC-610 after 324 cycles, x40.

The precipitate loss in capacity in electrolyte-starved designs with a relatively large amount of separation seems to be connected with electrolytic pumping of water or electrolyte out of the negative plate compartment and, perhaps, a gradual soaking up of the electrolyte by the separator. At any rate failure examination discloses that the negative plates of such cells are "dry," not bone dry, but dry enough for the active material, largely zinc oxide, to be crumbly.

## A FAILED CELL EXAMINATION

Figures 189 through 194 (R425) show a cell tear-down for examination of a failed cell. Figure 189 shows the electrochemical element extracted from the cell. A growth of mossy zinc outside the 3-layer wrap around the zinc plate shows up as a blue color near the top of the element. Figure 190 shows zinc plates extracted from the element with their separator (fibrous sausage casing) intact. The brown-black discoloration is caused by silver and palladium which have come through one cellulosic layer next to the positive plate and deposited in the third layer out from the negative (four layers of cellulosic, in total). The blue-white areas are metallic zinc (with a light film of oxide) which has come through the three layers of cellulosic around the negatives. Figure 191 shows the

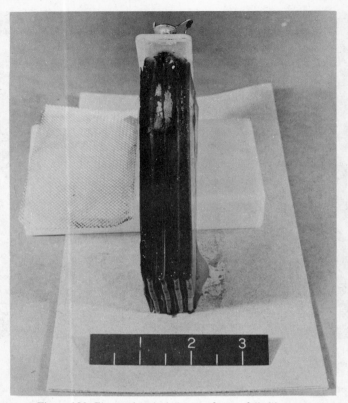

**Figure 189** Electrochemical element from a failed battery.

positive plate group extracted from the element and unfolded with separator intact. The surface of a single layer of sausage casing is presented. The blue-white color shows that metallic zinc has penetrated to the outside of the last layer away from the negative plate. The positive group with the sausage casing and the single interseparator layer of Dynel-370 turned back is shown in Figure 192. The eight plates lie at the bottom. The whitish surface discoloration results from a light coating of zinc oxide on the surface of the plate. Presumably this precipitates from supersaturated solution during discharge. Sometimes this oxide will show up as a bright pink. Figure 193 shows a pair of negative plates with the sausage casing unwrapped. The blue coloration means that considerable metallic zinc is present. The silver-palladium discoloration has yet to penetrate (in appreciable quantity) to the second layer around the negatives. The separator is physically in good shape, that is, there are no holes or tears or obviously degraded areas. Massive penetration of metallic zinc to the third layer has occurred. In Figure 194, a negative plate is shown with backlighting. Loss of

**Figure 190** Fibrous sausage casing, wrapped negative plates, U-fold.

active material from the top and one side is clearly evident. Weights of washed and dried plates show very little weight loss, rather the active material has redistributed toward the center of the plate. Thickness measurements showed that the negative plates grow in thickness in the central area to 2-3 times their original thickness. This particular cell had gone 108 cycles at 40% DOD at room temperature.

**Figure 191** Positive plate group with single sausage casing layer, unfolded.

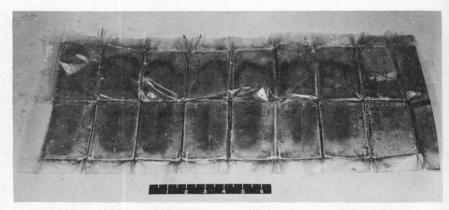

Figure 192  Positive plates displayed with separator and interseparator, turned back.

Figure 193  Negative plates with separator layers unwrapped.

## SEALED SECONDARY CELLS

In addition to the usual failure determinants for secondary cells on a cycling routine, sealed cells can exhibit other failure determinants. The mode can be an abrupt loss in capacity resulting from an exploded cell, or, if internal pressure is monitored, the mode will be an out-of-tolerance internal pressure, which is substantially equivalent to an exploded cell because cycling would have to be

Figure 194 Negative plate photographed with back lighting.

shut down in order to avoid an explosion—or a ruptured cell. In a vacuum environment, another determinant can be seal leakage which results in capacity loss because of evaporation of water from the electrolyte.

Pressure can increase in cells on a cycling routine because the charging process is not quite 100% efficient. In cells on a routine which does not require short charge times or high DODs, the cell can be overdesigned in ZnO/Ag ratio in the unformed cell, so that on charge the cell will gas oxygen, which is readily taken up by the negative plate (R376). As cycling progresses, however, and the zinc active material loses surface area and accumulates silver and palladium (if palladium is used in the positive plate), the recharge process at the negative plate becomes progressively less efficient, and the hydrogen concentration in the head space increases. This process is much accelerated in cells subjected to deep discharges, especially if available recharge times are short and recharge has to be done at a high rate.

## STRESS LEVELS IN CYCLE LIFE TESTING

In the ordinary course of cycle-life testing of secondary batteries, it is desirable to determine cycle-life capability over a range of DODs and over a range of temperatures in order to establish performance capability.

If cycle life of cells of a given design is examined as a function of DOD when sufficient separation is provided so that short-circuiting is not the failure determinant, it is found that cycle life falls off as DOD increases, as shown in Table 54. Not only does cycle life fall off, but also the total ampere-hour yield and, of course, the energy yield also fall off sharply. Based on this result, it might be thought that some stress factor is operating at different levels for different DODs, increasing as DOD increases. At least a portion of the decreased energy yield results from the requirement for various DODs. If a 25-Ah cell is used as the test cell, at 60% DOD the cell delivers just under 15 Ah when it fails but, at 25% DOD, the cell performance can degrade until it delivers only 6.25 Ah. It is certain that, once a cell fails at 60% DOD, it can be continued on cycle at lower DODs, but whether or not it would yield the same total capacities as for the lower DODs is uncertain because no reliable data exist to enable a judgment. If the discharges are made at different rates, the situation is additionally complicated.

When one thinks of increasing stress levels, temperature increase usually comes to mind. On a 2-hour cycle routine, the cell cycle life increases roughly linearly with temperature in the range 5-35° (R96). At temperatures around 5°, the failure determinant is different than it is at room temperature. Failed cells at 5° can be brought back to room temperature and cycled to failure as though they had started at room temperature in the first place. Failure is associated with lack of charge acceptance. At higher temperatures (50°) failure results from an increased inefficiency of charge. Thus several failure determinants are exhibited for zinc-silver cells as a function of temperature.

The reason for attempting to assess whether variations in cycle life with DOD and temperature result from accelerated zinc plate deterioration is because, if it were, then variations in stress levels could be involved, and, if so, determination of the stress factors might provide additional insight into failure mechanisms and provide clues for development work leading to improved performance.

It might be more fruitful to consider cycling failure as a fatigue phenomenon in which the system cycle by cycle, doesn't quite get returned to its original condition after each charge. In this case, "strain" in the system, would be measured directly by DOD, that is, the extent of chemical change occuring, and attention could be focused on such factors as electrolyte concentration, mercury content, or amount of overcharge, which are known to have appreciable effects on cycle life. Rates of change, both ways, could also be factors for consideration.

## DISCUSSION

It seems evident, as cycle life data pile up for empirical variations in cell design and zinc plate formulations, that improvements in cell performance are more and more likely to come about as a result of incisive investigations into failure processes. In this regard, a fair backlog of empirical data exists which could be screened in terms of what the individual experiments add to our knowledge and whether they were properly conducted and might be worth repeating.

Design of good diagnostic tests based on, hopefully, sound theoretical considerations or good intuitive guesses at cause-and-effect relationships should do much toward disclosing failure mechanisms and how they might be overcome. Zinc plate behavior is strongly reminiscent of behavior of the lead plate in the lead-acid battery when it is cycled without benefit of expander materials. The zinc plate may yet prove to be amenable to treatment based on our rather limited understanding of expander component functions. Investigators should not, however, confine approaches to this direction only because it may turn out to be too limited a viewpoint.

# 34.

## Orbiting Vehicle Batteries

### M. G. Gandel

Batteries are providing and will continue to provide the prime source of electrical power for many spacecraft for some time to come. The types are varied but are limited to a few practical electrochemical systems of which zinc-silver oxide has found greatest application. Nickel-cadmium and silver-cadmium batteries are widely used in rechargeable power systems; however, rechargeable silver-zinc batteries are coming into wider usage.

The spacecraft power system design engineer is responsible for providing a power system that will meet the mission requirements—usually stated in terms of power level, orbital parameters, and environmental conditions.

The intent of this paper is to present the logic of the spacecraft battery user to individuals representing suppliers and customers. Figure 195 schematically depicts the sequence of tasks necessary to deliver a battery for a flight vehicle.

The power system designer may begin by parametrically analyzing the available power system options which include zinc-silver, nickel-cadmium, and silver-cadmium batteries, fuel cells, solar arrays, and, in special cases, isotope power systems. Recent advances permit systems to be designed in which solar arrays provide charge to secondary batteries. After mission requirements, the ascent or booster environmental requirements must be considered. These requirements principally define static and dynamic loads of the forces of acceleration, shock, and vibration imposed on the equipment. A large part of present power system design effort involves the application of rechargeable battery solar array systems; in this regard, silver-zinc batteries are receiving increased attention. All of these related aspects are interesting; however, the prime purpose of this paper is to describe the engineering operations that a user might perform in obtaining and readying silver-zinc batteries for flight in an orbiting spacecraft.

The power system engineer digests the requirements and prepares a battery specification that can be satisfied by state-of-the-art or expected state-of-the-art capabilities. The design specification also reflects the tests to be performed to give assurance that all requirements will be satisfied in flight.

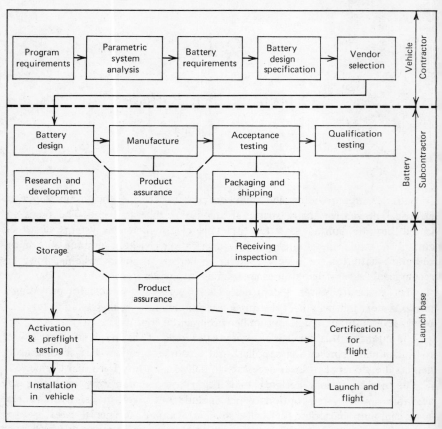

**Figure 195** Flow diagram of tasks necessary to deliver a battery for a flight vehicle.

The specifications state the design and qualification test requirements for each battery. By qualification testing is meant the testing of a limited number of production units to the maximum extremes the unit is expected to experience in flight. Once a unit has passed qualification testing, design is essentially frozen to guarantee that units above a minimum standard will be produced thereafter. When the requirements have been incorporated into a specification and the general configuration of the battery has been defined, quotations will be solicited for the development, design, and qualification testing of the battery to an established schedule.

Normal procurement practice, especially where any agency of the United States Government is involved, requires competitive bidding. The resulting quotations are analyzed from technical, financial, and management viewpoints, each normally weighted in that order.

By the time a contract and schedule are negotiated, the battery manufacturer must establish the design goals and a development plan. First to be evaluated is the ability of existing hardware (or modifications thereof) to meet the new requirements or the amount of development necessary. Most often pressure of time, budget, and the desire for highest possible reliability minimize innovation.

Any new cell design must be subjected to extensive development testing. Usually a development program will be organized into sets of discrete experiments in which only one or two variables can be varied simultaneously.

Following development and finalizing of design, units are manufactured for qualification testing. If qualifications tests are successful, the battery is certified for flight.

As part of the procurement of each production unit and prior to shipment to the launch base, the vendor, under customer product assurance cognizance, performs acceptance tests as prescribed by the design specification.

Shortly before vehicle installation, the manually activated silver-zinc batteries are activated and subjected to preflight testing and conditioning. In addition to testing formally prescribed by specification, there is testing necessary to characterize and define the functional peculiarities, reliability, and capacity expected under the environments defined. On the basis of the information developed, the power systems engineer predicts battery performance in space flight and, if necessary, analyzes for failure.

## POWER SYSTEM DEFINITION

The power system designer is an important member of the team responsible for the conceptual design of a spacecraft. A spacecraft may never leave the planning stage if an adequate electrical power system cannot be found. Much that has general validity has been written about optimizing power sources in terms of power level and life. However, primary Zn-Ag batteries are best suited for low power levels below a kilowatt, and life in terms of days or at most a few weeks. Its favorable characteristics include energy densities of 10 Wh/in$^3$. (610 Wh/dm$^3$) and 100 Wh/lb (220 Wh/kg), low cost compared to exotic systems, reliability, reproducible performance, and a capability of housing within the spacecraft.

The above points and perhaps others lead to a generally valid statement that, if weight permits, Zn-Ag batteries will be the selected power source. Thus the limitations of power level and time really are secondary to weight or booster capability. The facts that booster capability will increase and more efficient electrical power systems will be built will tend to keep silver-zinc a serious contender for a long time.

Consideration of the battery in a system sense is an important concept; it represents more than two terminals in the electrical system. As the battery discharges, it generates heat, and, at the same time, it serves as a heat sink so that one must evaluate the overall thermal balance of the battery during its life. Zinc-silver batteries find optimum application between temperatures of 10 and 21°, with the range usually a bit wider, going from 4 to 43°. A wide operating temperature range is essential when using passive thermal control in which the temperature of the battery is allowed to vary with respect to the overall thermal balance for the entire spacecraft.

Another interface is the structural mounting. This creates requirements for certain tolerances on the battery and, where cast magnesium cases have been used for batteries, these castings have had to be controlled within very close tolerances.

The spacecraft guidance system is also an interface to be considered. The battery, except for the sealed units, exhausts hydrogen gas and the upsetting impulses that the venting gas might cause must be considered. When the vented cells within sealed cast magnesium battery cases are used, relief valves are set to relieve below 1 atm and reseat above 0.3 atm. The reseating pressure is set high enough above the water vapor pressure over the KOH electrolyte to preclude loss of water. In normal operation, no problems have been experienced with the amount of hydrogen that is vented; however, a battery malfunction could account for abnormal gas generation. The vented cell must also be designed so that electrolyte is contained in a zero-g environment.

Still another interface is that of the telemetry signals from the spacecraft to ground command or receiving stations. Normally, the larger batteries are equipped with temperature and current sensors.

## SPECIFICATIONS

Specification requirements can be broken into two main categories: the functional requirements, which include voltage, current, operating temperature, and capacity, and the environmental requirements, which include shock, vibration, acceleration, and the static environments such as temperature, humidity, and vacuum. In addition to the functional and environmental requirements imposed by the battery user, there are the general requirements imposed by military specifications covering items such as wire, soldering, and connectors. The specification must also include sections calling out the necessary testing to give evidence of the capability of meeting requirements, that is, qualification testing. The hardware or the battery, in this case, is tested to levels exceeding those expected in actual flight. The units tested, while often being the first units produced following a development program, are typical of production. In fact, after a unit is in production, units may be selected at random and subjected to a requalification test to verify quality control. Normally, only two to six batteries are subjected to qualification testing. Therefore, if the expected limits are not exceeded by a reasonable margin, subsequent units could fall short of required capability. Also included within the body of the design specification is a section on acceptance testing; these tests are performed on all units produced and form the essential criteria for accepting delivery of the finished article.

## PROCUREMENT

The procurement of a battery is governed by a design specification, an outline drawing, and a work statement or purchase order to define the contractual relationship between the user and the battery supplier. Normal procurement practices require competitive bidding on battery requirements, and one function of the battery engineer is to evaluate proposals to develop new batteries.

Once a contract has been awarded for the development of a battery, there will be a meeting between the user and the battery manufacturer to establish the design considerations for the particular application and the approach the battery manufacturer will take in development and qualification of the battery. The amount of development testing necessary to verify a design is a direct function of how advanced the design may be and its critical design parameters.

The requirements of the work statement and the design specification are reflected in the battery vendor's design package. The vendor's design package will include process specifications that define items such as the preparation of electrolyte, electrodes, separator materials, and plating baths—virtually everything affecting the fabrication and assembly of a battery. The vendor will also have detailed engineering drawings of all parts and components of the battery and, depending upon the degree of control exercised by the work statement and specification, will employ various quality assurance control and auditing methods. Quality assurance audit check points at critical stages of the battery manufacture are enforced and therefore, during the entire course of the manufacture of a battery, there may be dozens of check points for the critical manufacturing steps. The quality controls enforced by the battery user may only be a small part of the total number of controls that the manufacturer may impose. Records are maintained and, in the event that a failure is found or suspected in a battery, an exhaustive search of the production records would follow in the attempt to search out any anomalies. During the progress of a battery-development program, design reviews are usually held and, at these meetings, progress in design and development of the battery are reviewed by representatives of the prime contractor, manufacturer, and customer.

Following qualification testing, acceptance testing and inspection are performed as the last steps before the buyer accepts the article. Customarily, acceptance testing will be performed at the vendor's facility just before shipment.

## BATTERY DEVELOPMENT

The unique characteristics of each battery development program, coupled with the value and need of individual technical creativity, results in a task that the author would not try to systematize. To the contrary, it would seem that technical direction and coordination, if excessive, will stifle successful development. Generally, the user is less adept in battery technology than the manufacturer, and further burden of improvement lies with the manufacturer. One point, however, should be emphasized—the value in setting the basis for sound development of a clear statement of requirements and goals.

The development program will in most cases be bound by budget and schedule, which demand that a plan be followed and milestones be met. The user can be more objective in evaluating progress than the manufacturer can, and he is funding the work; hence the user should decide between continued funding and termination. The general association of battery development with flight application and the requirement for success tend to limit the development goals of the aerospace user to the certain successes.

## QUALIFICATION TESTING

Qualification of batteries for flight involves testing to assure that the unit will function as required while being exposed to the specified environments. The functional characteristics of batteries need not be elaborated, but the major environments are dynamic shock, vibration, acceleration, temperature extremes, high vacuum, thermal shock, humidity, dust, and even fungus. Since the number of qualification test batteries may be limited from two to six, a single unit may be exposed to a series of functional or environmental tests, or both, in which failure of any unit in any test is grounds for rejection. The consequences of failing a qualification test are severe and costly; therefore care must be exercised in planning and executing these tests. Before the initiation of qualification testing, the manufacturer will supply a qualification test plan for review. It is absolutely essential that all methods of testing be reviewed and agreed upon before initiation of these tests.

A good example of the critical nature of test planning is in temperature measurement and control; ascertaining a true battery or battery cell temperature has been a persistent and plaguing problem. All too often, the battery is treated as a passive article in a poorly controlled thermal environment. It is easy to neglect the fact that the battery itself is generating heat with some proportionality to rate of discharge or charge, as the case may be. In the case of low-temperature testing, the manufacturer may take advantage of the self-heating of the battery to obtain a higher than obvious operating temperature. When dynamic testing a battery, a point is made to discharge the battery at some low rate so as to detect any momentary discontinuities in voltage that might be caused by an intermittent short. Generally batteries do not fall apart in dynamic testing; instead the functional problems or running slightly short on capacity are encountered, rather than the more dramatic kinds of failures. The results of a qualification test program will be very carefully reviewed and, if satisfactory, qualification test certification will follow.

## POSTDEVELOPMENT TESTING

A design specification is usually written as simply as possible so as not to confuse the designer with secondary objectives. Battery specifications are quite cumbersome as they are, and the state of the art is such that not many variables can be handled simultaneously. Therefore, subsequent to development and qualification testing, additional testing beyond the scope of the specification will be conducted, including: heat dissipation; voltage, temperature, current, and capacity relationships; load-sharing characteristics; wet-stand degradation; over-stress tests to induce failure; and charge-discharge life testing.

Knowledge of the heat dissipated by batteries as a function of discharge rate and state of charge is necessary for the calculation of vehicle thermal balances and expected battery temperatures. The author has employed both an oil bath calorimeter and a vacuum chamber for such testing (R234). The vacuum-chamber method determines heat generation by adiabatic increase in tempera-ture in a thermally isolated environment. These methods have been found to correlate closely. The effect of state of charge has been found to be minimal, and heat dissipation is quite linear with respect to rate of discharge up to rates exceeding $C/10$. Some self-discharge effects have been observed above $21°$, and measurements have been carried out to $53°$. The results for low-rate batteries indicate that approximately 15% of the electrical capacity delivered is manifested in heat. This information is used by the thermodynamicist in the following steps: (1) the power demand profile for the spacecraft is programmed to give battery heat dissipation as a function of time and (2) this information is then programmed into the vehicle thermal-balance calculation.

The thermal balance problem becomes further complicated when two or more batteries are wired in parallel. In this case, load sharing between batteries must be included in terms of respective temperatures, capacities, states of charge, and total load current. Load sharing between identical batteries wired in parallel is a manifestation of passive control as opposed to active thermal control wherein batteries are heated or cooled as required to maintain a narrow temperature band.

Wet-stand capability is a necessary consideration for any silver-zinc battery, and tests are conducted to determine this capability as a function of stand and operating temperatures.

## ACCEPTANCE TESTING AND INSPECTION

Acceptance is the legal action of transfer of ownership from the seller to the buyer. The buyer agrees to buy the batteries ordered if they comply to design specifications and drawings. The procurement of a primary battery means that capacity must be verified indirectly, and, in batteries stored without electrolyte, no functional tests can be performed. There must also be indirect verification of structural integrity of the unit since it would be unacceptable to fly a battery after it has seen high stress levels. Therefore, testing of cells representing the lot from which the battery was built is stipulated within the design specification which, if satisfied, forms the basis for acceptance of the battery by the customer. The lot sample cells are discharged at a given rate to verify minimum capacity. The cell capacities are accumulated and form the basis for determining statistical variations in production and for predicting battery capacity in flight.

Dielectric tests are conducted to ascertain if there are any shorts between either positive or negative terminals and the battery case. Humidity or moisture can account for excessive dielectric leakage. Conformance to the following is checked and verified during production and as part of acceptance testing: design drawings, design specifications, process specifications, and quality assurance documentation.

In many cases, customer representatives are assigned to the vendor's facility to assure adherence to contractual requirements; if acceptance testing and inspection do not occur at the vendor's facility, they will be performed upon receipt by the buyer.

## ACTIVATION AND PREFLIGHT HANDLING OF BATTERIES

Most of the silver-zinc batteries used for orbiting vehicles are manually activated. Activation methods have progressed from the syringe to drip to vacuum activation. The syringe method is suitable for small units in which a few batteries are activated at a time and the cells readily accept electrolyte.

For large units requiring close control on electrolyte volume, syringe activation proved cumbersome, and drip activators were used. The drip activator uses a syringe needle connected to the bottom of an electrolyte reservoir so that flow rate is a function of needle size and hydraulic head. This method was satisfactory though time consuming until cells using less absorbent separators and tighter cell packaging were developed. Consequently, vacuum activation became necessary. This method provides the following advantages: activation time is decreased from 1 to 6 hours, depending on the unit; electrolyte spillage and back-bubbling is reduced; and a large number of batteries may be activated sequentially rather than simultaneously.

After addition of electrolyte, the open-circuit voltage of each cell is checked at regular intervals for any indication of voltage decay indicating a short circuit. Voltage is measured under surge current, and pin-to-case voltage checks are made to assure that there are no electrolyte paths to the battery case. After a defined soak time has elapsed, the unit is inverted (cell vents or relief valves installed). Any evidence of electrolyte leakage is cause for rejection of the battery. To preclude problems with the electrolyte's shorting paths between cells, cell terminals, intercell connectors, and inside battery case, walls are insulated by coating or potting materials.

Each type of battery developed is tested to ascertain wet-stand limitations as a function of thermal environmental history. Any time the allowable wet-stand-temperature condition has been exceeded, that unit will not be used for flight. Consequently, maximum life is achieved by refrigerating activated batteries following activation and before vehicle installation. Care should be exercised in the refrigeration of batteries to avoid condensation and current leakage paths. Battery cases should be sealed, and connectors should be covered. Batteries should be allowed to approach room temperature before being opened or before the connector cover is removed.

The subject of preflight testing of batteries to ascertain battery "health" has received much thought, but one must still rely on statistical predictions based on experience. Perhaps the most useful check for battery integrity is simply allowing the battery to stand in an activated condition for a few days and taking cell open-circuit readings at set intervals. There is a good probability that, if a defective cell is not evidenced by a drop in open circuit voltage within a few days, normal life expectancy can be achieved.

## BATTERY INSTRUMENTATION

Instrumentation requirements are deduced from the major battery parameters and the need for diagnostic tools.

Voltage is usually measured on the electrical bus and not precisely at the battery connector, since a single bus voltage will be common (except for line-loss differences) for many pieces of equipment.

Current delivered is, in some cases, read from a current sensor located within the battery case. An ampere-hour meter may be used with a single battery; however, it would more commonly be used on a system level.

Temperature is the most significant parameter of all, since the available capacity of the batteries is quite temperature dependent below $15°$. The term *available capacity* is significant since capacity may not be lost, but yet not be available at about the minimum voltage limit. Thermistors may be located either near the center of the battery or on the outside wall, although the more central temperature should be more representative.

## CONCLUSION

Silver-zinc batteries, in large numbers, have been successfully applied in orbiting vehicles. Reliability has been established through careful attention to all details of manufacture, the application of extensive handling and testing procedures, and conservative power system design.

Battery manufacture is controlled to rigid specifications that, in turn, are enforced through quality control procedures common to aerospace and military standards. Procurement of the manufactured article is contingent upon satisfaction of specified requirements and tests. It is also a necessary condition that the battery be certified by test to be qualified for flight.

# 35.

## Quality Control Procedures
## for Zinc–Silver Oxide Batteries

### Fred E. Robbins

This paper presents the quality control program and basic procedures used for Minuteman battery production, together with more detailed explanations of certain operational procedures felt to have contributed significantly to attainment of a demonstrated 99.82% reliability.

Batteries for the Minuteman guidance system required two models, one, a 28 V, 300–Wh unit, the other, a dual-section 28 V battery with sections delivering approximately 100 and 200 Wh. Both batteries are of the automatically activated primary reserve type.

### QUALITY CONTROL DEFINED

Quality is that property which defines the degree of excellence attributable to a product. Reliability, as currently used, is defined as the mathematical probability that a product will perform its designed function for a desired length of time under expected environmental conditions.

Quality control (q.c.) will be defined here to be that discipline necessary to assure maintenance of product excellence at a level sufficiently high to equal or exceed a specified reliability goal. It is not simply a matter of developing a good set of q.c. procedures. Of themselves, these will not assure quality. However, when they are stringently, but practically, applied at all levels from bench operation to management decision, then a high quality, high reliability product can be made.

### ESSENTIALS FOR A QUALITY CONTROL PROGRAM

#### MANAGEMENT ORGANISATION

The first requisite in establishing an effective quality control program for a silver-zinc battery manufacturing facility is to establish a knowledgeable, responsible, and cooperative project management organization which is set up to

481

receive and weigh all necessary information when making decisions affecting product quality. For the Minuteman project a management team concept as shown in the organization chart of Figure 196 was employed. The project manager had direct responsibility for program administration, reporting to the general manager and his staff, who functioned to furnish general guidance and policy.

The project manager's staff members directed activities of their respective groups. Intergroup activities and problems were discussed, courses of action laid

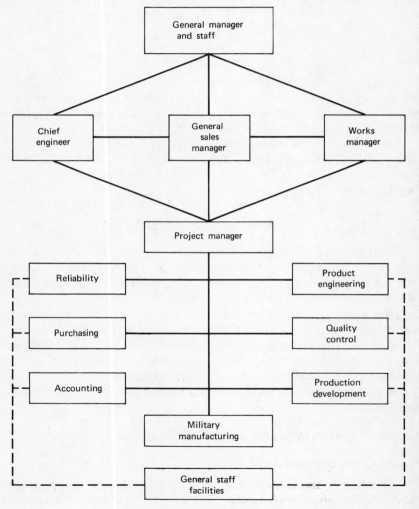

**Figure 196** Management team concept.

out, or problem solutions arrived at in project staff meetings. From the quality control viewpoint, the organisation functioned very well because most decisions affecting product quality were made by management members most familiar with the product, its application, and its operational requirements.

PERSONNEL

For an effective program a second requisite was to obtain people, screened and selected for capability and service, people who: (1) were capable of reading and understanding job operations, instructions, and specifications; (2) were capable of learning new and complex operations; (3) were careful and exacting workers, taking pride in work well done; and (4) had sufficient length of service to prevent displacement because of schedule fluctuations in other plant operations.

These personnel had to be taught to work to the exacting, precise methodology of silver-zinc production techniques required for this high reliability product, that is, work to instructions, follow them without deviation, and report anything preventing conformance to instructions. Indoctrination and education of personnel in the standards of perfection which had to be maintained for this product were done on a continuing basis with the goal of continuously improved product reliability through conscientious effort of every employee.

Increasing the product reliability was the responsibility of the reliability group and was accomplished through showing films pertaining to quality and reliability, distribution of informative material concerning application, and lectures by staff and technical management personnel emphasizing employee role in production of a quality-high reliability product. On the theory that a well-informed employee maintains a better morale and is a better worker, practice was to keep personnel informed of program progress, as far as possible.

PROCEDURES AND METHODS

The third requisite was to develop an extensive system of q.c. procedures and methods designed to: (1) ensure that the product would meet designed performance; (2) ensure that adequate quality control would be maintained through all manufacturing functions from receiving through shipping; and (3) provide objective evidence of such control.

The program was originally developed to conform to system requirements of MIL–Q–5923, revised to meet MIL–Q–9858, and upgraded in 1965 to conform to the rigorous quality program requirements of MIL–Q–9858A and NASA–NPC 200–2. These government specifications pretty well establish basic minimum q.c. procedures which must be used. They encompass broad general procedures relating to suppliers, materials (incoming and inplant), processing, manufacturing inspection, nonconforming materials and corrective action, records and documentation, and many others. Because these procedures are broadly defined, the manufacturer has the responsibility of developing and administering operating procedures within his system for the particular product to assure conformity.

## PROGRAM OBJECTIVES

Realization of a specified reliability goal of 99.7% at 90% confidence, and a five-year shelf life required setting up objectives and program design to fulfill the objectives. Minimum objectives established were:

1. Attain positive control of all materials and purchased subassemblies, with special attention to the remotely activated gas generator.
2. Develop a documentation system adequate to provide materials traceability and provide evidence of effective controls.
3. Provide specific and detailed instructions for all quality-control operations.
4. Assure leakproof reservoirs for electrolyte storage within the battery case.
5. Provide adequate surveillance of plate fabrication and processing to ensure conformance to specifications.
6. Provide a monitoring system for plate weight control and a testing system for continuous audit of plate performance.
7. Constantly control cell assembly and processing.
8. Provide an acceptance test program for cell-block assemblies to assess probable performance before assembly into a battery and to enable prediction of battery lot acceptance.
9. Provide a surveillance system for battery assembly operations with intermediate inspection of features which would be concealed by next operations.

## MATERIALS CONTROL

To attain the first objective, procedures had to be developed for control of suppliers, incoming material, and in-plant material.

### CONTROL OF SUPPLIERS

The first rule is that no material other than that from an approved source may be used in a production battery. An approval committee composed of representatives from purchasing, quality control, and engineering was set up to evaluate and approve suppliers through supplier facility visitation and evaluation, performance evaluation of qualification orders, or by review of past performance records. Critical materials suppliers required visitation.

The second rule is that the supplier must be fully informed of quality requirements for his materials. Quality Control reviewed all purchase orders to be sure the necessary requirements were included and set up interplant visits as necessary to establish full communication.

The third rule is that no supplier whose product falls below acceptable standards is to be retained on an approved listing.

## INCOMING MATERIALS CONTROL

All incoming materials were impounded by Receiving Inspection and acceptance and release to manufacturing was withheld until conformance was ascertained through specified inspections or tests. Each shipment was assigned a material lot number on arrival and records set up to include such information as purchase order number, receiving reports, and manufacturer. The number was placed on all documentation relating to a given shipment and on each unit of material. Accepted materials were released to stockrooms for storage and disbursement.

Receiving Inspection records provided a continuous picture of each supplier's quality performance.

## INTERNAL MATERIALS CONTROL

Segregation in stock was accomplished by storing by lot number with no lot mixing. Lot numbers automatically designated oldest stock in storage and oldest lots were disbursed first. Stock record cards for each item recorded such information as name, part number, specification, receipts, disbursments, and pertinent dates. Between stock identification through lot numbers and stock records, tight internal control and traceability through the manufacturing process were provided.

## RECORDS AND DOCUMENTATION

Eighty new forms applicable to silver-zinc battery production plus others adapted from commercial production were designed to provide traceability and evidence of product control. In addition to their record value, their maintenance also had a beneficial psychological effect on operators. Traceability to specific employees made them far less prone to error.

## OPERATING INSTRUCTIONS

To meet the third objective, each drawing, specification, and production flow plan was studied to determine where, what kind, and how much q.c. should be applied. An inspection operation instruction was written for every inspection station in the flow plan to ensure adequate inspection of each item or process. Instructions incorporated such controls as characteristics to be inspected, frequency sampling plan, tools or equipment used in inspection, pertinent specifications, definition of defects, accompanying documentation, and other information.

## MANUFACTURING CONTROL

The remaining objectives were attained in the manufacturing area through rigorous application of surveillance and control at specific points in the production flow of one or more of the following types of control exercised by skilled q.c. personnel:

Process control was maintained through such procedures as: equipment, instrumentation and personnel observation; laboratory analyses or tests; sampling inspections or tests on processed parts; and documentation for performance and trend evaluation:

2. Machine set-up and operational control was maintained by initial part-run approvals after any set-up or tool adjustment before production runs and by periodic examinations during runs.

3. Nondestructive test inspections were made as available, applicable, or necessary for adequate monitoring, for example, X-ray.

4. Parts and assemblies were inspected at points delineated on flow charts utilizing instructions applicable to control functions.

5. Test inspection, was provided through appropriate testing operations at control points where it would be economical to eliminate test defectives or where the test parameter first becomes available or where further processing would make the test ineffective or impossible;

6. Statistical control was provided where applicable, using sampling plans usually derived from Military Sampling Standards;

7. Calibration was performed for measuring and test equipment, periodically made on such equipment against known standards with adequate record-keeping through equipment life, conforming to MIL—C—45662 specification, Calibration Systems Requirements.

Points of inspection were established by development of q.c. flow charts designating all points in process where control was required to be applied.

## OPERATIONAL PROCEDURES ON MINUTEMAN BATTERIES

Construction of batteries was divided into five significant phases of manufacturing and a quality control flow diagram developed for each. These diagrams delineate points in the flow at which q.c. is exercised and the type or types of control applied.

PHASE I, PARTS AND SUBASSEMBLIES

Purchased parts, raw materials and subassemblies were generally handled as explained previously under incoming materials control. After material lot number assignment, materials, parts or subassemblies were sampled, usually according to the MIL–STD–105D sampling plans, and submitted to gage crib for dimensional checks and, where applicable, to engineering or process laboratories for testing. Normally, certified materials analyses are required to be furnished with purchased raw materials and parts.

Internal parts manufacturing and processing quality was maintained by: (1) first-piece approval and periodic reexamination and reapproval during production runs; (2) periodic surveillance of all processing; and (3) inspection of significant characteristics on finished parts.

Control of quality in subassembly fabrication employed rigid testing of elements in various stages of assembly and as finished components. Emphasis was placed particularly on elements of the electrolyte reservoir assembly at various points in its fabrication; because of its welded construction and cost, it was imperative that leaks be detected and eliminated before subsequent operations made it nonrepairable. Reservoirs were held to a leak rate no greater than $1.0 \times 10^{-6} \, cm^3$ (NTP)/sec. Both halogen-type sniffer and helium-detecting mass spectrometers were used.

PHASE II, NEGATIVE PLATES

Manufacturing was by the electrodeposition process onto expanded metal pure silver grids. (This book, Chapter 13.)

Tab-and-grid-strip assembly was effected by first-piece inspection and destructive testing of five weld samples at irregular intervals. Strip assemblies were inspected and coded to enable identification of each strip with plater, plating tank, and plating cycle. Every fifth strip was weighed and recorded for weight gain control.

Plating solutions were analyzed and approved before production runs and analyzed for zinc oxide content at specific points in the run. Temperature was monitored. Uniform weight was deposited on each strip through automatic control of the ampere-second input per cycle by integrating timers. Washing was controlled by pH measurement of wash water.

Uniform density of pressed plates was obtained by use of foolproof hydraulic-press pressure selector controls and by thickness measurements in each pressed strip.

After drying, each plate strip was weighed; the weight was recorded and weight gain was calculated using the weight of the fifth control strip in each cycle. Records completed at this station establish plate group identity and contain a complete process history of groups processed from each plating solution, including raw materials identification, and plate group identity was maintained through all subsequent operations. Die-cutting of individual plates from zinc-plated strips was controlled by first-piece approval and periodic appraisal of die condition throughout runs. Dies were rejected as soon as they began pulling burrs or breaking out material along edges.

Final inspection consisted of weighing each plate group, sampling for thickness, wrapping in cellulose paper and labeling with its identity. Plates to make one cell identical to endproduct cells were selected at random from the group, assembled into a cell by production methods, and discharged to the endproduct load profile as a basis for acceptance of the group. Records identified the cell, plate groups, date, operator, and disposition of plate groups. Finished plates were stored in continuously monitored, automatically controlled temperature-humidity cabinets.

PHASE III, POSITIVE PLATES

Positive-plate material mixing was regulated by milling the blend under controlled roll speeds and temperatures. Mixes were granulated, packed in sealed containers, and held by quality control for analysis for conformance to chemical composition. On release of material, it was remilled, calendered, and cut into strips of uniform thickness and weight. Control of the latter parameters was maintained by random sampling of five pieces of every hundred, recording both values.

Grid cutting, tab-and-grid assembly, and die-cut controls were applied as for the negative plate.

Hot pressing of strips on tab-grid strip assemblies was controlled by periodic check and recording of platen temperatures and dwell times. Oven temperatures and dwell times were periodically checked and recorded during the sintering operation. Cold pressing after sintering was controlled by thickness check on first piece and periodic sampling and charting of results.

Plates, after being divided into 80 plate groups for formation, were assigned a number which thereafter identified the group through all operations, assemblies, and records. Groups were weighed and a log of group numbers, material batch identity, weights, and inspection verification was originated.

Formation control in electrolyte of controlled specific gravity was accomplished by regulated constant-current, constant-time charge-control circuits monitored by q.c. personnel. Washing to constant pH and drying in automatically controlled temperature-humidity chambers and wrapping and storing were controlled and recorded. Positive plate tabs were dyed to provide a color code.

Acceptance testing was done by constructing a test cell using four plates per positive plate group of 80. Cells were pressure activated with the load on. The application could not tolerate high voltages and had a number of heavy load pulses, so that criteria for satisfactory performance became highest initial voltage on activation, lowest pulse voltage, and run time to a specified lower voltage limit.

The cell-test procedure not only served as a means of plate acceptance but also provided a continuous audit for effective control of plate fabrication, formation, and washing and drying processes. Test results plotted on a chart form afforded a running record over extended time periods for detection and evaluation of trends.

## PHASE IV, CELL AND CELL-BLOCK ASSEMBLY

Figure 197 shows the q.c. flow diagram for cell and cell-block assembly. It illustrates the manufacturing flow, the points at which q.c. is applied, and the types of control exercised.

Positive-plate wrapping, plate stacking, and cell-case edge sealing operations were all under q.c. surveillance. Inspection emphasis was on positive-plate edges during wrapping and on negatives during stacking. Detection of roughness required edge rolling to reduce the possibility of cell shorting.

To minimize effects of plate groups on a battery, usually one plate but never more than two plates from any group were permitted per cell. Enough cells in addition to those required for a cell block were assembled to provide block-acceptance test cells. A serial number was assigned for each block at this point, and a record of all plate groups used was established. This took the form of a history card which travels with the tray of cell build-ups. Cell build-ups were audited for correct plate numbers and assembly, and any loosened material or dust was vacuumed off before sealing.

After cell-case edge sealing, cells were short-checked then cured for 20 hours to allow the seal to harden. Cured cells were placed in a cell-group fixture, torqued to a specified pressure, and tested for shorts by measuring resistance between plate groups. Block-acceptance test cells were selected at random at this time from the total build-up for one block.

After assembly of intercell connectors to cell tabs, the plate tabs were cut off flush with the connector and soldered after an audit of the cut tabs to prevent reverse connecting of cells. Visual inspection of each soldered connection was followed by a short check of the cell group before presealing of cell-case openings around tabs and manifold connections. The presealed block was leak checked then wrapped, potted, and cured. Then the block was visually and dimensionally inspected and pressure tested against a specification. The history card which accompanied each block afforded at this time a complete record of each operation and inspection, by whom made and when accomplished.

Battery requirements specified performance at two operating temperature extremes. The same cell acceptance criteria previously discussed were used again for two cells at each temperature, individually discharged to the required load profile. Failure of one cell on any single criterion rejected the block. Tracings of

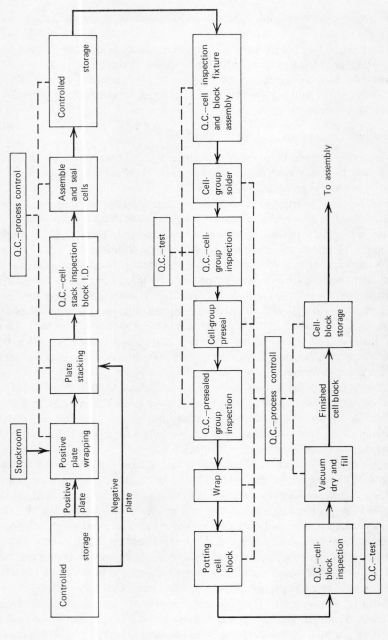

**Figure 197** Quality control flow diagram of silver-zinc battery cell and cell-block assembly.

load currents and cell voltages for all four cells were made on the same sheet of graph paper on $X-Y$ recorders. Copies of this sheet for each block incorporated into a battery became a part of the battery-lot acceptance-test data package.

With 25 cells assembled for each block, the random selection of four test cells represented a sample size of 12 (3 voltage check points per cell) parameters from a lot size of 75; acceptance on zero failures and rejection on one represents a 1.0% AQL sampling plan for block acceptance. In addition to accomplishing the purpose of cell-block acceptance, the test procedure continually audited effectiveness of cell fabrication and control. That assembly operators exercised due care and that the processes were properly controlled was evidenced by the fact that more than 36,000 such cell tests were made without a single failure attributable to cell construction.

Batteries were presented for acceptance in lots of 9 consecutively assembled units consisting of 225 individual cells from which 36 were tested for block acceptance. Because zero failures were permitted, this could be considered as a lot sampling plan whereby a sample size of 108 parameters from a lot size of 675 were tested or a 0.15% AQL plan for the 9 blocks in a battery lot. Thus, when viewed accumulatively, the results of cell-block acceptance testing demonstrate an appreciable increase in the probability of satisfactory performance of two randomly selected batteries during battery-lot acceptance testing.

PHASE V, BATTERY ASSEMBLY

The preparatory operations of reservoir filling and electrolyte trap packing were performed under q.c. surveillance. Electrolyte content of reservoirs was controlled by filling with an accurately measured volume and verified by measuring weight gain. Sealed reservoirs were thermally shocked 5 times with inspection for electrolyte leakage after each test. Reservoirs were serialized, and all operations on and information relative to assembly and q.c. were recorded on a history card which became a part of the battery history.

Before start of assembly the battery was assigned a serial number, and a battery history record was started with serial numbers of all components recorded. This document accompanied the battery through all operations and became a part of the permanent battery data package.

Final assembly operations such as soldering, potting, welding, foaming, and degassing were continuously monitored by q.c. personnel, and control functions were primarily of the surveillance, inspection, and testing types.

Nondestructive battery tests entailing visual examination, dimensional measurements, activator-circuit resistance measurement, hermetic seal and insulation resistance were performed by q.c. personnel on all finished batteries and results recorded.

Under the philosophy that the best gage of adequate control is operational results, representative samples selected from each manufacturing lot were activated and discharged under application-load conditions. On this test, a destructive test, the following parameters were monitored: activation-current pulse duration, gas-generator ignition time, battery-activation time, battery voltage under all load conditions, load currents, run time, current leakage to canister, and battery dimensional changes. Satisfactory performance on this test plus the cell results previously mentioned provided the basis for acceptance of the remaining batteries in the lot.


## THE ELECTRICALLY ACTIVATED GAS GENERATOR

Given a highly reliable cell-block, electrolyte-reservoir combination, reliability of the remotely activated total assembly requires high reliability of the electrically fired pyrotechnic gas generator. Reliability was achieved by redundancy and development of a high reliability gas generator.

In the battery designs concerned, redundancy was provided by using dual gas generators rather than dual electric matches as is sometimes done. A new generator was developed also for this program by the supplier to overcome design deficiencies existing in previous models. In excess of 18,200 pieces have been fired in generator-acceptance tests, battery tests, and flight tests without a single generator failure. This represents a demonstrated reliability of 99.98% at 90% confidence.

Controls applied to this item were probably the most severe of any in the battery. Generators were assembled in manufacturing lots, each lot having electric matches containing the same mixes of ignition powder and flash charge, with grains from the same lot of propellant, and with the same mixture of propellant ignitor. Process inspections consisting of combinations of first approvals, sampling plans, process verification, and 100% inspection were performed at 10 points in the assembly operations. Completed assemblies were subjected 100% to visual inspection, bridge-wire resistance, insulation resistance, helium leak test, and physical measurements. Nonconformance to any non-destructive test exceeding 4% was cause for lot rejection.

Lot acceptance was also dependent on satisfactory performance of a destructively tested sample consisting of a high percentage of the lot. All samples were subjected to a no-fire test followed by discharge of 80% of the sample to determine pressure characteristics and 20% for displacement capability. Ignition time was also measured. Customer representation at all acceptance tests was required, and one failure on destructive testing was cause for lot rejection.

## PROGRAM RESULTS

Development, begun in midyear 1960, ended with operational qualification early in 1962. Production was started in March of 1962. First batteries of these designs were flight tested on September 17, 1962, and they have been used on all subsequent flights. Batteries discharged for acceptance testing, life testing, and in actual flights totaled 2182 at the time of preparation of this paper. One failure was experienced, for a demonstrated reliability of 99.82% at 90% confidence.

# 36.

## Reliability Programs and Results

### Nicholas T. Wilburn

The remotely activated missile battery is a highly complex device consisting of interrelated mechanical, pyrotechnic, hydraulics, and electrochemical systems, each required to react instantly within narrow operational tolerances after several years of exposure to thermal and dynamic environmental stresses of varying degree. Following activation on command, the battery must then perform its electrical functions within a generally quite restricted voltage-time envelope while being subjected to severe and interacting environments of acceleration, deceleration, shock, vibration, spin, roll, yaw, pitch, and, occasionally, nuclear fields to which the missile is exposed during its mission.

Since it is a one-shot device to which no nondestructive tests of consequence may be applied, it is possible to achieve and maintain the required level of perfection for the missile battery only by proper initial design, by extreme quality control in production, and by effective surveillance programs in the field to detect performance degradation with prolonged storage.

Present missile battery developments rely primarily on designs, components. materials, and processes which have been proved in the past, particularly during the limited number of extensive reliability programs which have been conducted by the military. Two of these programs, each dating back about ten years, were of particular significance in their influence on subsequent battery developments. They were based on entirely different concepts in approach to reliability, each highly simple, straightforward, and effective. They were derived by considering the two aspects of the remotely activated battery: one as a largely mechanical assembly of many components and subassemblies of varying functions, each with potential weak points subject to failure from the effects of environments before and after activation; the other as an electrochemical power source which has to perform within the restricted time-voltage requirement envelope. To accomplish this, the emphasis had to be placed on the cell and battery components whose performance variability could markedly affect the statistical probability of a battery functioning or failing to function within the envelope.

Besides the difference in underlying concept, the two approaches have differed historically in their close connection with the two basic activation systems which are used in the vast majority of missile batteries. These are the copper-tube reservoir and the piston-cylinder reservoir activation systems. The basic structural mechanical reliability program established the reliability of the tube activator, and the basic statistical program was concerned with a battery employing the piston activator. To this date, no tube-activator battery has been subjected to an extensive statistical evaluation of its performance, and no piston-activated battery has been subjected to extreme environmental stresses. There is no likelihood of either possibility in the future. The reliability of batteries with each activation system has been established beyond doubt, and either system may be used with confidence in any future missile battery depending on design considerations indicating the use of one system or the other.

## RELIABILITY BASED ON ENVIRONMENTAL SAFETY FACTORS

The basic structural mechanical testing concepts were established in the reliability program for Battery BA-472/U from 1957 to 1960. This program has been reported extensively (R443, 636-637). It was based, in principle, on the test-to-failure concept which states, in general, that a high degree of engineering confidence will exist in a device if it can be demonstrated that the device has no weak components or weaknesses resulting from the assembly of components. Proof that no weaknesses exist results from testing all critical components, subassemblies, and assemblies to failure and then verifying that a sufficient safety factor exists between the failure points and all actual or anticipated environmental stresses.

The test-to-failure concept was modified since it was soon determined that batteries and components do not lend themselves readily to simple tests in which they will fail during the application of steadily increasing environmental stresses. The modification was the selection of acceptable stress levels above the known environments. These were determined to be four times the gravity force of the known requirements of shock, vibration, and acceleration. A temperature safety factor was established at a minimum of $7°$ above the anticipated maximum environmental temperature. Components and assemblies would then be tested at these levels. Any components failing at these levels would be redesigned until they would meet the standards. Successful completion of the component and assembly phase would be followed by battery tests with the same standards for redesign if necessary.

The environmental requirements for the program are given in Table 74. A further environment, perhaps the most severe of all, to which components and batteries were subjected was thermal shock, ten cycles, each consisting of

TABLE 74   ENVIRONMENTAL REQUIREMENTS,
BA-472/U RELIABILITY PROGRAM

| ENVIRONMENT | ACTUAL | RELIABILITY PROGRAM |
|---|---|---|
| SHOCK | 60G | 240G |
| VIBRATION | 10G (5-2000 cps) | 40G (5-2000 cps) |
| ACCELERATION | 60G | 240G |
| TEMPERATURE | -65°F +165°F | -80°F +212°F |

stabilization at $-62°$ followed by immediate entry into a $100°$ chamber. Following stabilization, the next cycle would begin by immediate exposure at $-62°$. This thermal shock was done prior to shock and vibration, unactivated, at both temperature extremes. This was done purposely so that any weakening of parts by the effect of thermal coefficients would be sure to reveal themselves during the shock and vibration.

As anticipated, several component limitations were discovered during the component evaluation phase. The first was the inability of the original 2½ mil (0.06 mm) thick copper foil reservoir diaphragms to withstand 40 g vibration. This was corrected by increasing the thickness to 4 mils (0.1 mm). Pressure checks showed that the heavier diaphragms would burst at 700 psi (49 kg/cm²), well below the 2,100 psi (148 kg/cm²) average burst pressure of the unsupported copper tube. The heavier diaphragms were also determined to have no adverse effect on activation time. The 40 g vibration test at $100°$ revealed a major design limitation in the pile type assembly of the cell stack. The heat-softened plastic caused a general deformation of the pile which permitted a considerable movement of the battery plates. This led to a breaking of the intercell connectors. This problem led to a complete redesign of the battery block.

Figure 198 shows the original pile design on the right and the redesigned battery block. This was a unit-cell construction assembled from pieces of stress-relieved sheet Lucite. This case design successfully met the high-temperature vibration requirement in all axes both unactivated and activated. In order to permit the required temperature stabilization at $100°$ for the thermal shock and dynamic environment tests, it was found necessary to first stabilize the battery components by a combined vacuum and high-temperature treatment. Otherwise oxygen given off by the positive plates would oxidize volatile organic

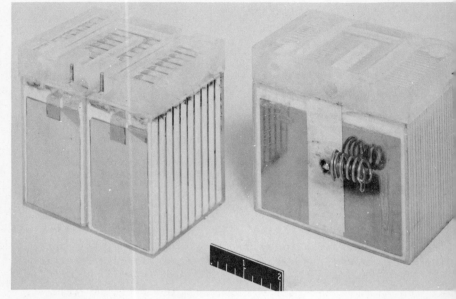

**Figure 198** Battery-block designs of BA-472: (left) unit-cell design; (right) pile-type design.

substances in the battery, generating heat which would cause the separators to oxidize.

Several bimetallic thermostats are used in the battery to maintain the temperature of the electrolyte in the reservoir at certain minimum levels prior to activation. These thermostats, controlling blanket electrical heaters around the reservoir, are subject to constant cycling over the years in which the battery may be installed in the missile. The design reliability of the thermostats was checked by repetitive cycling after exposure to thermal shock, shock, and vibration. Excessive drift in temperature setting was noted after 20,000 cycles with the original design with swedged contacts. A new design was obtained with welded contacts. Drift was found to be minimal during over 50,000 cycles.

Other component deficienies were detected and in each case were corrected by proper redesign. A program was then conducted with complete battery assemblies to determine if other weaknesses existed as a result of component assembly. The program for these batteries is shown in Table 75. They were subjected to thermal shock and to shock and vibration at the extreme temperatures. They were then stabilized at the actual requirement extreme temperatures, $-40°$ or $74°$, and then given either activated shock, vibration, or acceleration test, each at four times the required gravity level in all principle axes. The results of the tests are indicated by the service times in minutes to end

TABLE 75    ASSEMBLY TEST PROGRAM OF BA-472/U BATTERY

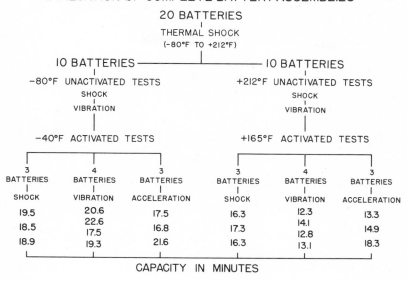

EVALUATION OF COMPLETE BATTERY ASSEMBLIES

20 BATTERIES
|
THERMAL SHOCK
(-80°F TO +212°F)

10 BATTERIES ——————————————— 10 BATTERIES
|                                                    |
-80°F  UNACTIVATED  TESTS            +212°F  UNACTIVATED  TESTS
SHOCK                                            SHOCK
|                                                    |
VIBRATION                                     VIBRATION

-40°F  ACTIVATED  TESTS                +165°F  ACTIVATED  TESTS

| 3 BATTERIES | 4 BATTERIES | 3 BATTERIES | 3 BATTERIES | 4 BATTERIES | 3 BATTERIES |
|:---:|:---:|:---:|:---:|:---:|:---:|
| SHOCK | VIBRATION | ACCELERATION | SHOCK | VIBRATION | ACCELERATION |
| 19.5 | 20.6 | 17.5 | 16.3 | 12.3 | 13.3 |
| 18.5 | 22.6 | 16.8 | 17.3 | 14.1 | 14.9 |
| 18.9 | 17.5 | 21.6 | 16.3 | 12.8 | 18.3 |
|  | 19.3 |  |  | 13.1 |  |

CAPACITY  IN  MINUTES

voltage, each substantially above the requirement of 8.5 minutes. The average activation time to minimum voltage, 95% of the nominal voltage, was one-half second, well under the requirement of a maximum of one second.

The final phase of the reliability program was an accelerated effort to determine shelf-life capability. Batteries were stored at four temperatures; 27° for controls, the maximum required field storage temperatures of 53° with peaks to 74° for four hours a day, 60° and 85°. The storage period was 18 months. At all temperatures other than 85°, no failures occurred during or after the 18 months' storage. At 85° failures were noted after 3 months as shown on Figure 199. The failures were the result of failure of the gas generators to activate. Batteries were stored the full 18 months at 85°. In all cases beyond three months, the generators were replaced with new units following storage. All batteries thus tested met the specified requirements, establishing the fundamental fact that all battery components other than the generator had an unlimited shelf-life capability under the maximum 53°-85° temperatures. Battery shelf-life capability was then limited only by the gas generator. Generator failure rate data at higher temperatures were then used to make the extrapolated prediction of a minimum capability of ten years as shown. This prediction has been substantiated by field data on batteries up to nine years old, none of which have failed to activate properly and give the required service.

The success of battery BA-472/U over the years and many other batteries

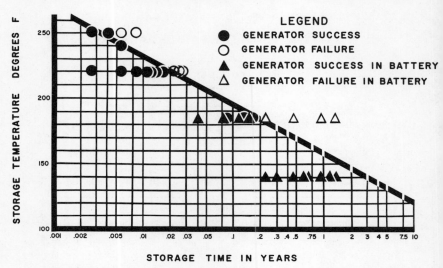

**Figure 199** Gas generator performance versus elevated temperature storage.

based on the design principles established by the reliability program has demonstrated the value of the environmental safety factor concept. The actual probabilities of successful operation were not determined, but extremely high engineering confidence was established in the basic battery design.

## STATISTICAL DETERMINATION OF PROBABILITY OF FAILURE

The second basic approach to missile battery reliability is based on a total statistical assessment of factors, both variables and attributes, entering into the production and performance of the battery. A design is established with precisely defined quality control procedures and limits. Sample batteries are produced and tested under the specified thermal and dynamic environmental conditions, and the results are assessed against the production processes involved, and the indicated quality control adjustments are made. Tests are done to determine that significant changes have been accomplished, and further battery fabrication and testing is pursued to verify that the performance parameters are within acceptable limits.

Predicted failure rates, adjusted for related error distributions, are combined as much as possible to arrive at a thorough reliability demonstration of the battery's performance capability with respect to the specified requirements.

The statistical approach to reliability was developed in its fullest extent for the Minuteman missile program. The major areas of the reliability program and the significant mathematical concepts employed are discussed in the literature

(R510) for an early phase of the program. A fairly thorough knowledge of statistical concepts and quality control procedures is necessary to properly understand the published discussion. In this paper, there will be attempted only the presentation of the major concepts relating to the determination and interpretation of probability of failure with respect to the critical electrical performance parameters of the battery.

Figure 200 illustrates a typical missile battery discharge profile defined within the envelope established by the voltage requirements, the activation time, and

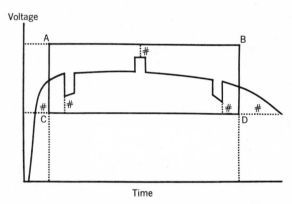

**Figure 200** Missile battery time-voltage envelope: ABCD, envelope, AB, maximum voltage; CD, minimum voltage; AC, activation time; BD, service time; #, critical intervals.

the service time requirements. The battery is required to reach minimum operating voltage within the specified activation time and to remain between minimum and maximum voltages for the prescribed service time. It must do this within the range of the specified minimum and maximum operating temperatures and while being exposed to any of the specified dynamic environments such as shock, vibration, and acceleration. It must also remain within the envelope during load shifts up or down from the steady-state load.

In the Minuteman program it was required that predictions based on test data demonstrate that there is a 99.7% probability of the battery's remaining within the voltage-time envelope under all thermal and dynamic environmental conditions. The specified reliability was thus 99.7% at a designated confidence level of 90%.

The demonstration of reliability for a specific set of test conditions was based on testing a moderate-sized sample, for example, 15 to 30 units, determining the performance of each unit with respect to the critical performance parameters (activation time, maximum and minimum voltages, and service time to minimum voltage), and then determining the mean and variability of the entire sample with respect to each parameter. This can be done with established statistical formulas, but the method used was to plot a frequency distribution on arithmetic probability paper. This method has the advantage that a linear plot will support the assumption that the particular sample distribution is from a normally distributed population.

Figure 201 gives a typical example of this method. Fifteen batteries have been tested and yield service times ranging from 154 to 196 seconds, to be evaluated against a service-time requirement of 120 seconds. A tally is made of the times in intervals of 2.5 seconds. The frequency is recorded and cumulative frequency percentages are determined. These are plotted on the probability paper and a best straight line is drawn through the points. The plot is clearly linear, supporting the assumption that the population from which the sample was drawn is normally distributed. The mean value is estimated at 178 seconds, from the intersection of the plot with the 50% line. The sample standard deviation is calculated from the intersections of the plot with the 16% (+S) and the 84% (−S) lines. This is 190.5 − 166.5 = 24, which is equivalent to 2S.

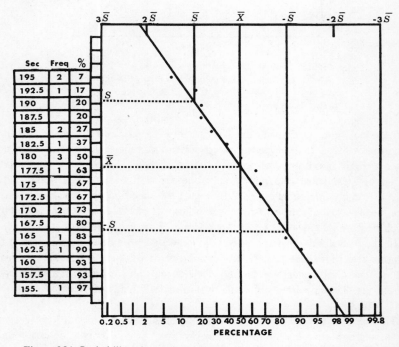

**Figure 201**  Probability plot to determine service-life performance reliability.

Therefore the sample standard deviation is estimated as 12. The sample mean, $\bar{X}$, is 58 sec above the 120-second requirement. The $K$ factor is the number of standard deviation units between the mean and the requirement, in this case $58/12 = 4.8$.

Statistical tables are then consulted to evaluate the sample $K$ factor of 4.8 against the required value of $K$ to demonstrate a reliability of 99.7% at a confidence level of 90% for a sample size of 15. This is determined to be 4.0$S$. Therefore the sample has demonstrated the required reliability. Actually a $K$ factor of 4.8$S$ for a sample size of 15 will result in prediction of a reliability of 99.95% at a confidence level of 90%. The effect of sample size is important in such calculations. Instead of the required 4.0$S$ for a sample size of 15, a factor of only 3.6$S$ is required for a sample size of 25.

This example illustrates the method of predicting the probability of success or reliability for each of the critical performance parameters of the battery under a given set of test conditions. Limitations under particular environmental conditions are corrected by modifying the process controls involved or by redesign if necessary.

The extremely reliable Minuteman missile battery design was achieved in this manner and by many extensive quality control measures. The highly critical gas generator, for example, is produced under rigid process controls and 50% of all units are destructively tested by the manufacturer. The same percentage of all batteries produced are similarly tested by the battery manufacturer. It is not surprising that no battery failures have ever occurred in missile flight.

## SUMMARY

A reasonable concluding statement might be that the two basic reliability programs as discussed are impressive, almost monumental, in scope, imagination, resourcefulness in the solution of problems, and for persistence in achieving a technological goal of immense importance. There have been many discussions in the past as to the relative merits of the two approaches, so different in concept. The success of each over the years would tend to make such arguments superfluous.

One should not be left with the impression that one approach under-emphasizes performance parameters and the other, environmental stresses. The BA-472 in Nike Hercules operates well within its prescribed time-voltage envelope; the Minuteman battery does not fail under any combination of its specified environments. The same basic objectives have been achieved in each case.

# References

*Editor's note.* Reports issued on government research and development contracts have been identified by AD- and N- numbers from Technical Abstract Bulletin (TAB), Scientific and Technical Aerospace Reports (STAR), and U. S. Government Research and Development Reports (USGRDR); such reports except where restricted or classified are available from the Clearinghouse for Federal Scientific and Technical Information (CFSTI). Chemical Abstract references have been added to those of Russian papers and to some from less common journals. The complete pagination has also been shown in most cases.

1. L. T. Abrahamson, "Silver-Zinc Batteries as a Source of Power for Pilotless Aircraft," in *AIEE Paper* No. DP-57-491 (1951).
2. L. M. Adams, W. W. Harlowe, and G. C. Lawrason, "Fabrication and Testing of Battery Separator Material from Modified Polyethylene," Final report, 67 pp., June 1966, N66-29543.
3. L. M. Adams, W. W. Harlowe, and G. C. Lawrason, "Development of Battery Separator Material," Mid-program report, 15 Feb. 1967, N67-25679.
4. L. M. Adams, W. W. Harlowe, and G. C. Lawrason, Monthly reports, Mar. 1967-Aug. 1968, on contract JPL951718, "Develop Battery Separator Material Process." (not generally available).
5. G. C. Akerlof and P. Bender, *J. Am. Chem. Soc.,* **63,** 1085-1088 (1941).
6. J. A. Allen, *Australian J. Chem.,* **13,** 431-442 (1960).
7. J. A. Allen, *Australian J. Chem.,* **14,** 20-32 (1961).
8. Allied Chemical Co., Solvay Technical and Engineering Service, *Caustic Potash,* Bulletin No. 15 and private communication from Solvay Technical Service Dept.
9. K. R. Ameln, U.S. Pat. 2,677,006 (Apr. 27, 1954).
10. American Chemical Society, *Reagent Chemicals,* Specifications, Fourth Edition (1968), American Chemical Society Publications; (A) "Zinc Oxide," 639-641; (B) "Mercuric Oxide," 364-365.
11. R. F. Amlie, J. B. Ockerman, and P. Ruetschi, *J. Electrochem. Soc.,* **108,** 377-383 (1961).
12. R. F. Amlie and P. Ruetschi, *J. Electrochem. Soc.,* **108,** 813-819 (1961).
13. R. E. Amsterdam, *Proc. 19th Annual Power Sources Conf.,* 66-69 (1965).
14. R. E. Amsterdam, *Proc. 20th Annual Power Sources Conf.,* 105-109 (1966).
15. H. Andre, *Bull. Soc. Franc. Electriciens* (6th Series), **1,** 132-146 (1941).
16. H. Andre, U.S. Pat. 2,594,712 (Apr. 29, 1952).
17. H. Andre, U.S. Pat. 2,317,711 (Apr. 27, 1943).
18. P. J. Antikainen, S. Heitanen, and L. G. Sillen, PB-145613, and *Acta Chem. Scand.,* **14,** 95-101 (1960).
19. G. M. Arcand, "Reactions pertaining to Zn/Ag and Cd/Ag Batteries," Final report (April 1, 1967), N67-25332, 45 pp.

20. R. D. Armstrong, M. Fleischmann, and H. R. Thirsk, *J. Electroanal. Chem.*, 11, 208-223 (1966).

21. S. Arouete and K. F. Blurton, Final report (Oct. 31, 1966), 54 pp., N67-39054.

22. S. Arouete, K. F. Blurton, and H. G. Oswin, "Controlled Current Deposition of Zinc from Alkaline Solution," *Extended Abstracts, Battery Division, The Electrochemical Society, Inc.,* 12, 85-87 (1967); also *J. Electrochem. Soc.,* 116, 166-169 (1969).

23. F. C. Arrance, R. Greve, and A. Rosa, Final report (Feb. 1968), 177 pp., N68-16321.

24. G. Babarovsky and G. Kuzma, *Z. physik. Chem.,* 67, 48-63 (1909).

25. Yu. V. Baimakov, *Electrolytic Production of Metal Coatings,* Leningrad: Nautschnoechimikotechnitsch, Isdatelswo, 1925, 188 pp., p. 163; *CA* 20, 553.

26. C. T. Baker and I. Trachtenberg, *J. Electrochem. Soc.,* 144, 1045-1046 (1967).

27. C. C. Balke, U.S. Pat. 2,672,415 (Mar. 16, 1954).

28. A. D. Bangham, M. M. Standish, and J. C. Watkins, *J. Mol. Biol.,* 13 (1), 238-252 (1965).

29. G. A. Barbieri, *Atti Reale Accad. dei Lincei,* Ser [5] 16, 72-79 (1907); *Chem. Centralblatt,* II, 1224, (1907).

30. G. A. Barbieri, *Ber.,* 60B, 2427-2428 (1927).

31. G. Barker, "Square Wave Polarography and Some Related Techniques," 118-131, in *Modern Electroanalytical Methods, Proc. Internatl. Symposium on Modern Electrochemical Methods,* Paris (1957), G. Charlot, editor, Elsevier Publishing Co., London, 1958, p. 122 cited.

32. C. T. Baroch, R. V. Hilliard, and R. S. Lang, *J. Electrochem. Soc.,* 100, 165-172 (1953).

33. R. G. Barradas and G. H. Fraser, *Can. J. Chem.,* 42, 2488-2495 (1964).

34. R. G. Barradas and G. H. Fraser, *Can. J. Chem.,* 43, 446-449 (1965).

35. C. S. Barrett, *Structure of Metals. Crystallographic Methods, Principles, and Data,* McGraw-Hill Book Co, New York, 1952, Appendix 9, p. 648.

36. E. L. Barrett, U.S. Pat. 2,739,179 (Mar. 20, 1956).

37. J. L. Barton and J. O'M. Bockris, *Proc. Roy. Soc.* London, Ser. A, 268, 485-505 (1962).

38. R. Yu. Bek and N. T. Kudryavtsev, *Zh. Prikl. Khim.,* 34, 2013-2020 (1961); *CA* 56, 1284*d*.

39. R. Yu. Bek and N. T. Kudryavtsev, *Zh. Prikl. Khim.,* 34, 2020-2027 (1961); *CA* 56, 1284*g*.

40. A. F. Benton and L. C. Drake, *J. Am. Chem. Soc.,* 54, 2186-2194 (1932).

41. N. P. Berezina and N. V. Nikolaeva-Fedorovich, *Elektrokhim.,* 3 (1), 3-7 (1967); *CA* 66, 11290.

42. C. Berger, M. P. Streier, and H. Frank, Final report (June 30, 1966), 78 pp., N67-12969.

43. C. Berger and F. C. Arrance, (A) U.S. Pat. 3,379,569 (Apr. 23, 1968); (B) U.S. Pat. 3,379,570 (Apr. 23, 1968).

44. F. Bergsma and C. A. Kruissink, *Advan. Polymer Sci.,* 2, 307 (1961).

45. G. Bianchi, G. Mazza, and S. Trasatti, *Intern. Congr. Metal. Corrosion,* Second, New York City, 1963. 905-915 (Pub. 1966); *CA* 65, 11755.

46. L. P. Bicelli and G. Poli, *Electrochim. Acta* 11, 289-305 (1966).

47. G. Biedermann and L. G. Sillen, *Acta Chem. Scand.,* 14, 717-725 (1960).

48. H. Blades and J. R. White, U.S. Pat. 3,081,519 (Mar. 19, 1963); U.S. Pat. 3,227,664 (Jan. 4, 1966).

49. J. O'M. Bockris and G. A. Razumney, *Fundamental Aspects of Electrocrystallization,* Plenum Press, New York, 1967, Chap. 6.

50. G. W. Bodamer, editor, "Heat Sterilizable Impact-Resistant Cell Development," Interim Summary Report (Sept. 1967), 196 pp., N68-17342.

51. G. W. Bodamer, editor, Quarterly Report (15 July 1968), 37 pp., N68-31518.

52. H. Bode, "Cadmium-Silver(II) Oxide Secondary Batteries," in *Performance Forecast of Selected Energy Conversion Devices*, 29th Meeting of AGARD Propulsion and Energetics Panel, Liege (June 12-16, 1967) 125-142, AD-671685.

53. H. Bode and A. Oliapuram, *Electrochim. Acta*, **13**, 71-80 (1968).

54. E. Boettcher, German Pat. 57188 (Mar. 27, 1890).

55. R. S. Bogner, "Heat Sterilizable Silver-Zinc Investigation," Final report (Mar. 15, 1965), 152 pp., N65-27367.

56. J. F. Bonk and A. B. Garrett, *J. Electrochem. Soc.*, **106**, 612-615 (1959).

57. T. I. Borisova and V. I. Veselovskii, *Zh. Fiz. Khim.*, **27**, 1195-1207 (1953); *CA* **48**, 1174h.

58. F. M. Bowers, R. D. Wagner, N. R. Berlat, and G. L. Cohen, "The Decomposition of Argentic and Argentous Oxides in Concentrated KOH Electrolyte," Report NOL-TR 62-187 (Apr. 1963) AD-403777.

59. H. Braekken, *Kgl. Norske Videnskab. Selskabs. Forh.*, **7**, 143-146 (1935), *CA* **29**, 4647.

60. G. W. D. Briggs, *Electrochim. Acta*, **1**, 297-299 (1959).

61. G. W. D. Briggs, I. Dugdale, and W. F. K. Wynne-Jones, *Electrochim. Acta* **4**, 55-61 (1961).

62. G. W. D. Briggs and M. Fleischmann, *Trans. Faraday Soc.*, **62**, 3217-3228 (1966).

63. G. W. Briggs, M. Fleischmann, D. J. Lax, and H. R. Thirsk, *Trans. Faraday Soc.*, **64**, 3120-3127 (1968).

64. O. H. Brill and F. Solomon, U.S. Pat. 2,654,795 (Oct. 6, 1953).

65. O. H. Brill and K. H. Brown, U.S. Pat. 2,773,924 (Dec. 11, 1956).

66. H. Brintzinger and J. Wallach, *Angew. Chem.*, **47**, 61-63 (1934).

67. H. T. S. Britton, *J. Chem. Soc.*, **127**, 2110-2120, 2120-2141 (1925); *CA* **20**, 26.

68. D. S. Brown, J. P. G. Farr, N. A. Hampson, D. Larkin, and C. Lewis, *J. Electroanal. Chem.*, **17**, (3-4), 421-424 (1968).

69. J. Burbank and C. P. Wales, *J. Electrochem. Soc.*, **112**, 13-16 (1965).

70. J. Burbank, "Identification and Characterization of Electrochemical Reaction Products by X-ray," NRL Report 6626, 20 pp., (Dec. 20, 1967) AD-664945.

71. W. K. Burton, N. Cabrera, and T. C. Frank, *Trans. Royal Soc.* (London), *A243*, 299-358 (1951); *CA* **46**, 16b.

72. E. A Butler, "Discharge Behavior of the AgO/Ag Electrode," Report JPL-TR-32-535, 13 pp., (Dec. 22, 1963) N64-13285.

73. E. A Butler and A. U. Blackburn, Final report, 43 pp., (Apr. 10, 1967), N67-37220.

74. A. Butts and C. D. Coxe, editors, *Silver, Economics, Metallurgy and Use*, D. Van Nostrand Co. Inc., Princeton, N.J., 1967, 93-94; *cf.* also U.S. Military specification, MIL-S-13282A, "Silver and Silver Alloys."

75. B. D. Cahan, J. B. Ockerman, R. F. Amlie, and P. Ruetschi, *J. Electrochem. Soc.*, **107**, 725-731 (1960).

76. B. D. Cahan, U.S. Pat. 3,017, 448 (Jan. 16, 1962).

77. E. Calvet, "Microcalorimetry of Slow Phenomenon," in *Experimental Thermochemistry*, F. O. Rossini, editor, Vol. I, Chap. 12, Interscience, John Wiley & Sons, New York, 1956, p. 237.

78. E. Calvet, "Recent Progress in Microcalorimetry," in *Experimental Thermochemistry*, H. A. Skinner, editor, Vol. II, Chap. 17. Interscience, John Wiley & Sons, Inc., New York, 1962, p. 385.

79. W. N. Carson, Jr., and J. M. McQuade, "The Use of Auxiliary Electrodes in Sealed Cells," *Extended Abstracts*, The Electrochemical Society, Inc., Battery Division, **8**, 32-33 (1963).
80. W. N. Carson, Jr., "Auxiliary Electrode for Charge Control," *Proc. 18th Annual Power Sources Conf.*, 59-61 (1964).
81. W. N. Carson, Jr., Final report, (June 1964), 45 pp., N64-12025.
82. W. N. Carson, Jr., J. A. Consiglio, and J. W. Marr, Final report, (Mar. 1966), 41 pp., N66-18430.
83. W. N. Carson, Jr., and R. L. Hadley, "Rapid Charging of Secondary Cells," in *Space Power Systems Engineering*, G. C. Szego and J. E. Taylor, editors, **16**, 1145-1157 (1966), Academic Press, New York.
84. W. N. Carson, Jr., Final report, (April 1966), 111 pp., X66-16153. (This report on contract NAS5-9193, also identified as NASA-CR-62029, is available only to U.S. government agencies and their contractors).
85. W. N. Carson, Jr., "Charge Control Methods for Nickel-Cadmium Batteries," *Proc. 20th Annual Power Sources Conf.*, 103-105, (1966).
86. W. N. Carson, Jr., *NASA Technical Brief* 66-10340 (July 1966).
87. W. N. Carson, Jr., U.S. Pat. 3,356,533 (Dec. 5, 1967).
88. W. N. Carson, Jr., and R. L. Hadley, "Rapid Charging of Secondary Cells," preprint, *Proc. 6th International Battery Symposium*, Brighton (1968).
89. E. J. Casey and W. J. Moroz, *Can. J. Chem.*, **43** (5) 1199-1214 (1965).
90. H. G. Cassidy, M. Ezrin, and T. H. Updegraff, *J. Am. Chem. Soc.*, **75**, 1615-1617 (1953).
91. A. J. Catotti and M. D. Read, "Charge Control of Nickel-Cadmium Batteries," *Proc. 19th Annual Power Sources Conf.*, 63-66 (1965).
92. C. L. Chapman, "Technical Problems Associated with the Silver Zinc Battery," *Proc. First International Symposium on Batteries* (21-23 Oct., 1958) Paper f, 11 pp., 4 fig.
93. A. Charkey and G. A. Dalin, "Research and Development Study of the Silver-Cadmium Couple for Space Application," 2d Quarterly Progress Report (1 Oct. 31 Dec. 1963) NASA-CR-53286.
94. P. Chartier, *Thesis*, Science Dept., University of Strasbourg, 1968, p. 116.
95. A. M. Chreitzberg, "Effect of Test Temperature on Tensile Strength of Polystyrene and ABS Catalyzed Cements," Internal Report, ESB Inc., (Sept. 28, 1966).
96. W. W. Clark, I. F. Luke, and E. A. Roeger, Jr., Final report, AFAPL-TR-65-42 (May 1965) 225 pp., AD-465888.
97. C. L. Clarke, British Pat. 1932 (Apr. 17, 1883).
98. S. G. Clarke and J. F. Andrew, *Trans. Inst. Met. Finishing*, **32**, 262 (1955).
99. F. J. Cocca, "Stabistors," *Proc. 18th Annual Power Sources Conf.*, 65-67 (1964).
100. G. L. Cohen and G. Atkinson, *Inorg. Chem.*, **3**, 1741-1743 (1964).
101. G. L. Cohen and E. H. Ostrander, "On the Valence of Silver in Argentic Oxide," NOLTR64-13 (June 1964) AD-601196, 10 pp.
102. G. L. Cohen and G. Atkinson, "The Formation of a Soluble Silver(III) Species in Aqueous KOH Solution," *Extended Abstracts*, The Electrochemical Soc., Battery Division, **10**, 25-28 (1965).
103. G. L. Cohen and G. Atkinson, "The Characterization of the Tetrahydroxyargen-tate(III) Species in Alkaline Solution," The Electrochemical Soc., Battery Division, **12**, 106-109 (1967).
104. G. L. Cohen and G. Atkinson, *J. Electrochem. Soc.*, **115**, 1236-1242 (1968).
105. D. B. Colbeck, "Low Rate Sealed Zinc-Silver Oxide Batteries," *Proc. 17th Annual Power Sources Conf.*, 135-138 (1963).

106. J. R. Coleman and T. E. King, "Synthetic Silver Oxide in the Preparation of Battery Electrodes," in *Power Sources 1966, Research & Development in Non-Mechanical Electrical Power Sources,* D. H. Collins, editor, Pergamon Press, New York, 1967, pp. 193-205.

107. J. E. Cooper and A. Fleischer, editors, *Characteristics of Separators for Alkaline Silver Oxide Zinc Secondary Batteries,* (1965), AD-447301; (A) Appendix B, 147-155; (B) "Silver Diffusion," Chap. 10, 103-114; (C) "Zinc Penetration," Chap. 12, 129-145; (D) "Electrolyte Absorption and Retention," Chap. 4, 21-29; (E) "Degradation by Soluble Silver," Chap. 7, 77-83; (F) "Electrical Resistance AC Method," Chap. 6b, 69-75; "Electrolyte Diffusion," Chap. 9, 93-102.

108. R. L. Corbin, Delco-Remy Division, General Motors Corp., Private Communication (5 August 1968). See also R378.

109. J. T. Crennell and F. M. Lea, *Alkaline Accumulators,* Longmans, Green & Co., London, 1928; *CA* 22, 3846.

110. G. T. Croft, *J. Electrochem. Soc.,* 106, 278-284(1959).

111. G. A. Dalin and Z. Stachurski, "Aging of Silver Oxide-Zinc Primary Batteries," *Extended Abstracts,* The Electrochemical Society, Inc., Battery Division, 8, 146-147 (1963).

112. G. A. Dalin, M. Sulkes, and Z. Stachurski, "Sealed Silver Zinc Batteries," *Proc. 18th Annual Power Sources Conf.,* 54-58 (1964).

113. G. A. Dalin and M. Sulkes, "Sealed Silver Zinc Batteries, I," *Proc. 19th Annual Power Sources Conf.,* 69-73 (1965).

114. G. A. Dalin and M. Sulkes, "Design of Sealed Silver-Zinc Cells," *Proc. 20th Annual Power Sources Conf.,* 120-123 (1966).

115. G. A. Dalin and Z. O. J. Stachurski, "Absorption and Diffusion of Zincate Ions in Cellulose Membranes," *Power Sources 1966, Proc. 5th International Symp.,* Brighton, 21-37.

116. G. A. Dalin, "Performance and Economics of the Silver-Zinc Battery in Electric Vehicles," in *Power Systems for Electric Vehicles,* H. L. Linford and H. P. Gregor, editors, U.S. Dept. HEW, Public Health Service Publication 999-AP-37 (1967).

117. A. Damjanovic, T. H. V. Setty, and J. O'M. Bockris, *J. Electrochem. Soc.,* 113, 429-440 (1966).

118. A. Damjanovic, A. Dey, and J. O'M. Bockris, *Electrochim. Acta,* 11, 791-814 (1966).

119. J. A. Darbyshire, *Trans. Faraday Soc.,* 27, 675-678 (1931).

120. M. A. Dasoyan, *Chemical Current Sources,* translated by Foreign Technology Division, Wright-Patterson AFB, FTD-TT-66-7/1+2, AD-638888, p. 121.

121. A. Dassler, U.S. Pat. 2,104,973 (Jan. 11, 1938).

122. W. P. Davey, *Phys. Rev.,* 19, 248-251 (1922).

123. J. A. Davies, B. Domeij, P. S. Pringle, and F. Brown, *J. Electrochem. Soc.,* 112 (7), 675-680 (1965).

124. R. E. Davis, G. L. Horvath, and C. W. Tobias, *Electrochim. Acta,* 12, 287-297 (1967).

125. O. K. Davtyan and E. G. Missyuk, *Deuxièmes Journées Internationales d'Etude des Piles à Combustible,* Brussells (1967), pp. 226-235.

126. A. J. DeBethune, *Standard Aqueous Electrode Potentials and Temperature Coefficients at 25°C,* C. A. Hampel, Skokie, Ill., 1964.

127. A. J. DeBethune, "Electrode Potentials Table," p. 424 in *The Encyclopedia of Electrochemistry,* C. A. Hampel, editor, Reinhold Publishing Corp., New York, 1964.

128. Defense Dept., U.S.A., "Aircraft Storage Batteries," in *Digest of U.S. Naval Aviation Weapon Systems,* Avionics Edition, July 1969, pp. 18-19; also in the Aeronautics Edition, Apr. 1969, pp. 18-19. (Order from Commanding Officer, Code 43, Naval Aviation Engineering Service Unit, Philadelphia, Pa. 19112.)

129. Defense Dept. (DOD), T.O. IF-84F-2-1 "Technical Manual, General Airplane, USAF Series, F-84F," Changed 1 June 1968.

130. – – –, T.O. IF-84(R)F-2-1 "Technical Manual, General Airplane, USAF Series RF-84F," changed 30 May 1968.

131. – – –, T.O. IF-105B-2-1 "Technical Manual, General Airplane, USAF Series F-105B," Change Number 9, 20 March 1968.

132. – – –, T.O. IF-105D-2-1 "Technical Manual, General Airplane, USAF Series F-105D," Change Number 13, 15 February 1968.

133. – – –, T.O. IF-105F-2-1 "Technical Manual, General Airplane, USAF Series F-105F," Change Number 14, 15 February 1968.

134. – – –, T.O. IF-106A-2-1 "Technical Manual, General Airplane, USAF Series F-106A and F-106B," Changed 5 October 1967.

135. – – –, MIL-E-5272C(ASG), "Environmental Testing, Aeronautical and Associated Equipment, General Specification For," 13 April 1959 and Amendment-1, 20 January 1960.

136. – – –, PD-SANE-83A/1 "Purchase Description, Battery, Emergency for F-106, 15 Ampere-Hour, Silver Zinc," dated 27 April 1967.

137. – – –, PD-SANE-77A/1 "Purchase Description, Battery, Canopy for F-106, 3 Ampere-Hour, Silver Zinc."

138. – – –, MIL-B-8565E(AS) "Military Specification, Batteries, Storage, Aeronautical," 2 March 1967 and Amendment-2, 1 December 1967.

139. P. Delahay, M. Pourbaix, and P. Van Rysselberghe, *J. Electrochem. Soc.*, **98**, 65-67 (1951).

140. P. Delahay, *Double Layer and Electrode Kinetics,* Interscience, New York, 1965.

141. I. A. Denison, *Trans. Electrochem. Soc.*, **90**, 387-403 (1946).

142. I. A. Denison, W. J. Pauli, and G. R. Snyder, "Silver Oxide-Zinc Battery; Low Temperature Operation without Auxiliary Heating," DOFL-TR-560, 29 pp., (Dec. 13, 1957); PB-131781.

143. V. V. Deshpande and M. B. Kabadi, *J. Univ. Bombay,* **20A**, 28-38 (1952); **21A**, 14-21 (1952); **22A**, 42-46 (1953); *CA* **47**, 949, 11063; **48**, 11231, resp.

144. A. R. Despic, J. Diggle, and J. O'M. Bockris, "Semi-Annual Progress Report #11, 127 pp., (31 Dec. 1967); N68-18217.

145. A. R. Despic, J. Diggle, and J. O'M. Bockris, *J. Electrochem. Soc.,* **115** (5), 507-508 (1968).

146. T. W. DeWitt, "A Sodium-Sulfur Secondary Battery," in *Power Systems for Electric Vehicles,* 277-287, *cf.* Ref. #116.

147. H. G. Dietrich and J. Johnston, *J. Am. Chem. Soc.,* **49**, 1419-1431 (1927).

148. M. J. Dignam, H. M. Barrett, and G. D. Nagy, *Can. J. Chem.,* **47**, 4253-4266 (1969).

149. R. DiPasquale, U.S. Pat. 3,002,834 (Oct. 3, 1961).

150. T. P. Dirkse, *J. Electrochem. Soc.,* **101**, 328-331 (1954).

151. T. P. Dirkse, *J. Electrochem. Soc.,* **101**, 638 (1954).

152. T. P. Dirkse, C. Postmus, Jr., and R. Vandenbosch, *J. Am. Chem. Soc.,* **76**, 6022-6024 (1954).

153. T. P. Dirkse, *J. Electrochem. Soc.,* **102**, 497-501 (1955).

154. T. P. Dirkse, *J. Electrochem. Soc.,* **106**, 154-155 (1959).

155. T. P. Dirkse and D. B. DeVries, *J. Phys. Chem.,* **63**, 107-110 (1959).

156. T. P. Dirkse and G. J. Werkema, *J. Electrochem. Soc.,* **106**, 88-90 (1959).

157. T. P. Dirkse and B. Wiers, *J. Electrochem. Soc.,* **106**, 284-287 (1959).

158. T. P. Dirkse, *J. Electrochem. Soc.,* **106**, 453-457 (1959).

159. T. P. Dirkse, *J. Electrochem. Soc.,* **106**, 920-925 (1959).

160. T. P. Dirkse, *J. Electrochem. Soc.,* **107**, 859-864 (1960).

161. T. P. Dirkse and L. A. Vander Lugt, "Oxygen Adsorption by the Silver Electrode," 15 pp. (Oct. 1, 1961); AD-265383.
162. T. P. Dirkse, *J. Electrochem. Soc.,* **109**, 173-177 (1962).
163. T. P. Dirkse, L. A. Vander Lugt, and H. Schnyders, *J. Inorg. Nucl. Chem.* **25**, (7), 859-865 (1963).
164. T. P. Dirkse and J. B. DeRoos, *Z. Phys. Chem.,* (Frankfurt) **41**, (1/2) 1-7 (1964).
165. T. P. Dirkse, "Low Temperature Cycling Behavior of the Silver Electrode," 13 pp. (1 Nov. 1964); AD-609876.
166. T. P. Dirkse, "Silver Migration and Transport Mechanism Studies in Silver Oxide-Zinc Batteries," Final report, 65 pp., AF-APL-TR-64-144 (Dec. 1964); AD-610563.
167. T. P. Dirkse, D. Vander Hart, and J. Vriesenga, *J. Inorg. Nucl. Chem.,* **27**, 1779-1786 (1965).
168. T. P. Dirkse, "Investigation of the Transport and Reaction Processes Occurring within Silver Oxide-Zinc Batteries," AFAPL-TR-66-5, 53 pp. (Mar. 1966); AD-478674.
169. T. P. Dirkse, *J. Electrochem. Tech.,* **4** (3/4) 163-165 (1966).
170. T. P. Dirkse, *J. Electrochem. Tech.,* **5** (1/2) 18-21 (1967).
171. T. P. Dirkse, "Electrode Migration and Reaction Processes Occurring Within Alkaline Zinc Batteries," 6th Quarterly Report (15 June 1967); AD-816527.
172. T. P. Dirkse, "Electrode Migration and Reaction Processes Occurring Within Alkaline Zinc Batteries," 7th Quarterly Report (Sept. 1967); AD-820273.
173. T. P. Dirkse, U.S. Pat. 3,348,973 (Oct. 24, 1967).
174. T. P. Dirkse, D. DeWit, and R. Shoemaker, *J. Electrochem. Soc.,* **114** (12), 1196-1200 (1967).
175. T. P. Dirkse, "Electrode Migration and Reaction Processes," 9th Quarterly Report (Mar. 1968); AD-829092.
176. T. P. Dirkse, D. DeWit, and R. Shoemaker, *J. Electrochem. Soc.,* **115**, 442-444 (1968); (A) p. 443, fig. 2.
177. T. P. Dirkse and R. Shoemaker, *J. Electrochem. Soc.,* **115** (8), 784-786 (1968).
178. T. P. Dirkse, "Chemistry of the Zinc/Zinc Oxide Electrode," *Extended Abstracts,* The Electrochemical Society, Inc., Battery Division, **13**, 33-34 (1968).
179. S. Dmitriev, *Ref. Zh. Khim.,* **9**, Pt. 1 (1966) Abstract No. 9B1007.
180. J. F. Donohue, "Silver Batteries," in *Silver, Economics, Metallurgy, and Use,* A. Butts and C. D. Coxe, editors, Chap. 10, 153-179, D. Van Nostrand and Co. Inc., Princeton, N.J., 1967.
181. J. Doyen, U.S. Pat. 2,867,678 (Jan. 6, 1959).
182. C. Drucker and A. Finkelstein, *Galvanische Elemente und Akkumulatoren,* Akademische Verlagsges. M. B. H., Leipzig, 1932.
183. J. C. Duddy, U.S. Pat. 2,881,237 (Apr. 7, 1959).
184. J. C. Duddy, U.S. Pat. 3,007,991 (Nov. 7, 1961).
185. I. Dugdale, M. Fleischmann, and W. F. K. Wynne-Jones, *Electrochim. Acta,* **5**, 229-239 (1961).
186. A. Dun and F. Hasslacher, British Pat. 1862 (Feb. 5, 1887).
187. T. A. Edison, U.S. Pat. 684,204-5 (Oct. 8, 1901).
188. T. A. Edison, U.S. Pat. 1,016,874 (Feb. 6, 1912).
189. Edison Storage Battery Co., West Orange, N.J., Under U.S. Navy's BuShips contracts NXss-29252 and 48435, the following reports were issued: Low-rate Primary, A and B types (Dec. 16, 1942); Thin-plate high rate types, C and D (Aug. 18, 1943); Thin-plate high rate type Z (Dec. 28, 1943); Patentability (Jan. 24, 1944); and Cell Construction, Types F and G; and under Nord contract 5070 (1943-44), a production contract, 60 and 300 kW torpedo batteries, primary type were to be built.

190. M. Eisenberg, H. F. Baumann, and D. M. Brettner, *J. Electrochem. Soc.*, **108**, 909-915 (1961).

191. M. Eisenberg, *J. Electrochem. Tech.*, **3** (11/12), 340 (1965).

192. V. Engelhardt, editor, *Electrometallurgy of Aqueous Solutions*, Leningrad, 1937, p. 273; cf. *Handbuch der technische Elektrochemie*, 6 volumes, *CA* **38**, 2884.

193. ESB Incorporated, *see* Ref. 50.

194. K. J. Euler and A. Fleischer, "Primary and Secondary Galvanic Cells; History and State-of-the-Art Summary," pp. 3-56 in *Performance Forecast of Selected Static Energy Conversion Devices*, AD-671685; cf. Ref. 52.

195. F. F. Faizullin and L. K. Yuldasheva, *Uch. Zap. Kazan. Gos. Univ.* Obshcheuniv. Sbornik, **116**, (5), 82-85 (1956); *CA* **51**, 17517*a*.

196. R. Faivre, *Ann. Chim.* (Paris), **19**, 58-101 (1944); *CA* **39**, 1341.

197. H. F. Farmery and W. A. Smith, "Some Material Problems in the Silver-Zinc Secondary Battery," in *Batteries. Research and Development in Non-Mechanical Electrical Power Sources, Proc. 3rd International Symposium, Bournemouth* (Oct. 1962), D. H. Collins, editor, The Macmillan Co., pp. 179-205, 1963.

198. J. P. G. Farr and N. A. Hampson, *Trans. Faraday Soc.*, **62**, 3493-3501 (1966).

199. J. P. G. Farr and N. A. Hampson, *J. Electroanal. Chem.*, **13**, 433-441 (1967).

200. J. P. G. Farr and N. A. Hampson, *J. Electroanal. Chem.*, **18**, 407-411 (1968).

201. J. W. Faust and H. F. John, *J. Electrochem. Soc.*, **108**, 109-110 (1961).

202. L. V. Favroskaya and E. I. Stolyarova, *Izvest. Akad. Nauk Kaz. S.S.R., Ser. Gorn. Dela, Met. i Stroimaterialov*, No. 6, 92-103 (1956); Referat. *Zh. Khim.* (1957), Abstract No. 27432; cf. *CA* **53**, 8555*e*.

203. N. P. Fedot'ev and G. G. Khad'mash, *Zh. Prikl. Khim.*, **28**, 1104-1112 (1955); *CA* **50**, 6219*c*.

204. W. Feitknecht, *Helv. Chim. Acta*, **13**, 314-345 (1930).

205. W. Feitknecht and H. Weidmann, *Helv. Chim. Acta*, **26**, 1911-1930 (1943).

206. W. Feitknecht, *Helv. Chim. Acta*, **32**, 2294-2298 (1949).

207. W. Feitknecht and W. Haeberli, *Helv. Chim. Acta*, **33**, 922-936 (1950).

208. M. Feller-Kniepmeier, H. G. Feller, and T. Titzenthaler, *Ber. Bunsenges. Physik. Chem.*, **71**, 606-612 (1967).

209. A. Fischbach, U.S. Pat. 2,700,693 (Jan. 25, 1955).

210. A. Fischbach and A. L. Almerini, U.S. Pat. 2,811,572 (Oct. 29, 1957).

211. H. Fischer and N. Budiloff, *Z. Metallk.*, **32**, 100-105 (1940).

212. A. Fleischer, *J. Electrochem. Soc.*, **115**, 816 (1968).

213. M. Fleischmann, H. R. Thirsk, and J. Tordesillas, *Trans. Faraday Soc.*, **58**, 1865-1877 (1962).

214. M. Fleischmann, D. J. Lax, and H. R. Thirsk, *Trans. Faraday Soc.*, **64**, 3137-3146 (1968).

215. M. Fleischmann and H. R. Thirsk, *Advances in Electrochemistry and Electrochemical Engineering*, P. Delahay, editor, John Wiley & Sons, Inc., New York, Chap. 3, pp. 123-210, 1963.

216. M. Fleischmann, D. J. Lax, and H. R. Thirsk, *Trans. Faraday Soc.*, **64**, 3128-3136 (1968).

217. M. Fleischmann, private communication.

218. V. N. Flerov, *Zh. Prikl. Khim.*, **29**, 1779-1785 (1956); *CA* **51**, 7211*a*.

219. V. N. Flerov, *Zh. Fiz. Khim.*, **31**, 49-54 (1957); *CA* **51**, 15220*f*.

220. V. N. Flerov, *Zh. Prikl. Khim.*, **32**, 132-137 (1959); *CA* 8876*i*.

221. V. N. Flerov, *Izv. Vysshykh Vchebn. Zavedenii, Khim. i Khim. Tekhnol.*, **6** (2), 280-285 (1963); *CA* **59**, 8354*c*.

222. F. Foerster and O. Guenther, *Z. Elektrochem.*, **5**, 16-23 (1898); **6**, 301-303 (1899).

223. Ford Motor Co., Aeronutronics Div. of Philco Corp., JPL contract N-21453, 4th Technical Progress Report (Jan. 1961).
224. J. S. Fordyce and R. L. Baum, *J. Chem. Phys.*, **43**, 843-846 (1965).
225. H. T. Francis, Space Batteries, NASA-SP-5004 (1964), N64-18052.
226. T. C. Frank, *Discussions Faraday Soc.*, **5**, 48-54 (1949).
227. E. Frankland, *Proc. Roy. Soc.* London, **46**, 304-308 (1899).
228. R. Fricke and T. A Ahrndts, *Z. Anorg. Allgem. Chem.*, **134**, 344-356 (1924).
229. R. Fricke and B. Wullhorst, *Z. Anorg. Allgem. Chem.*, **205**, 127-144 (1932).
230. A. N. Frumkin, V. S. Bagotskii, Z. A. Iofa, and B. N. Kabanov, *Kinetika Elektrodynkh Protsessov*, Izd. Moskov. Gos. Univ., 1952. Izdatel'stvo Moskovs Kogo Universiteta; translated volume, *Kinetics of Electrode Processes*, 545 pp., FTD-HT-67-13; AD-668917.
231. A. N. Frumkin and L. N. Nekrasov, *Dokl. Akad. Nauk SSSR*, **126**, 115 (1959).
232. H. Fry and M. Whitaker, *J. Electrochem. Soc.*, **106**, 606-611 (1959).
233. V. P. Galushko and E. F. Zavgorodnyaya, *Tr. Soveshch. Elektrokhim. Akad. Nauk SSSR, Otd. Khim. Nauk*, (1950); 482-487 (1953); *CA* **49**, 12159*b*.
234. M. G. Gandel and R. H. Kinsey, *J. Spacecraft and Rockets*, **2** (6), 996-998 (1965); *CA* **64**, 7671*e*.
235. P. Garine, U.S. Pat. 2,636,059 (Apr. 21, 1953).
236. W. E. Garner and L. W. Reeves, *Trans. Faraday Soc.*, **50**, 254-260 (1954).
237. R. Gaspar and K. Molnar-Ivanecsko, *Acta Phys. Acad. Sci. Hung.*, **6**, 105-118 (1956); *CA* **50**, 12569*b*.
238. H. Gerischer, *Z. Physik. Chem.*, **202**, 302-317 (1953).
239. H. Gerischer and M. Krause, *Z. Physik. Chem.* (Frankfurt), **10**, 264-269 (1957).
240. P. T. Gilbert, *J. Electrochem. Soc.*, **99**, 16-21(1952).
241. G. Gilmont and R. F. Walton, *J. Electrochem. Soc.*, **103**, 549-552 (1956).
242. R. Giovanoli, H. R. Oswald, and W. Feitknecht, *Helv. Chim. Acta*, **49** (7), 1971-1983 (1966).
243. J. H. Gladstone and A. Tribe, *Nature*, **25**, 221, 461.
244. S. Glasstone, K. J. Leidler, and H. Eyring, *The Theory of Rate Processes*, McGraw-Hill Book Co. Inc., New York, 1941, p. 402.
245. W. E. Gloor et al., cf. Ref. 412.
246. Gmelin's *Handbuch d. Anorg. Chem., Syst.*, No. 32, Zinc Ergaenzungsband, 1956, p. 251.
247. H. Goehr and E. Lange, *Z. Physik. Chem.* (Frankfurt), **17**, 100-119 (1958).
248. H. Goehr and H. Reinhard, *Z. Elektrochem.*, **64**, 414-421 (1960).
249. D. H. Gold and H. P. Gregor, *J. Phys. Chem.*, **64**, 1464-1467 (1960).
250. V. M. Goldschmidt, "The Laws of Crystal Chemistry," *Skrifter Norske Videnskaps-Akad. Oslo, I: Mat.-Naturv. Kl.*, No. 2 (1926); cf. *The Structure of Crystals*, R. W. G. Wyckoff, Second Edition, The Chemical Catalog Co. Inc., p. 428 (Ref. 106).
251. J. Goodkin and F. Solomon, "A Zinc-Silver Oxide Cell for Extreme Temperature Application," in *Batteries 2, Research and Development in Non-Mechanical Electrical Power Sources, Proc. 4th Internatl. Symposium*, D. H. Collins, editor, Pergamon Press, Elmsford, New York, Sept. 1964, pp. 475-487.
252. J. Goodkin and A Charkey, "High Rate Charge and Low Rate Discharge of the Silver Electrode, *Extended Abstracts*, The Electrochemical Society, Inc., Battery Division, **10**, 29-31 (1965).
253. J. Goodkin, "Long Life Stable Zinc Electrodes for Alkaline Secondary Batteries," 1st Quarterly Report, 40 pp., (March 1967); AD-655285.
254. J. Goodkin, "Long Life Stable Zinc Electrodes for Alkaline Secondary Batteries," 2nd Quarterly Report, 41 pp. (Dec. 1967); AD-824719.

255. J. Goodkin, "Long Life Stable Zinc Electrodes for Alkaline Secondary Batteries," 3rd Quarterly Report, 42 pp. (Jan. 1968); AD-828224.

256. J. Goodkin, see Ref. 255.

257. Yu. S. Gorodetskii, *Elektrokhimiya*, **1** (6), 681-685 (1965); *CA* **64**, 290h.

258. K. Gossner and H. Polle, *Z. Phys. Chem.* (Frankfurt/Main), **54** (1-2), 93-100 (1967).

259. D. F. Goudriaan, *Proc. Acad. Sci. Amsterdam*, **22**, 179-189 (1919); *CA* **14**, 664-665.

260. W. S. Graff and H. H. Stadelmaier, *J. Electrochem. Soc.*, **105**, 446-449 (1958).

261. H. P. Gregor, "Ion Exchange Membrane Electrolytes," Final report (Aug. 1963); AD-434245.

262. H. P. Gregor, D. Dolar, and G. K. Hoeschele, *J. Am. Chem. Soc.* **77**, 3675 (1955).

263. H. P. Gregor and M. Beltzer, *J. Polymer Sci.* **53**, 125-129 (1961).

264. F. P. Greiger, "Third Electrode Charge Control," *Proc. 20th Annual Power Sources Conf.*, 99-102 (1966).

265. R. C. Griffin, *Technical Methods of Analysis*, McGraw-Hill Book Co., New York, 1927, p. 134.

266. W. L. Grube and S. R. Rouze, *Proc. Amer. Soc. Testing Mat.* **52**, 573 (1952).

267. R. S Gurnick and R. T. Joy, U.S. Pat. 2,709,651 (May 31, 1955).

268. W. J. Hamer and D. N. Craig, *J. Electrochem. Soc.*, **104**, 206-211 (1957).

269. N. A. Hampson and M. J. Tarbox, *J. Electrochem. Soc.*, **110**, 95-98 (1962).

270. N. A. Hampson and M. J. Tarbox, J. T. Lilley, and J. P. G. Farr, *Electrochem. Tech.*, **2**, 309-313 (1964).

271. N. A. Hampson and J. P. G. Farr, *Electrochem. Tech.*, **3**, 340-341 (1965).

272. N. A. Hampson, P. C. Jones, and R. F. Phillips, *Can. J. Chem.*, **45**, 2039-2044 (1967).

273. N. A. Hampson and D. Larkin, *J. Electroanal. Chem.*, **18**, 401-406 (1968).

274. N. A. Hampson, unpublished data.

275. J. D. Hanawalt, H. W. Rinn, and L. K. Frevel, *Ind. Eng. Chem., Anal. Ed.*, **10**, 457-512 (1938).

276. *Handbook of Chemistry and Physics*, 39th edition, (1957-1958).

277. A. Hantzsch, *Z. Anorg. Chem.*, **30**, 338-341 (1902).

278. V. H. Haskell, see Ref. 412.

279. O. M. Hechter, A. Polleri, G. Lester, and H. P. Gregor, *J. Am. Chem. Soc.*, **81**, 3798-3799 (1959).

280. L. Helmholtz and R. Levine, *J. Am. Chem. Soc.*, **64**, 354-358 (1942).

281. T. J. Hennigan and R. D. Sizemore, "Charge Control of Silver-Cadmium and Silver-Zinc Cells," *Proc. 20th Annual Power Sources Conf.*, 113-116 (1966).

282. T. J. Hennigan, private communication.

283. J. Heyrovsky, *Trans. Faraday Soc.*, **19**, 692-702 (1924).

284. A. Hickling and S. Hill, *Discussions Faraday Soc.*, **1**, 236-246 (1947).

285. A. Hickling and D. Taylor, *Discussions Faraday Soc.*, **1**, 277-285 (1947).

286. T. W. Higgins and P. F. Bruins, "The Cause and Prevention of Dendritic Growth on Secondary Zinc Electrodes," *Extended Abstracts*, The Electrochemical Society, Inc., Battery Division, **8**, 133-134 (1963).

287. T. W. Higgins, "The Causes and Prevention of Dendritic Growth in Zinc Electrodeposition," Ph.D. thesis, Polytechnic Inst. of Brooklyn, 193 pp., Univ. Microfilms, Order No. 62-5641, *CA* **58**, 9876c.

288. J. H. Hildebrand and W. G. Bowers, *J. Am. Chem. Soc.*, **38**, 785-788 (1916).

289. S. Hills, *J. Electrochem. Soc.*, **108**, 810-811 (1961).

290. A. Himy, "Development of a One Ah Sterilizable Silver-Zinc Cell," 2d Quarterly Report, (Jan. 1967), N67-30802.

291. A. Himy, "Improved Zinc Electrode," Final Report, 200 pp., (July 1967), N67-31486.

292.  J. J. Holochek, U.S. Pat. 3,276,975 (Oct. 4, 1966).
293.  P. L. Howard, *J. Electrochem. Soc.*, **99**, 200C-201C (1952).
294.  P. L. Howard, U.S. Pat. 2,724,734 (Nov. 22, 1955).
295.  P. L. Howard, "Zinc-Silver Oxide Batteries," *Proc. 10th Annual Battery Research and Development Conf.*, 1956, pp., 41-44.
296.  P. L. Howard and F. Solomon, "Silver Oxide Secondary Batteries," *Proc. 13th Annual Power Sources Conf.*, 1959, pp. 92-96.
297.  P. L. Howard, "Wet Cell Batteries for Power," *Product Engineering* (15 Feb. 1960), p. 75.
298.  P. L. Howard, "Battery Performance," *Product Engineering* (26 Oct. 1964), p. 85.
299.  P. L. Howard, "The Deep Sea," *Product Engineering* (Mar. 14, 1966), p. 37.
300.  H. C. Hubbell, U.S. Pat. 897,833 (1 Sept. 1908).
301.  K. Huber, *Z. Elektrochem.*, **48**, 26-29 (1942).
302.  K. Huber, *Helv. Chim. Acta*, **26**, 1037-1054, 1253-1281 (1943); **27**, 1443 (1944).
303.  K. Huber and B. Bieri, *Helv. Phys. Acta*, **21**, 375-378 (1948).
304.  K. Huber, *J. Electrochem. Soc.*, **100**, 376-382 (1953).
305.  G. F. Huettig and B. Steiner, *Z. Anorg. Allgem. Chem.*, **199**, 149-164 (1931).
306.  G. F. Huettig and K. Toischer, *Z. Anorg. Allgem. Chem.*, **207**, 273-288 (1932).
307.  G. F. Huettig and H. Moeldner, *Z. Anorg. Allgem. Chem.*, **211**, 368-378 (1933).
308.  N. S. Hush and J. Blackledge, *J. Electroanal. Chem.*, **5**, (6) 420-434, 435-449 (1963).
309.  N. Ibl, "Applications of Mass Transfer Theory; The Formation of Powdered Metal Deposits," in *Advances in Electrochemistry and Electrochemical Engineering*, C. W. Tobias, editor, Vol. 2, Interscience Publishers, 1962, pp. 49-143.
310.  T. Inoue, M. Sato, and F. Ishii, *J. Electrochem. Soc. Japan*, **22**, 679-684 (1954); *CA* **49**, 9346c.
311.  Z. A. Iofa, S. Ya. Mirlina, and N. B. Moiseeva, *Zh. Prikl. Khim.*, **22**, 983-984 (1949); *CA* **46**, 4397e.
312.  Z. A. Iofa, L. V. Komlev, and V. S. Bagotskii, *Zh. Fiz. Khim.*, **35**, 1571-1577 (1961); *CA* **55**, 25540g.
313.  S. S. Jaffe, U.S. Pat. 3,005,943 (Oct. 24, 1961).
314.  W. W. Jakobi, "Alkaline Secondary Batteries," in Kirk-Othmer's *Encyclopedia of Chem. Tech.*, second edition, Interscience, New York, 1964, Vol. 3, pp. 161-249.
315.  G. Jayme and K. Balser, *Papier* **18** (12) 746-758 (1964); *CA* **62**, 4205e.
316.  G. Jayme and K. Balser, *Papier* **21** (10A) 678-688 (1967); *CA* **68**, 40845g.
317.  K. Jellinek and H. Gordon, *Z. Physik. Chem.*, **112**, 207-249 (1924).
318.  Jet Propulsion Laboratory, JPL Specification XSO-32075-TST-A.
319.  F. Jirsa, *Z. Elektrochem.*, **33**, 129-134 (1927).
320.  F. Jirsa, *Chem. Listy*, **19**, 3-9 (1925); *CA* **19**, 2460.
321.  F. Jirsa, *Z. Anorg. Allgem. Chem.*, **148**, 130-155 (1925).
322.  F. Jirsa, J. Jelinek, and J. Srbek, *Z. Anorg. Allgem. Chem.*, **158**, 33-60 (1926).
323.  E. Joensson and S. U. Falk, U.S. Pat. 3,049,578 (Aug. 14, 1962).
324.  H. L. Johnston, F. Cuta, and A. B. Garrett, *J. Am. Chem. Soc.*, **55**, 2311-2325 (1933).
325.  F. Jolas, *Electrochim. Acta*, **13**, 2207–2221 (1968).
326.  P. Jones and H. R. Thirsk, *Trans. Faraday Soc.*, **50**, 732-739 (1954).
327.  P. Jones, H. R. Thirsk, and W. F. K. Wynne-Jones, *Trans. Faraday Soc.*, **52**, 1003-1011 (1956).
328.  L. Jumau, *Etude Resumée des Accumulateurs Electriques*, Dunod, third edition, 1928, pp. 125-129.
329.  W. Jungner, Swedish Pat. 11,132 (1899).
330.  W. Jungner, U.S. Pat. 670,024 (Mar. 19, 1901).

331. W. Jungner, U.S. Pat. 692,298 (Feb. 4, 1902).
332. S. S. Kabalkina, S. V. Popova, N. R. Serebryanaya, and L. V. Vereshchagin, *Dokl. Akad. Nauk SSSR,* **152** (4), 853-854 (1963); *CA* **60**, 2400*d*.
333. B. N. Kabanov and D. I. Leikis, *Z. Elektrochem.,* **62** 660-663 (1958).
334. B. N. Kabanov, *Electrochim. Acta,* **6** (1-4), 253-257 (1962).
335. B. N. Kabanov, *J. Phys. Chem. (USSR)* **36,** 1432-9 (1962).
336. J. Kaplan, *Z. Physik,* **52,** 883 (1928).
337. H. Kawabe, H. Jacobson, I. F. Miller, and H. P. Gregor, *J. Colloid and Interface Sci.,* **21,** 79-93 (1966).
338. D. Kay, *Techniques for Electron Microscopy,* Oxford University Press, New York, 1961, Chap. 5.
339. G. Z. Kazakevich, I. E. Yablokova, and V. S. Bagotskii, *Elektrokhimiya,* **2** (9), 1055-1060 (1966), *CA* **65**, 19681*e*.
340. J. M. Keen and J. P. G. Farr, *J. Electrochem. Soc.,* **109**, 668-678 (1962).
341. J. J. Kelley and J. Szymborski, "Alkaline Battery Separator Characterization Studies," contract NAS5-10418, Second Quarterly Report, (23 Aug-23 Nov. 1967), 67 pp.
342. J. J. Kelley and J. Szymborski, "Alkaline Battery Separator Characterization Studies," Third Quarterly Report, (23 Nov. 1967-23 Feb. 1968), 31 pp. (Editor's note: NASA-CR or N numbers for this report and the previous one could not be located although the First Quarterly and Fourth Quarterly Reports are indexed as N68-16914 and N69-15810, resp.)
343. J. A. Keralla and J. J. Lander, "Secondary Silver Oxide-Zinc Battery Studies," Delco Remy Internal Report on Project 4260-K (July 5, 1961).
344. J. A. Keralla and J. J. Lander, "Effect of Surfactants Additions to the Zinc Plate on Cycle Life Performance in Secondary Silver Oxide-Zinc Cells," AFAPL-TR-67-107, 113 pp., (31 Aug. 1967), AD-819967.
345. J. A. Keralla and J. J. Lander, "Effect of Surfactants Additions to the Zinc Plate on Cycle Life Performance in Secondary Silver Oxide-Zinc Cell," *Extended Abstracts,* The Electrochemical Society, Inc., Battery Division, **12,** 94-95 (1967).
346. J. A. Keralla and J. J. Lander, *Electrochem. Tech.,* **6** (5/6), 202-205 (1968), cf. p. 204.
347. F. G. Keyes and H. Hara, *J. Am. Chem. Soc.,* **44**, 479-485 (1922).
348. E. M. Khairy and A. E. S. Mahgoub, *J. Chem. U.A.R.,* **9** (1) 31-46 (1966).
349. N. E. Khomatov and M. F. Sorokina, *Zh. Fiz. Khim.,* **40** (1), 44-8 (1966); *CA* **64**, 10764*g*.
350. M. Kiliani, *Berg- Huettenmaenn. Ztg.,* **251** (1883).
351. K. Kinoshita, *Bull. Chem. Soc. Japan,* **12**, 164-72, 366-376 (1937).
352. W. Klemm, *Z. Anorg. Allgem. Chem.,* **201**, 32-33 (1931).
353. O. Klein, *Z. Anorg. Chem.,* **74**, 157-169 (1912).
354. F. Kober and H. West, "The Anodic Oxidation of Zinc in Alkaline Solutions," *Extended Abstracts,* The Electrochemical Society, Inc., Battery Division, **12**, 66-69 (1967).
355. L. D. Kovba and N. A. Balashova, *Zh. Neorg. Khim.,* **4**, 225-226 (1959); *CA* **53**, 11952*i*.
356. T. R. E. Kressman, P. A. Stanbridge, and F. L. Tye, *Trans. Faraday Soc.,* **59**, 2129-2149 (1963).
357. T. A. Kryukova, "The Growth of Zinc Dendrites in Some Swelling Polymers," in *Soviet Electrochemistry, Proc. 4th Conference on Electrochemistry,* Consultants Bureau, N.Y. 1961, Vol. III, pp. 147-151. Applied.

358. O. K. Kudra, *Zh. Fiz. Khim.* **5**, 121 (1935); but see O. K. Kudra and K. N. Ivanov, *J. Phys. Chem.* **40**, 769-777 (1936).

359. N. T. Kudryavtsev and A. A. Nikiforova, *Zh. Prikl. Khim.*, **22**, 4 (1949); 367-376 (1949); *CA* **43**, 6521.

360. N. T. Kudryavtsev, *Dokl. Akad. Nauk SSSR*, **72**, 93-95 (1950); *CA* **44**, 7673g.

361. N. T. Kudryavtsev, *Zh. Fiz. Khim.*, **26**, 270-281 (1952); *CA* **47**, 4765c.

362. N. T. Kudryavtsev, A. I. Lipovetskaya, and K. N. Knarlamova, *J. Appl. Chem. USSR*, **25**, 459-62 (1952); *CA* **48**, 8680b.

363. N. T. Kudryavtsev, *Tr. Soveshchaniyapo Elektrokhim. Akad. Nauk SSSR, Otd. Khim. Nauk*, (1950) 258-275 (1953); *CA* **49**, 12156i.

364. N. T. Kudryavtsev, R. Yu. Bek, and I. F. Kushevich, *Tr. Moskov. Khim. Tekhnol. Inst. D. I. Mendeleeva*, (1956), No. 22, 137-42; *CA* **51**, 16142i.

365. N. T. Kudryavtsev, R. Yu. Bek, and I. F. Kushevich, *Zh. Prikl. Khim.*, **30**, 1093-1096 (1957); *CA* **51**, 17519b.

366. F. I. Kukoz, G. V. Mikhailenko, and M. F. Skalozubov, *Izv. Vysshikh Uchebn. Zavedenii, Khim, Khim. Teckhnol.*, **8** (3), 448-452 (1965); *CA* **64**, 4580-4581.

367. F. Kunschert, *Z. Anorg. Chem.*, **41**, 337-358 (1904).

368. N. Kuska and N. A. Cronander, "Selection and Utilization of Batteries for Deep Submergence Vehicles." AIAA Paper 67-138 (1967), 8 pp.; AIAA Advanced Marine Vehicle Meeting, Norfolk, Va., May 22-24 (1967).

369. P. E. Lake and E. J. Casey, *J. Electrochem. Soc.*, **106**, 913-919 (1959).

370. J. J. Lander, "Sealed Zinc-Silver Oxide Batteries," *Proc. 15th Annual Power Sources Conf.*, 77-80 (1961).

371. J. J. Lander, "Long Life Silver-Zinc Batteries," Final Report, ASD-TDR-62-668 (Oct. 1962), AD-291646.

372. J. J. Lander and J. A. Keralla, "Silver-Zinc Battery Investigation," RTD-TDR-63-4029 (Oct. 1963), 76 pp., *sec* p. 6, AD-422077.

373. J. J. Lander, L. M. Cooke, and P. A. Scardaville, "Applied Research Investigation of Sealed Silver-Zinc Batteries," Final report, AFAPL-TDR-64-85, 363 pp. (1 Aug. 1964), AD-603023.

374. J. J. Lander, J. A. Keralla, and P. A. Scardaville, "Sealed Silver-Zinc Batteries, Part 3," *Proc. 19th Annual Power Sources Conf.*, 77-80 (1965).

375. J. J. Lander, "Sealed Silver-Zinc Batteries," in *Space Power Systems Engineering*, Vol. 16, G. C. Szego and J. E. Taylor, editors, Academic Press, New York, 1966, pp. 1101-1109.

376. J. J. Lander, J. A. Keralla, and R. S. Bogner, "Zinc Electrode Investigation," AFAPL-TR-66-79, 203 pp. (31 Aug. 1966), AD-489498.

377. J. J. Lander and J. A. Keralla, "Silver-Zinc Electrodes and Separator Research," AFAPL-TR-67-107, 138 pp. (31 Aug. 1967), AD-819967.

378. J. J. Lander, private communication, R. L. Corbin, Delco-Remy, (5 Aug. 1968). See also R108.

379. R. Landsberg, *Z. Physik. Chem.* (Leipzig) **206**, 291-298 (1957).

380. R. Landsberg and H. Bartelt, *Z. Elektrochem.*, **61**, 1162-1168 (1957).

381. R. Landsberg and L. Mueller, *Wiss. Z. Tech. Hochsch. Chem. Leuna-Merseburg*, **2**, 103-104 (1959); *CA* **54**, 11647.

382. R. Landsberg, H. Fuertig, and L. Mueller, *Wiss. Z. Tech. Hochsch. Chem. Leuna-Merseburg*, **2**, 453-560 (1960); *CA* **54**, 17175d.

383. M. Lang, "Production of Silver Oxide Electrodes by Direct Rolling Techniques," *Proc. 23rd Annual Power Sources Conf.*, 84-86 (1969).

384. A. Langer and J. T. Patton, *J. Electrochem. Soc.* **114** (2), 113-117 (1967).

385. A. Langer and M. Scala, "Separator Development for a Heat Sterilizable Battery," JPL contract 951525 (Sept. 1966-June 1968), Quarterly Reports N67-13212, N67-17973, N67-35534, (No N number found for Fourth Quarter), N68-11225, N68-17493, N63-25105, and N69-26324).

386. A. Langer and E. A. Pantier, *J. Electrochem. Soc.*, **115** (10), 990-993 (1968).

387. W. M. Latimer and J. H. Hildebrand, *Reference Book of Inorganic Chemistry,* The Macmillan Company, New York, third edition (1951); (A) p. 103, (B) p. 163.

388. W. H. Latimer, *Oxidation States of the Elements and Their Potentials in Aqueous Solutions,* Prentice-Hall, Englewood Cliffs, N.J., second edition, 1952.

389. E. Laue, *Z. Anorg. Allgem. Chem.* **165**, 325-363 (1927).

390. K. R. Lawless and G. T. Miller, *Acta Cryst.*, **12**, 594-600 (1959).

391. D. J. Lax, Ph.D. Thesis, University of Durham, U.K., 1963.

392. M. LeBlanc and H. Sachse, *Physik Z.*, **32**, 887-889 (1931).

393. G. L. Leclanche, French Pat. 71865 (Jun. 8, 1866).

394. H. F. Leibecki, "Argentic Oxysalt Electrodes," NASA-TN-D-3208, 15 pp., N66-14759.

395. D. I. Leikis and G. L. Vidovich, *Elektrokhimiya* 1, 477-478 (1965); *CA* **63**, 11012*b*.

396. D. I. Leikis, G. L. Vidovich, L. L. Knoz, and B. N. Kabanov, *Z. Physik. Chem.* (Leipzig), **214**, 334-342 (1960).

397. G. R. Levi and A. Quilico, *Gazz. Chim. Ital.*, **54**, 598-604 (1924); *CA* **19**, 942.

398. H. B. Linford, H. P. Gregor, and B. J. Steigerwald, *Power Systems for Electric Vehicles. A Symposium* (April 6-8, 1967), Public Health Service Publication No. 999-AP-37, 323 pp.

399. K. J. Liu and H. P. Gregor, *J. Phys. Chem.*, **69** (4), 1252-1259 (1965).

400. M. A. Loshkarev, A. M. Ozerov, and N. T. Kudryavtsev, *Zh. Prikl. Khim.*, **22**, 294-306 (1949); *CA* **43**, 5674*h*.

401. R. Luther and F. Pokorny, *Z. Anorg. Allgem. Chem.*, **57**, 290-310 (1908).

402. F. T. Magg and D. Sutton, *Trans. Faraday Soc.*, **54**, 1861-70 (1958); **55**, 974-980 (1959).

403. C. G. Maier, G. S. Parks, and C. T. Anderson, *J. Am. Chem. Soc.*, **48**, 2564-2576 (1926).

404. P. A. Malachevsky and R. Jasinski, *J. Electrochem. Soc.*, **114**, 1239-1242 (1967).

405. P. A. Malachevsky and R. Jasinski, *J. Electrochem. Soc.*, **114**, 1258-1259 (1967).

406. P. A. Malachevsky and R. Jasinski, "Potentiostatic Oxidation of Silver," *Extended Abstracts,* The Electrochemical Soc. Inc., Theor. Electrochem. Div. **6**, 419-421 (1968).

407. L. Malaprade, *Compt. Rend.*, **210**, 504-505 (1940).

408. L. Malatesta, *Gazz. Chim. Ital.*, **71**, 467-474 (1941); *CA* **36**, 6929.

409. H. J. Mandel, U.S. Pat. 2,941,022 (June 14, 1960).

410. R. Marc, *Z. Physik. Chem.*, **73**, 685-723 (1910).

411. H. F. Mark *et al.*, editors, *Encyclopedia of Polymer Science and Technology,* John Wiley & Sons, Inc., New York, 1965; (A) V. C. Haskell, Vol. 3 p. 60, (B) Vol. 3, p. 131, and (C) A. H. Nissan and G. K. Hunger, Vol. 3, p. 181.

412. H. Mark *et al.*, editors, *High Polymers,* Vol. 5, Interscience Publishers, New York, (1954); (A) W. E. Gloor and E. D. Klug, p. 1480, (B) p. 1482, (C) J. A. Howsman, p. 393, (D) p. 434, (E) W. D. Nicoll, N. L. Cox, and R. F. Conaway, p. 825, (F) p. 837, and (G) H. M. Spurlin, p. 1057.

413. H. Mark *et al.*, editors, *Encyclopedia of Chemical Technology,* second edition, John Wiley & Sons, Inc., New York, 1968; (A) R. L. Mitchell and G. C. Daul, Vol. 17, p. 168.

414. R. P. P. Mathur and N. R. Dhar, *Z. Anorg. Allgem. Chem.*, **199**, 387-391 (1931); *CA* **25**, 5821.

415. P. B. Mathur and R. Gangadharan, *Indian J. Technol.*, **2**, 331-334 (1964); *CA* **62**, 7374b.

416. Yu. A. Mazitov, K. J. Rozental, and V. I. Veselovskii, *Zh. Fiz. Khim.*, **38** (2) 449-455 (1964).

For authors' names beginning with Mc, see at the end of the M entries, numbers 438 and following, also number 418 B.

417. S. M. Mehta and M. B. Kabadi, *J. Univ. Bombay*, **18A**, 38-44, 45-49, 50-53, 54-57 (1949); *CA* **44**, 5191h, 5192a, d, 5250b.

418. A. Meller, *Tappi*, **48**, 231-238 (1965).

418B. D. M. McDonald, *Tappi*, **48**, 708-713 (1965).

419. J. W. Mellor, *A Comprehensive Treatise on Inorganic and Theoretical Chemistry*, Longmans, Green & Co., London, U.K., Vol. III, p. 457.

420. M. Mendelsohn, U.S. Pat. 2,785,106 (Mar. 12, 1957).

421. M. Mendelsohn, U.S. Pat. 3,013,099 (Dec. 12, 1961).

422. E. Menzel and C. Menzel-Kopp, *Surface Sci.*, **2**, 376-380 (1964).

423. I. A. Menzies, D. L. Hill, G. J. Hill, L. Young, and J. O'M. Bockris, *J. Electroanal. Chem.*, **1**, 161-170 (1959/60).

424. G. H. Miller, Tech. Memorandum AFAPL-APIP-TM-68-2, Aerospace Power Div., AF Aero Propulsion Lab., Wright-Patterson AFB, Ohio, (obtainable only on specific request).

425. G. H. Miller and C. T. Napier, Unreported Failure Examination Work at the AF Aero Propulsion Lab., Wright-Patterson AFB.

426. R. L. Mitchell and G. C. Daul, *see* Ref. 413.

427. J. Moir, *Proc. Chem. Soc.*, **21**, 310-311 (1905).

428. H. Morawetz, *Macromolecules in Solution*, Interscience Publishers, New York. 1967.

429. H. J. Morgan and O. C. Ralston, *Trans. Electrochem. Soc.*, **30**, 229-239, (1916).

430. H. J. Morgan and J. D. Gray, *Eng. Min. J.* **151** (3) 72-75 (1950); *CA* **44**, J. *Electrochem. Soc.*, **100**, 590-591 (1953).

431. P. W. Morgan, U.S. Pat. 2,999,788 (Sept. 12, 1961).

432. D. H. Morrell and D. W. Smith, "The Fabrication of Battery Plates from Metal Powders," in *Power Sources 1966, Research and Development in Non-Mechanical Electrical Power Sources, Proc. 5th Internatl. Symposium, Brighton,* (Sept. 1966), D. H. Collins, editor, Pergamon Press, New York, 1967, pp. 207-225.

433. W. Morrison, U.S. Pat. 916,575 (Mar. 30, 1909); 975,885 (Nov. 15, 1910); 975,980-1 (Nov. 15, 1910); 976,092 (Nov. 15, 1910); 976,277-9 (Nov. 22, 1910).

434. L. H. Mott, U. S. Pat. 2,792,302 (May 14, 1957).

435. J. D. Moulton and R. F. Enters, U. S. Pat. 2,615,930 (Oct. 26, 1952).

436. E. Mueller, J. Mueller, and A. Fauvel, *Z. Elektrochem.*, **33**, 134-144 (1927); *CA* **21**, 2214.

437. R. H. Muller, *J. Electrochem. Soc.*, **113**, 943-947 (1966).

438. J. McBreen, "Investigation of Zinc Shape Change and Low Rate Capacity Loss in Silver Positives," Yardney Electric Co., Internal Report (May 1966).

439. J. McBreen and G. A. Dalin, "The Mechanism of Zinc Shape Change in Secondary Batteries," *Extended Abstracts,* The Electrochemical Society, Inc., Battery Divsion, **11**, 123-124 (1966).

440. J. McBreen, "Study to Investigate and Improve the Zinc Electrode for Spacecraft Electrochemical Cells," Final report, 48 pp. (Oct. 1966), N68-15716.

441. J. McCallum and C. L. Faust, "Failure Mechanisms in Sealed Batteries, Part II," AFAPL-TR-67-48 (Pt II), (Oct. 1967), AD-821895.

442. D. H. McClelland and M. Shaw, "Study of the Silver Oxide Electrode by Potentiostatic Sweep," *Extended Abstracts,* The Electrochemical Society, Inc., Battery Division, **10**, 32-34 (1965).

443. R. H. McCutcheon, "Reliability Program," *Proc. 12th Annual Battery Research & Development Conf.,* 51-57 (1958).

For McDonald entry, see number 418 B with *Tappi* entries.

444. A. S. McKie and D. Clark, "Crystallographic and Chemical Studies of the Oxides of Silver," in *Batteries,* D. H. Collins, editor, *see* Ref. 197, pp. 285-296.

445. J. A. McMillan, *Acta Cryst.,* **7**, 640 (1954).

446. J. A. McMillan, *J. Inorg. Nucl. Chem.,* **13**, 28-31 (1960).

447. J. A. McMillan, *Chem. Rev.,* **62**, 65-80 (1962).

448. J. A. McMillan, *Nature,* **195**, 594-595 (1962).

449. K. Nagel, R. Ohse, and E. Lange, *Z. Elektrochem.,* **61**, 795-803 (1957).

450. G. D. Nagy, "The Anodic Behavior of Silver in Alkaline Solutions," Ph.D. Thesis, University of Toronto (1964); *CA* **63**, 7895*a*, *Univ. Microfilms* (Ann Arbor, Mich.) Order No. 65-3154, 143 pp., *Dissertation Abstr.* **25** (11) 6258 (1965).

451. G. D. Nagy, W. J. Moroz, and E. J. Casey, "Studies on the Stability of Oxides of Silver," *Proc. 19th Power Sources Conf.,* 80-85 (1965), also AD-626229.

452. G. D. Nagy, unpublished results.

453. N. Nakajima, "Role of Dissolved Silver in Silver Peroxide-Zinc-Alkaline Battery," M.S. Thesis, 121 pp., Polytechnic Institute of Brooklyn (1955).

454. K. Nakamoto, *Infrared Spectra of Inorganic and Coordination Compounds,* John Wiley & Sons, Inc., New York, 1963, p. 107.

455. I. Naray-Szabo and K. Popp, *Z. Anorg. Allgem. Chem.,* **322**, 286-296 (1963).

456. I. Naray-Szabo, G. Argay, and P. Szabo, *Acta Cryst.,* **19** (2) 180-184 (1965).

457. R. Näsänen, *Suomen Kemistilehti,* **16B**, 1-3 (1943); *CA* **39**, 1796.

458. R. D. Naybour, "A Report on Morphologies of Zinc Electrodeposited from Zinc-Saturated Aqueous Alkaline Solution," Report ECRC/R26, Electricity Research Council, Cheshire, U.K. (Aug. 1967).

459. R. D. Naybour, *Electrochim. Acta,* **13**, 763-769 (1968).

460. A. B. Neiding and I. A. Kazarnovski, *Dokl. Akad. Nauk SSSR,* **78**, 713-716 (1951); *CA* **45**, 8385*h.*

461. G. Neumann, U.S. Pat. 2,571,927 (Oct. 16, 1951) and 2,636,058 (Apr. 21, 1953).

462. P. Niggli, *Z. Krist.,* **57**, 253-299 (1922); *CA* **17**, 2525.

463. Z. Ya. Nikitina, *Zh. Prikl. Khim.,* **32**, 132 (1959).

464. A. A. Noyes, K. S. Pitzer, and C. L. Dunn, *J. Am. Chem. Soc,* **57**, 1229-1242 (1935).

465. J. J. O'Connell and E. A. McElhill, "Separator Development for a Heat Sterilizable Battery," JPL contract 951524; (A) Final Summary Progress Report, 64 pp. (July 1967), N67-30918; (B) Supplement (12 June 1968), N69-13084.

466. R. Ohse, *Z. Elektrochem.,* **63**, 1063-1068 (1959)

466A L. E. Orgel and J. D. Dunitz, *Nature* 179, 462-465 (1957); *J. Phys. Chem. Solids* 3, 20-29 (1957).

467. J. Osterwald, *Z. Elektrochem.,* **66**, 492-496 (1962).

468. E. M. Otto, *J. Electrochem. Soc.,* **113**, 643-645 (1966).

469. E. M. Otto, *J. Electrochem. Soc.,* **115**, 878-881 (1968).

470. C. G. Overberger and A. Lebovits, *J. Am. Chem. Soc.,* **77**, 3675-3676 (1955).

471. J. E. Oxley, "Improvement of Zinc Electrodes for Electrochemical Cells," contract NAS5-3908, First Quarterly Report, 24 pp., (Sept. 1964), N64-33804.

472. J. E. Oxley, "Improvement of Zinc Electrodes for Electrochemical Cells," Second Quarterly Report, 28 pp. (to 12 Dec. 1964), N65-16741.

473. J. E. Oxley, "Improvement of Zinc Electrodes for Electrochemical Cells," Final Report, 40 pp. (12 Jun. 1964 to 24 Mar. 1965), N67-30067.

474. J. E. Oxley and C. W. Fleischmann, "Improvement of Zinc Electrodes for Electrochemical Cells," contract NAS5-9591, First Quarterly Report, 21 pp. (Sept. 1965), N66-13568.

475. J. E. Oxley and C. W. Fleischmann, "Improvement of Zinc Electrodes for Electrochemical Cells," Second Quarterly Report, 28 pp., (Dec. 1965), N66-19656.

476. J. E. Oxley, G. K. Johnson, and H. G. Oswin, "The Growth of Dendrites from Alkaline Zincate Solutions," *Extended Abstracts,* The Electrochemical Society, Inc., Battery Division, 10, 3-5 (1965).

477. J. E. Oxley and C. W. Fleischmann, "Improvement of Zinc Electrodes for Electrochemical Cells," Third Quarterly Report, 30 pp., (March 1966), N66-26870.

478. J. E. Oxley, C. W. Fleischmann, and H. G. Oswin, "Silver Batteries 2, Improved Zinc Electrodes for Secondary Batteries," *Proc. 20th Annual Power Sources Conf.,* 123-126 (1966).

479. J. E. Oxley, G. K. Johnson, and H. G. Oswin, unpublished paper.

480. R. K. Packer, "The Thermal Decomposition of Silver Dioxide, I," Report No. AUWE-TN-110/63, 13 pp. (Jan. 1963), AD-405846.

481. R. K. Packer and G. W. Allvey, "The Thermal Decomposition of Silver Dioxide, II," Report No. AUWE-TN-110/63-Pt-2, AD-478639.

482. T. Z. Palagyi, *J. Electrochem. Soc.,* 106, 846 (1959).

483. T. Z. Palagyi, *J. Electrochem. Soc.,* 108, 201-203 (1961).

484. T. Palagyi and I. Naray-Szabo, *Acta Chim. Acad. Sci. Hung.,* 30, 1-9 (1962), *CA* 57, 5550*i*.

485. N. B. Palmquist, "Evolution of the Low Rate Silver Peroxide/Zinc Primary Battery as a Basic Power Source for Space Flight," Aerospace Corp. Report TDR469-5102-1, 70 pp. (10 Nov. 1964), AD-451692L.

486. D. A. Payne, H. Tachikawa, and A. J. Bard, "Adsorption of Organic Materials on Zinc Electrodes," AFAPL-TR-68-115, Appendix III, 122-136, AD-839883.

487. D. A. Payne, H. Tachikawa, and A. J. Bard, Final Progress Report, AFAPL-TR-69-57, Appendix 2 (30 June 1969), AD-860365.

488. Z. G. Pinsker and L. I. Tatarinova, *Electron Diffraction,* Butterworth & Co., London, U.K., 1953, p. 262.

489. R. Piontelli, L. P. Bicelli, and C. Romagnani, *Atti Accad. Naz. Lincei, Rend., Classe Sci. Fis. Mat. Nat.,* 36 (3), 281-290 (1964); *CA* 62, 3666*e*.

490. K. S. Pitzer and J. H. Hildebrand, *J. Am. Chem. Soc.,* 63, 2472-2475 (1941).

491. G. Planté, *Recherches sur l'Electricité.* Bureaux de la Revue La Lumière Electrique, Paris (1883).

492. Yu. V. Pleskov and B. N. Kabanov, *Zh. Neorg. Khim.,* 2, 1807-1811 (1957); *CA* 52, 9842*b*.

493. Yu. V. Pleskov, *Dokl. Akad. Nauk SSSR,* 117, 645-647 (1957); *CA* 52, 12617.

494. G. Poli, Z. Siedlecka, and B. Rivolta, *Rend. Ist. Lombardo Sci. Lettere,* A97, 631-638 (1963); *CA* 65, 6715*h*.

495. T. I. Popova and V. I. Bagotskii, *Papers of the Mendeleev Congress,* Section 13, Academy of Sciences, SSSR, 1958 p. 42.

496. T. I. Popova, V. S. Bagotskii, and B. N. Kabanov, *Zh. Fiz. Khim.,* 36, 766, 770 (1962).

497. I. N. Pospelova, A. A. Rakov, and S. Ya. Pshezhetskii, *Zh. Fiz. Khim.,* 30, 1433-1437 (1956); *CA* 51, 6400*h*.

498. A. Pouchain, British Pat. 282,449 (Dec. 20, 1926); 290,665 (May 19, 1927); 303,825 (Jan. 10, 1928); 347,010 (May 8, 1929); 379,468 (Sept. 1, 1932); U.S. Pat. 1,895,397 (Jan. 24, 1933), 1,900,616 (Mar. 7, 1933).

499. M. Pourbaix, *Atlas of Electrochemical Equilibria* Pergamon Press, New York, 1966.

500. R. W. Powers, (A) *Electrochem. Tech.*, **5**, 429-440 (1967); (B) Plating from Alkaline Solutions, Progress Report no. 1, ILZRO Project No. ZE120 (Oct. 1966); (C) Progress Report No. 2, (Feb. 1967).

501. M. E. Prostakov, A. I. Levin, and V. P. Kochergin, *Zh. Fiz. Khim.*, **35**, 420-425 (1961); *CA* **55**, 11133*b*.

502. T. H. Purcell, Jr., U.S. Pat. 3,223,558 (Dec. 14, 1965).

503. M. Quintin, L. Kervajan, and M. Collier, *Compt. Rend.*, **260**, (17) (Groupe 7), 4510-4513 (1965); *CA* **63**, 9450*c*.

504. G. Raedlein, *Z. Elecktrochem.*, **61**, 727-733 (1957).

505. See Ref. 609, V. I. Veselovskii.

506. G. Rampel, U.S. Pat. 3,258,362 (June 28, 1966).

507. J. E. B. Randles, *Discussions Faraday Soc.*, **1**, 11-19 (1947).

508. P. Ray and K. Chakravarty, *J. Indian Chem. Soc.*, **21**, 47-50 (1944); cf. *CA* **39**, 36.

509. Republic Aviation Corp., (A) Product Specification, RE-B-7, Revision F, Amendment 3, Silver-Zinc Battery, 5 Ah, 26.25 V, (5 June 1958); (B) RE-B-22A, Silver-Zinc Battery, 100 Ah (Sept. 1, 1960).

510. A. R. Riedel, "Quality Control for Missile Batteries," *Proc. 15th Annual Power Sources Conf.*, 86-89, (1961), AD-421601.

511. R. A. Robinson and R. H. Stokes, *Electrolyte Solutions,* Butterworths London, U.K., second edition, 1965, p. 31.

512. V. I. Radionova, *Uch. Zap. Mosk. Gos. Ped. Inst.* **99**, 221-226 (1957); *CA* **54**, 48*c*.

513. V. V. Romanov, *Zh. Prikl. Khim.*, **33**, 2071-2078 (1960); *CA* **55**, 17294*h*.

514. V. V. Romanov, U.S.S.R. Pat. 141,187 (1961).

515. V. V. Romanov, *Zh. Prikl. Khim.*, **34**, 2692-2699 (1961); *CA* **56**, 9866*c*.

516. V. V. Romanov, *Zh. Prikl. Khim.*, **35** (6), 1293-1302 (1962); *CA* **57**, 10908*d*.

517. V. V. Romanov, *Zh. Prikl. Khim.*, **36** (5), 1050–1056 (1963); *CA* **59**, 14891*f*.

518. V. V. Romanov, *Zh. Prikl. Khim.*, **36** (5), 1057-1063 (1963); *CA* **59**, 14891*g*.

519. F. D. Rossini, D. D. Wagman, W. H. Evans, S. Levine, and I. Jaffe, Natl. Bur. Standards (U.S.) Circular 500, 1266 pp. (1952).

520. W. A. Roth and P. Chall, *Z. Elektrochem.*, **34**, 185-199. (1928).

521. A. L. Rotinyan, N. P. Fedot'ev, and Sok-Li Un, *Zh. Fiz. Khim.*, **31**, 1295-1303 (1957); *CA* **52**, 2613*b*.

522. J. J. Rowlette, "Heat Generation in the Surveyor Main Battery," *Extended Abstracts,* The Electrical Society, Inc., Battery Division, **12**, 120-123 (1967).

523. N. D. Rozenblyum, N. C. Bubyreva, and C. Z. Kazakevich, *Zh. Fiz. Khim.*, **40** (10), 2464-2467 (1966); *CA* **66**, 14239*j*.

524. K. I. Rozental and V. I. Veselovskii, *Zh. Fiz. Khim.*, **35**, 2670-2675 (1961); *CA* **56**, 13954*a*.

525. G. F. Rublee, U.S. Pat. 2,269,040 (Jan. 6, 1942).

526. P. Ruetschi, U.S. Pat. 2,951,106 (Aug. 30, 1960).

527. P. Ruetschi, B. D. Cahan, and W. S. Herbert, U.S. Pat. 3,057,944 (Oct. 9, 1962).

528. P. Ruetschi and J. B. Ockermann, *Electrochem. Tech.* **4**, 383-393 (1966).

529. P. Ruetschi, U.S. Pat. 3,311,501 (Mar. 28, 1967)

529B. P. Ruetschi and B. D. Cahan, U.S. Pat. 3,080,440 (Mar. 5, 1963).

530. A. J. Salkind, *Electrochem. Tech.*, **2**, 54-55 (1964).

531. A. J. Salkind, personal communication; Salkind noted the description of the apparatus and the process for his laboratory preparation of cellophane sheet in one of Andre's notebooks when reviewing them for his book, *Alkaline Storage Batteries,* John Wiley & Sons, Inc., 1969.

532. C. W. Saltonstall *et al.,* O. S. W. Report, Contract 14-01-001-338, (Oct. 1966). Office of Saline Water, Dept. of the Interior.

533. I. O. Salyer, E. V. Kirkland, and P. H. Wilken, "Silver-Zinc Battery Separator Material Development," Final Report, 64 pp., (July 1968), N68-36585.
534. A. I. Samokhotskii and A. P. Byudov, *Vestnik Metalloprom,* **15** (12), 122-124, (1935); *Chim. Ind.* (Paris), **36**, 936; *CA* **31**, 2155.
535. S. Sampath, R. Viswanathan, H. Udupa, and B. B. Dey, *J. Sci. Ind. Res.* (India), **15B**, 669-670 (1956); *CA* **51**, 7197c.
536. H. J. S. Sand, *Phil. Mag.,* **1**, 45. (1901).
537. I. Sanghi and W. F. K. Wynne-Jones, *Proc. Indian Acad. Sci.,* **47A**, 46-64 (1958); CA **52**, 16090g.
538. I. Sanghi and M. Fleischmann, *Proc. Indian. Acad. Sci.,* **49A**, 6-24 (1959).
539. I. Sanghi and M. Fleischmann, *Electrochim. Acta,* **1**, 161-176 (1959).
540. P. A. Scardaville and T. J. Weatherell, "Fabrication and Test of Battery Separator Material Resistant to Thermal Sterilization," Final Report, 298 pp. (Dec. 1965), N66-16190.
541. V. Scatturin, P. Bellon, and R. Zannetti, *Ric. Sci.,* **27**, 2163-2172 (1957); *CA* **52**, 35b.
542. V. Scatturin, P. Bellon, and R. Zannetti, *J. Inorg. Nucl. Chem.,* **8**, 462-467 (1958).
543. V. Scatturin, P. Bellon, and A. Salkind, *Ric. Sci.,* **30**, 1034-1044 (1960).
544. V. Scatturin, P. Bellon, and A. Salkind, *J. Electrochem. Soc.,* **108**, 819-822 (1961).
545. P. Schindler, H. Althaus, and W. Feitknect, *Helv. Chim. Acta,* **47**, 982-991 (1964).
546. W. J. Schlotter, U. S. Pat. 2,738,375 (Mar. 13, 1956).
547. A. X. Schmidt and C. A. Marlies, *Principles of High-Polymer Theory and Practice: Fibers, Plastics, Rubbers, Coatings, and Adhesives,* McGraw-Hill Book Co., New York, 1948, 755 pp.
548. R. Scholder and H. Weber, *Z. Anorg. Chem.,* **215**, 355-366 (1933).
549. R. Scholder, and G. Hendrich, *Z. Anorg. Allgem. Chem.,* **241**, 76-92 (1939).
550. P. Schoop, *Z. Elektrotech. Elektrochem.* pp. 131-134 (1894).
551. G. Schorsch, *Bull. Soc. Chim. France,* (7) 1456-1461 (1964).
552. R. W. Schult and W. T. Stafford, *Electro-Technology* 84-90 (1961).
553. G. M. Schwab and G. Hartmann, *Z. Anorg. Allgem. Chem.,* **281**, 183-186 (1955).
554. W. S. Sebborn, *Trans. Faraday Soc.,* **29**, 825-829 (1933); *CA* **27**, 5649.
555. H. N. Seiger, R. C. Shair, and P. F. Ritterman, "The Adhydrode in Charge Control," *Proc. 18th Annual Power Sources Conf.,* 61-64 (1964).
556. T. Sekine, *Acta Chem. Scand.,* **19**, 1526-1538 (1965).
557. J. Selbin and M. Usategui, *J. Inorg. Nucl. Chem.,* **20**, 91-99 (1961).
558. C. M. Shepherd, "The Silver Oxide-Zinc Alkaline Primary Cell, I," NRL-C-3478 (July 1, 1949), PB-109794.
559. C. M. Shepherd and H. C. Langelan, *J. Electrochem. Soc.,* **114**, 8-13 (1967).
560. P. R. Shipps, "Secondary Metal-Air System," *Proc. 20th Annual Power Sources Conf.,* 86-88 (1966).
561. N. A. Shumilova, G. V. Zhutaeva, and M. P. Tarasevich, *Electrochim. Acta,* **11**, 967-974 (1966).
562. N. A. Shumilova and V. S. Bagotzky, *Electrochim. Acta,* **13**, 285-293 (1968).
563. R. Shuttleworth, R. King, and B. Chalmers, *Metal Treat.,* **14**, 161 (1957).
564. J. H. Simons, *J. Phys. Chem.,* **36**, 652-657 (1932).
565. M. F. Skalozubov, *Novocherk. Politekhn. Inst.,* **134**, 3-18 (1962).
566. A. Skogseid, Dissertation, Oslo (1948).
567. R. N. Snyder and J. J. Lander, *Electrochem. Tech.,* **3** (5/6), 161-165 (1965).
568. F. Solomon and K. N. Brown, U.S. Pat. 3,055,964 (Sept. 25, 1962).
569. L. Spencer, *J. Chem. Soc.,* **123**, 2124-2128 (1923).
570. W. D. Spencer and B. Topley, *J. Chem. Soc.,* **26**, 2633-2650 (1929).

571.  Z. Stachurski, "Investigation and Improvements of Zinc Electrodes for Electro-chemical Cells," Final Report, 150 pp., (Dec. 1965), N67-26278.

572.  Z. Stachurski, First Quarterly Report, 38 pp. (June 1967) Contract NAS5-10231; continuation of Ref. 571, N67-39303.

573.  Z. Stachurski, and G. A. Dalin, unpublished work.

574.  M. v. Stackelberg and H. v. Freyhold, *Z. Elektrochem.,* **46**, 120-129 (1940).

575.  M. v. Stackelberg, M. Pilgram, and V. Toome, *Z. Elecktrochem.,* **57**, 342-350 (1953).

576.  B. Stehlik and P. Weidenthaler, *Chem. Listy,* **52**, 402-404 (1958); *CA* **52**, 19399e.

577.  B. Stehlik, P. Weidenthaler, and J. Vlach, *Chem Listy,* **52**, 2230-2236 (1958); *CA* **53**, 5809.

578.  B. Stehlik and P. Weidenthaler, *Collection Czech. Chem. Commun.* **24**, 1416-1419 (1957).

579.  B. Stehlik, *Chem. Zvesti,* **15**, 469-473, 474-478, (1961); *CA* **55**, 25437c.

580.  V. V. Stender and M. D. Zholudev, *Zh. Prikl. Khim.,* **32**, 1296-1299 (1959); *CA* **53** 18683g.

581.  M. Sulkes and G. A. Dalin, "Development of the Sealed Zinc-Silver Oxide Secondary Battery System," Second Quarterly Report, 83 pp. (Dec. 1963); AD-600395.

582.  M. Sulkes and G. A. Dalin, Third Quarterly Report, *see* Ref. 581, 78 pp. (to 31 Mar. 1964); AD-444798.

583.  H. E. Swanson, R. K. Fuyat, and G. M. Ugrinie, *NBS Circular 539,* Vol. IV, p. 61 (1955).

584.  H. E. Swanson, M. C. Morris, R. P. Stinchfield, and E. H. Evans, *NBS Monograph 25,* Section 1 (1961).

585.  N. Tanaka and R. Tamamushi, *Electrochim. Acta,* **9** (7), 963-989 (1964).

586.  M. R. Tarasevich, N. A. Shumilova, and R. Kh. Burshtein, *Izv. Akad. Nauk SSSR,* Ser. Khim., **1**, 17-26 (1964); *CA* **61**, 306c.

587.  U. B. Thomas, "Kinetic Basis for the Operating Characteristics of Sealed Nickel-Cadmium Cells," in *Batteries,* D. H. Collins, editor, *see* Ref. 197, pp. 117-128.

588.  J. J. Thomson, *Proc. Phys. Soc.,* **40,** 79 (1928).

589.  R. L. Tichenor, U.S. Pat. 2,578,027 (Dec. 11, 1951).

590.  W. A. Tiller, *Science,* **146** (3646), 871-879 (1964).

591.  P. S. Titov and L. Baldanova, *Jubilee Collection of Scientific Works, Moscow Inst. of Nonferrous Metals and Gold,* No. 9, pp. 580-589. (1940): *CA* **37,** 5660.

592.  P. S. Titov and E. N. Paleolog, *Jubilee Collection of Scientific Works, Moscow Inst. of Nonferrous Metals and Gold,* No. 9, pp. 602-611 (1940).

593.  P. S. Titov and E. N. Paleolog, *Jubilee Collection of Scientific Works, Moscow Inst. of Nonferrous Metals and Gold,* No. 9, p. 603 (1940).

594.  F. D. Trischler, "Separator Development Phase of Heat Sterilizable Battery Development Program," Final Report, 36 pp. (Oct. 1966); N67-15399.

595.  A. F. Turback, U.S. Pat. 3,072,618 (Jan. 8, 1963).

596.  S. B. Tuwiner, *Diffusion and Membrane Technology,* Reinhold Book Corp., New York, 1962. p. 153.

597.  A. Tvarusko, *J. Electrochem. Soc.,* **115** (11), 1105-1110 (1968).

598.  R. N. Tweedy, "The Drumm Battery—Is It a Revolution?" *The Electrical Review,* cvii, 290-292 (Aug. 22, 1930).

599.  C. S. Tsa and Z. A. Iofa, *Dokl. Akad. Nauk SSSR,* **131**, 137-140 (1960).

600.  U.S. Department of Interior, Test Manual for Permselective Membranes, Office of Saline Water Report No. 77 (1964).

601.  J. Valcha and B. Stehlik, *Collection Czech. Chem. Commun.,* **25**, 676-681 (1960); *CA* **54**, 19116h.

602.  W. J. van der Grinten, "Secondary Zinc Electrodes with Zincate Trapping," *Extended Abstracts,* The Electrochemical Society, Inc., Battery Division, **12**, 96-97 (1967).

603. Q. Van Winkle, "Electrokinetic Phenomena," in *Encyclopedia of Science and Technology*, McGraw-Hill Publishing Co., New York, 1966, Vol. 4, pp. 4-79.
604. A. E. Varker, "Development of Bonding Processes for Plastic to Plastic and Plastic to Metal Seals," ESB Incorporated, internal report, (June 25, 1965).
605. T. B. Vaughan and H. J. Pick, *Electrochim. Acta,* 2, 179-194 (1960).
606. J. B. Vergette, N. C. Wright, and T. E. King, "Studies of Silver Oxide," DCBRL Report No. 501 (Oct. 1966); AD-806861.
607. D. A. Vermilyea, "Anodic Films," in *Advances in Electrochemistry and Electrochemical Engineering*, John Wiley & Sons, Inc., 1963, Vol. 3, chap. 4, pp. 211-286.
608. E. J. W. Verwey and J. H. deBoer, *Rec. Trav. Chim.,* 55, 531-540 (1936); *CA* 30, 7948.
609. V. I. Veselovskii, "Mechanism of Electrochemical Oxidation," *Soviet Electrochemistry, Proc. of the Fourth Conference on Electrochemistry (Oct. 1-6, 1956)*, Academy of Science, U.S.S.R., Vol. II pp. 163-70, Consultants' Bureau, New York, 1959; also, *CA* 54, 10590e, *Tr. Chetvertogo Soveshch. po Elektrokhim. Moscow, 1956*, pp. 241-251 (Pub. 1959).
610. V. I. Veselovskii, T. I. Borisova, A. A. Jakoleva, and S. O. Izidinov, *Electrochim. Acta,* 10 (3), 325-337 (1965).
611. K. J. Vetter, *Elektrochemische Kinetik*, Springer Verlag, Berlin, 1961; *CA* 55, 17308i.
612. G. L. Vidovich, D. I. Leikis, and B. N. Kabanov, *Dokl. Akad. Nauk SSSR,* 124, 855-857 (1959); *CA* 55, 8118d.
613. G. L. Vidovich, D. I. Leikis, and B. N. Kabanov, *Dokl. Akad. Nauk SSSR,* 142, 109-112 (1962); *CA* 57, 5700c.
614. G. W. Vinal, *Storage Batteries*, Third Edition, John Wiley & Sons, Inc., 1947.
615. G. W. Vinal, *Primary Batteries*, John Wiley and Sons, Inc., 1950; (A) pp. 235-254, (B) pp. 263-264.
616. H. Vogel, *J. Prakt. Chem.* 87, 288 (1862); *Pogg. Ann.,* 118, 145 (1863).
617. Yu. Ts. Vol and N. A. Shishakov, *Izv. Akad. Nauk SSSR. Otd. Khim. Nauk,* 586-591 (1962); *CA* 57, 4088e.
618. Yu. Ts. Vol and N. A. Shishakov, *Izv. Akad. Nauk SSSR, Ser. Khim.,* 11, 1920-1923 (1963); *CA* 60, 6465d.
619. A. Volta, *Phil. Trans. Roy. Soc.,* 90, 403 (1800).
620. G. S. Vozdvizhenskii and E. D. Kochman, *Zh. Fiz. Khim.,* 39 (3), 657-663 (1965); *CA* 63, 5235e.
621. I. L. Wadehra, R. St. J. Manley, and D. A. I. Goring, *J. Appl. Polymer Sci.,* 9, 2627-2630 (1965), *see* p. 2634.
622. O. C. Wagner and R. F. Enters, "Development of Manufacturing Methods and Techniques for the Production of Improved Alkaline Batteries," AFML-TR-66-236 (Sept. 1966); AD-802279.
623. R. D. Wagner, "A Method for the Quantitative Analysis of the Silver Oxide Cathode," NOLTR64-214, 23 pp. (1965); AD-618479.
624. C. P. Wales and J. Burbank, *J. Electrochem. Soc.* 106, 885-890 (1959).
625. C. P. Wales, *J. Electrochem. Soc.,* 108, 395-400 (1961).
626. C. P. Wales, *J. Electrochem. Soc.,* 109, 1119-1124 (1962).
627. C. P. Wales, *J. Electrochem. Soc.,* 111, 131-135 (1964).
628. C. P. Wales, and J. Burbank, *J. Electrochem. Soc.,* 111, 1002-1005 (1964).
629. C. P. Wales, and J. Burbank, *J. Electrochem. Soc.,* 112, 13-16 (1965).
630. C. P. Wales, *J. Electrochem. Soc.,* 113, 757-63 (1966).
631. C. P. Wales and A. C. Simon, "Feasibility of Microscopy for Investigating the Silver Oxide Electrode," NRL Report 6647, 26 pp. (Jan. 9, 1968); AD-664993.
631B. R. A. Wallace, M.Ch.E. Thesis, Polytechnic Inst. of Brooklyn, (June 1959).
632. P. Weidenthaler and J. Vlach, *Chem. Listy,* 52, 2230 (1959).

633. E. Wellisch, M. K. Gupta, L. Marker, and O. J. Sweeting, *J. Appl. Polymer Sci.*, 9, 2591-2606 (1965).

634. J. C. White, R. T. Pierce, and T. P. Dirkse, *Trans. Electrochem. Soc.*, 90, 467-473 (1946).

635. N. T. Wilburn, "Zinc-Silver Oxide Batteries," *Proc. 10th Annual Battery Research and Development Conf.* pp. 20-22 (1956); PB-125427.

636. N. T. Wilburn, "Concluding Results of the Reliability Program for Battery BA-472( )/U," *Proc. 13th Annual Power Sources Conf.* pp. 152-156 (1959); PB-145521.

637. N. T. Wilburn, "Reliability Program on High-Rate Zinc-Silver Oxide Batteries," *Proc. 15th Annual Power Sources Conf.* pp. 83-85 (1961); AD-421601.

638. N. T. Wilburn, K. E. Meade, and C. J. Bradley, "Dry Process Divalent Silver Oxide Electrodes," ECOM-2628, 69 pp., (1965); AD-622698.

639A. N. T. Wilburn, K. E. Meade, and C. J. Bradley, "Dry Process Divalent Silver Oxide Electrodes," ECOM-2749, 19 pp. (Sept. 1966); AD-640053.

639B. N. T. Wilburn, private communication, April 2, 1968.

640. F. Will, *J. Electrochem. Soc.*, 111, 145-151, 152-160 (1963).

641. G. Wranglen, *J. Electrochem. Soc.*, 97, 353-360 (1950).

642. G. Wranglen, *Trans. Roy. Inst. Technol.* (Stockholm), No. 94, 41 pp., (1955); *CA* 50, 5429a.

643. R. W. G. Wyckoff, *Am. J. Sci.*, 3, 184-188 (1922).

644. R. W. G. Wyckoff, *Crystal Structures*, Second Edition, Interscience, New York., 1963, Vol. 1.

645. G. M. Wylie, "Investigation of AgO Primary Batteries," Final Report, 110 pp. (Aug. 1961); AD-267953.

646. G. M. Wylie, U.S. Pat. 3,282,740 (Nov. 1, 1966).

646B. G. M. Wylie, "Sealed Zinc-Silver Oxide Batteries, Part 2," *Proc. 19th Annual Power Sources Conf.*, 73-76 (1965).

647. A. A. Yakovleva, T. I. Borisova, and V. I. Veselovskii, *Zh. Fiz. Khim.*, 36, 1426-1431 (1962); *CA* 57, 13523f.

648. M. N. Yardney, U.S. Pat. 2,983,777 (May 9, 1956).

649. Yardney Electric Corp., "Production of Dry Process Silver Oxide Electrodes by Continuous Rolling," Quarterly and Final Reports, Contract DAAB-05-68-C-1801, R. F. Enters, Fourth Quarterly Report, 33 pp., AD-853455 (1969).

650. T. Yawataya, T. Hani, Y. Oda, and A. Nishihara, *Dechema Monograph.*, 74, No. 781-804, pp. 501-514 (1962); *CA* 58, 14209e.

651. S. Yoshizawa and Z. Takehara, *J. Electrochem. Soc. Japan*, 31 (3), 91-104 (1963); *CA* 61, 1503d.

652. S. Yoshizawa and Z. Takehara, *Denki Kagaku*, 32 (1), 27-35 (1964); *CA* 61, 11607b.

653. S. Yoshizawa and Z. Takehara, *Denki Kagaku*, 32 (1), 35-39 (1964); *CA* 61, 11607b.

654. S. Yoshizawa and Z. Takehara, *Denki Kagaku*, 32 (3), 197-200 (1964); *CA* 62, 7374h.

655. M. D. Zholudev and V. V. Stender, *Bull. Acad. Sci. Kazakh SSR*, 7, 30 (1957).

656. M. D. Zholudev and V. V. Stender, *Zh. Prikl. Khim.*, 31, 1036-1039 (1958); *CA* 52, 19596h.

657. N. A. Zhulidov and F. I. Yefremov, *Vestnik Elektropromyshlennosti* 2, 74-75 (1963); *CA* 52, 19596g; "The New Nickel-Zinc Storage Battery," unedited rough draft translation, 8 pp., Wright-Patterson AFB, FTD-TT-64-605 (6 Jan. 1965), AD-610270.

658. G. V. Zhutaeva, N. A. Shumilova, and V. I. Luk'yanycheva, *Elektrokhimiya*, 4 (2), 196-198 (1968); *CA* 68, 92931h,

659. J. G. Zimmerman, *Trans. Electrochem. Soc.*, 68, 231-249 (1935).

# Index